AF393198

Compact Textbooks in Mathematics

This textbook series presents concise introductions to current topics in mathematics and mainly addresses advanced undergraduates and master students. The concept is to offer small books covering subject matter equivalent to 2- or 3-hour lectures or seminars which are also suitable for self-study. The books provide students and teachers with new perspectives and novel approaches. They may feature examples and exercises to illustrate key concepts and applications of the theoretical contents. The series also includes textbooks specifically speaking to the needs of students from other disciplines such as physics, computer science, engineering, life sciences, and finance.

- **compact**: small books presenting the relevant knowledge
- **learning made easy**: examples and exercises illustrate the application of the contents
- **useful for lecturers**: each title can serve as basis and guideline for a semester course/lecture/ seminar of 2–3 hours per week.

Gunther Leobacher ·
Friedrich Pillichshammer

Introduction to Quasi-Monte Carlo Integration and Applications

Second Edition

 Birkhäuser

Gunther Leobacher [iD]
Institute of Mathematics and Scientific
Computing
University of Graz
Graz, Austria

Friedrich Pillichshammer [iD]
Institute of Financial Mathematics
and Applied Number Theory
Johannes Kepler University
Linz, Austria

ISSN 2296-4568   ISSN 2296-455X   (electronic)
Compact Textbooks in Mathematics
ISBN 978-3-032-05445-6   ISBN 978-3-032-05446-3   (eBook)
https://doi.org/10.1007/978-3-032-05446-3

Mathematics Subject Classification: 65C05, 91G60, 11K45, 65Cxx

To Maria and Gisi

Preface to the Second Edition

The present second edition of *Introduction to Quasi-Monte Carlo Integration and Applications* represents a thorough revision and expansion of the first edition from 2014. In addition to correcting errors and small inaccuracies, the "Further Reading" sections and the bibliography have also been thoroughly revised and updated—but without any claim to completeness. The main innovation, however, is the addition of a few more detailed sections. While the first edition, with one exception, largely focused on the presentation of classical theory without taking weighted problems into account, this is now done in the update. In particular, the material about Koksma-Hlawka inequalities has been revised and complemented by a new section about the weighted Koksma-Hlawka inequality, which also comprises the notion of weighted discrepancy. Two new sections about weighted discrepancy of lattice point sets and tractability properties have been added to Chaps. 4 and 6, respectively. Also in Chap. 4 we added a new section about the CBC construction of lattice rules for the worst-case error in weighted Korobov spaces and an outlook on lattice rules for the integration of functions which are not necessarily periodic. Finally, in Chap. 5 we added a section about polynomial lattice point sets, which constitute an important construction method of digital nets.

We are grateful to many readers of the first edition of our book for their feedback and for making us aware of mistakes and inaccuracies. Special thank goes to Klaus Ritter and the students in his seminar at RPTU Kaiserslautern for reading the book carefully and for their many hints. Any remaining errors are the responsibility of the authors.

As for the first edition, we hope that the book and the newly added material will turn out to be useful for teaching, self-study and as a reference, and that it will encourage many people to study quasi-Monte Carlo methods and/or apply them to problems from Mathematical Finance or other areas.

<table>
<tr><td>Graz, Austria</td><td>Gunther Leobacher</td></tr>
<tr><td>Linz, Austria</td><td>Friedrich Pillichshammer</td></tr>
<tr><td>June 2025</td><td></td></tr>
</table>

Preface to the First Edition

While Fubini's theorem states that the integral of a function on the s-dimensional unit cube can be computed simply by computing iterated integrals, the attempt of doing so for a function for which those integrals cannot be given by closed formulas is in most cases doomed to fail, if s, say, is greater than 10. The reason for this is that by iteration of a one-dimensional integration rule the number of function evaluation needed for the corresponding product rule grows exponentially in s.

This constraint on the practical computation of integrals led to the development of probabilistic methods. Here, the integral is interpreted as the expected value of the integrand evaluated at a random variable that is uniformly distributed on the s-dimensional unit cube. These methods where first applied by E. Fermi, S. Ulam and J. von Neumann, the latter also being the originator of the name "Monte Carlo simulation".

In contrast to Monte Carlo integration rules, which sample the integrand at random points, so-called *quasi*-Monte Carlo rules use deterministic sample points. The relationship between Monte Carlo and quasi-Monte Carlo corresponds to the relationship of two notions of "uniform distribution" in mathematics. The first notion is the probabilistic concept of a random variable for which the probability of taking values in a given subset of the unit cube is precisely the volume of that set. The second notion is that of a sequence of points, for which the proportion of points of the sequence lying in a given s-dimensional sub-interval of the unit cube equals the volume of the sub-interval.

While the notion of uniformly distributed sequences and examples thereof had been coined earlier, the birth of the theory of "Uniform Distribution Modulo One" is marked by H. Weyl's seminal paper "Über die Gleichverteilung mod. Eins", first published in the year 1916. It was already known by then that, at least in principle, uniformly distributed sequences could be used to integrate Riemann-integrable functions. However, under the weak assumptions the convergence of the sample mean to the integral can be arbitrarily slow, making the "method" impractical.

As the starting point for the analysis of quasi-Monte Carlo methods for numerical integration one can consider the establishment of the Koksma-Hlawka inequality, which was shown by J. F. Koksma in 1942 for the one-dimensional case and in 1961 by E. Hlawka for arbitrary dimensions. Since then the Koksma-Hlawka inequality is the prototypical error estimate for quasi-Monte Carlo integration. Its main feature is that it bounds the integration error by the product of two terms, the variation of the function and the star discrepancy of the underlying sample nodes. The second notion is related to that of uniform distribution of a sequence, but while the latter is an asymptotic quality, the star discrepancy allows to assess the quality of uniformity of a finite number of points. Knowing how well the points can be chosen with respect to that measure means—thanks to the Koksma-Hlawka inequality—knowing the possible convergence of the integration error. This is where concepts from Discrepancy Theory enter the game.

From the early 1960s on several people, such as, e.g., N. M. Korobov, E. Hlawka, I. M. Sobol', J. Halton, H. Faure, H. Niederreiter and C. P. Xing provided constructions of point sets and sequences with excellent distribution properties, i.e. with low star discrepancy or related/alternative quality measures. The point sets and sequences constructed in this way are therefore suitable sample points for quasi-Monte Carlo rules. However a certain disadvantageous dependence of the discrepancy bounds on the dimension led to the believe that quasi-Monte Carlo rules can only be applied in very moderate dimensions. Contrary to these opinions, quasi-Monte Carlo rules are nowadays used for numerical integration of functions in hundreds or even thousands of dimensions and since recently there is also a stream of research which studies infinite-dimensional integration. The motivation for this paradigm change lies in results of numerical experiments published by S. H. Paskov and J. F. Traub in 1995 who studied quasi-Monte Carlo rules for functions in 360 dimensions coming from Mathematical Finance. But, despite their apparent effectivity even for those very high dimensional problems, the question of exactly *why* quasi-Monte Carlo rules should give these good results is still not completely resolved. In 2010, at the MCQMC meeting in Warsaw, I. H. Sloan spoke in this context about "The unreasonable effectiveness of quasi-Monte Carlo". Although in the meantime some partial answers come from the study of weighted function spaces and from tractability theory, the quest for an explanation of this unreasonable effectiveness of quasi-Monte Carlo is still a very active part of research.

As suggested by its title, this book is an introductory text into quasi-Monte Carlo methods and some of their applications, and it aims at giving a comprehensible treatment of the subject with detailed explanations of the basic concepts. Origination from a 2 h—one semester undergraduate course, it should be accessible to students in mathematics or computer science with basic knowledge in algebra, calculus, linear algebra and probability theory. Although the main focus is on the theory behind the concepts of quasi-Monte Carlo, several practical applications with an emphasis on financial problems are discussed.

The topics of the book roughly retrace the history of quasi-Monte Carlo methods as sketched above, but do so using up-to-date concepts and notations. Thus we start with the classical multi-dimensional integration problem and its first high-dimensional alternative, Monte Carlo integration. Chapter 2 is devoted to uniform distribution of sequences and several concepts of discrepancy. We give a discrepancy estimate for one of the oldest specimens of low discrepancy sequences, the Halton sequence. In Chap. 3 we introduce the modern framework of reproducing kernel Hilbert spaces for obtaining bounds on the integration error for functions in those spaces. The Koksma-Hlawka inequality, though not its most general form, appears as a special case of that theory. The next two chapters, are mostly devoted to constructions of low-discrepancy point sets and sequences, namely lattice point sets, (t, m, s)-nets and (t, s)-sequences. The chapter on lattice rules include a section on integration in weighted Korobov spaces. The concept of weighted spaces has some bearing on the question about effectiveness of quasi-Monte Carlo for very high-dimensional problems. Chapter 6 concludes the theoretical part by providing more information about the curse of dimensionality and tractability of discrepancy.

The last two chapters constitute the application part of the book. Chapter 7 gives a very condensed introduction to concepts from Mathematical Finance, in particular derivative pricing. We introduce some models and derivative that can serve as specimens for trying out the simulation methods provided in Chap. 8. This last chapter covers some of the basics of simulation, like generation of non-uniform random variables and generation of Brownian paths. An emphasis is on (fast-)orthogonal transforms for speeding up convergence. Several examples serve to illustrate the methods.

The compilation of a text book demands a great deal not only from the authors, but also from their families, colleagues and students, from which some of the common time has to be diverted to the project. We want to thank all of them for their support and understanding.

We appreciate valuable comments, suggestions and improvements from several colleagues which we would like to mention here: Josef Dick, Aicke Hinrichs, Peter Kritzer, Gerhard Larcher, Harald Niederreiter, Klaus Ritter and Wolfgang Ch. Schmid.

We hope that the book will turn out to be useful for teaching, self-study and as a reference that it will encourage many people to study quasi-Monte Carlo methods and/or apply them to problems from Mathematical Finance or other areas.

Linz, Austria   Gunther Leobacher
March 2014   Friedrich Pillichshammer

Competing Interests The authors have no competing interests to declare that are relevant to the content of this manuscript.

Contents

Notations

$\bar{a}$	Complex conjugate of a complex number a
$A(J, \mathcal{P}, N)$	Number of elements of $\mathcal{P}$ belonging to J
$\#B$	Number of elements of a finite set B
$\mathbb{C}$	The set of complex numbers
$C^\alpha([0, 1])$	The class of all f on $[0, 1]$ with continuous derivatives up to order α
χ_J	Characteristic function of a set J
$\Delta_{\mathcal{P},N}$	Local discrepancy, discrepancy function
D_N	Extreme discrepancy of N points
D_N^*	Star discrepancy of N points
$\mathbb{E}$	Expectation
$\exp$	Exponential function
$\mathbb{F}_b$	Finite field of order b
$\|f\|_{L_\infty}$	$\|f\|_{L_\infty} = \sup_{x \in D} |f(x)|$ for $f : D \to \mathbb{R}$
$f(x) \asymp g(x)$	Means that $cg(x) \le f(x) \le Cg(x)$ for some $C > c > 0$
$\Gamma_{m,s}$	Regular lattice
i	$\mathrm{i} = \sqrt{-1}$
λ_s	s-dimensional Lebesgue measure
$\log x$	Natural logarithm of x
$\log_2 x$	Dyadic (base 2) logarithm of x
$L_{p,N}$	L_p discrepancy of N points
$L_{2,N}$	L_2 discrepancy of N points
$\mathbb{N}$	The set of positive integers
$\mathbb{N}_0$	The set of non-negative integers
$O(f(x))$	$g(x) = O(f(x))$ if $|g(x)| \le Cf(x)$ for some $C > 0$
$\mathcal{P}$	Finite point set in $[0, 1)^s$
$\mathbb{P}$	Probability
ϕ_b	Radical inverse function
$Q_{N,s}(f)$	QMC rule
$\mathbb{R}$	The set of real numbers
$[s]$	$[s] = \{1, 2, \ldots, s\}$
$\mathcal{S}$	Infinite sequence in $[0, 1)^s$

$\top$	Transpose of a vector or matrix
Var	Variance
$\{x\}$	Fractional part of the real number x
$\lfloor x \rfloor$	Integer part of the real number x
$\lceil x \rceil$	Smallest integer larger than or equal to x
$\|\boldsymbol{x}\|_1$	$\|\boldsymbol{x}\|_1 = |x_1| + \cdots + |x_s|$ for $\boldsymbol{x} = (x_1, \ldots, x_s) \in \mathbb{R}^s$
$\|\boldsymbol{x}\|_\infty$	$\|\boldsymbol{x}\|_\infty = \max_{j=1,\ldots,s} |x_j|$ for $\boldsymbol{x} = (x_1, \ldots, x_s) \in \mathbb{R}^s$
$\mathbb{Z}$	The set of integers
$\mathbb{Z}_b$	Residue class ring modulo b (identified with $\{0, 1, \ldots, b-1\}$)

Introduction **1**

1.1 The Univariate and Multivariate Integration Problem

In this book we consider the problem of numerical integration over the s-dimensional unit cube $[0, 1]^s$,

$$\int_{[0,1]^s} f(\boldsymbol{x})\, \mathrm{d}\boldsymbol{x} = \int_0^1 \cdots \int_0^1 f(x_1, \ldots, x_s)\, \mathrm{d}x_1 \ldots \mathrm{d}x_s. \tag{1.1}$$

Here the dimension s may be large in practical applications. The restriction to integration problems over the unit cube $[0, 1]^s$ is mostly for simplicity and in many cases does not impose a big limitation, since most integrals over bounded or unbounded regions can be transformed into integrals over the unit cube (although one has to be careful in choosing suitable transformations which, of course, have influence on the behavior of the transformed integrand).

For most functions arising in practice the integral (1.1) cannot be solved analytically since, for instance, the given integrand does not have an antiderivative or its antiderivative, if existent, is not expressible as a finite combination of elementary functions. Hence we must solve such integrals numerically, that is, we want to find an algorithm which enables us to approximate the true value of the integral to any prescribed level of accuracy.

We will always assume that for a function f we can compute its function values at finitely many points in $[0, 1]^s$. However, we remark here already that the knowledge of finitely many function values is not enough to solve the integration problem since in this case the integral still could be any number. So we will need further "global" information, as, for example, the smoothness of f, which restricts the class of integrands to certain function classes.

© The Author(s), under exclusive license to Springer Nature Switzerland AG 2026
G. Leobacher and F. Pillichshammer, *Introduction to Quasi-Monte Carlo Integration and Applications*, Compact Textbooks in Mathematics,
https://doi.org/10.1007/978-3-032-05446-3_1 1

In the one-dimensional case, there are many classical quadrature rules available, as, for instance, the rectangle rule (midpoint rule), the trapezoidal rule, Simpson's rule, or the Gauss rule, which all have the general form

$$T_m(f) = \sum_{n=0}^{m} q_n f(x_n) \qquad (1.2)$$

with quadrature points $x_0, \ldots, x_m$ from $[0, 1]$ and with weights $q_0, \ldots, q_m \in \mathbb{R}$. As an example we mention the trapezoidal rule for which $q_0 = q_m = 1/(2m)$, $q_n = 1/m$ for $n = 1, \ldots, m - 1$ and $x_n = n/m$ for $n = 0, \ldots, m$. If $f \in C^2([0, 1])$ the error of the trapezoidal rule is of order $O(m^{-2})$ (cf. Exercise 1.2).

In the multi-dimensional case the classical methods use Cartesian products of one-dimensional quadrature rules which are then often called *product rules*. This means that one applies a one-dimensional quadrature rule of the form (1.2) to each one-dimensional integral in (1.1) which results in the product rule of the form

$$\sum_{n_1=0}^{m} \cdots \sum_{n_s=0}^{m} q_{n_1} \cdots q_{n_s} f(x_{n_1}, \ldots, x_{n_s}).$$

When writing this product rule again in the form

$$\sum_{n=0}^{M} w_n f(\boldsymbol{x}_n)$$

then the set of quadrature points $\{\boldsymbol{x}_0, \ldots, \boldsymbol{x}_M\}$ is just the s-fold product of the one-dimensional quadrature points $\{x_0, \ldots, x_m\}$. Hence the total number of nodes of a product rule is $N = M + 1 = (m + 1)^s$ which grows dramatically with the dimension s. For example, if $s = 30$ a product rule based on a one-dimensional rule with only two points involves already $N = 2^{30} \approx 10^9$ nodes. But if $s = 500$, then a product rule based on a 2-element one-dimensional rule already requires 2^{500} nodes and the time until the end of our universe will not suffice to compute that many function evaluations.

The error analysis of product rules follows immediately from that for the underlying one-dimensional quadrature rule. For instance, the error of approximating an s-dimensional integral with a product rule based on the trapezoidal rule is of order $O(m^{-2})$, where $m + 1$ is the number of nodes used in the underlying one-dimensional rule, provided that all partial derivatives of order two in each variable are continuous on $[0, 1]^s$. This looks quite promising on first sight, but in terms of the actual number $N = (m + 1)^s$ of integration nodes, this error is of order $O(N^{-2/s})$. For large dimensions, which might be in the hundreds for practical problems, such an error convergence is less than satisfying. This phenomenon is often called the *curse of dimensionality*. The phrase "curse of dimensionality" was coined already in 1957 by R. Bellman and means in our context that the minimal number of function values needed to compute an ε-approximation to multivariate integrals is an exponential function in s.

Example 1.1 Let $\mathcal{F}_{\mathrm{Lip}} := \{f : [0,1]^s \to \mathbb{R} : \mathrm{Lip}(f) \le L\}$ denote the class of Lipschitz continuous functions with Lipschitz constant of at most L, i.e., for $f \in \mathcal{F}_{\mathrm{Lip}}$ we have $|f(x) - f(y)| \le L\|x - y\|_\infty$, where $\|x\|_\infty = \max_{j=1,\dots,s} |x_j|$ for $x = (x_1, \dots, x_s)$. Assume we want to use the Cartesian product of the one-dimensional midpoint rule for the integration of functions $f \in \mathcal{F}_{\mathrm{Lip}}$. Then, for $m \in \mathbb{N}$, the set of integration nodes is

$$\Gamma_{m,s} := \left\{ x_k = \left(\frac{2k_1 + 1}{2m}, \dots, \frac{2k_s + 1}{2m} \right) : k = (k_1, \dots, k_s) \in \{0, 1, \dots, m-1\}^s \right\}.$$

Denote the integration error of the product midpoint rule for $f \in \mathcal{F}_{\mathrm{Lip}}$ by

$$e(f, \Gamma_{m,s}) := \int_{[0,1]^s} f(x)\, dx - \frac{1}{m^s} \sum_{k_1,\dots,k_s=0}^{m-1} f(x_k)$$

and put $Q_k = \prod_{j=1}^s \left[\frac{k_j}{m}, \frac{k_j+1}{m} \right)$ for $k = (k_1, \dots, k_s) \in \{0, 1, \dots, m-1\}^s$. Then we have

$$|e(f, \Gamma_{m,s})| = \left| \sum_{k_1,\dots,k_s=0}^{m-1} \int_{Q_k} (f(x) - f(x_k))\, dx \right| \le L \sum_{\substack{k \in \mathbb{N}_0^s \\ \|k\|_\infty < m}} \int_{Q_k} \|x - x_k\|_\infty\, dx.$$

For $x \in Q_k$ we have

$$\|x - x_k\|_\infty = \max_{j=1,\dots,s} \left| x_j - \frac{2k_j + 1}{2m} \right| \le \frac{1}{2m}$$

and hence we find that

$$|e(f, \Gamma_{m,s})| \le \frac{L}{2m} = \frac{L}{2} \frac{1}{N^{1/s}}$$

where $N = \#\Gamma_{m,s} = m^s$ is the number of employed integration nodes.

This result cannot be significantly improved. For example, consider the function

$$g(x_1, x_2, \dots, x_s) = \frac{L}{2\pi m}(1 + \cos(2\pi m x_1)).$$

It is easy to see that $g \in \mathcal{F}_{\mathrm{Lip}}$ and that

$$\int_{[0,1]^s} g(x)\, dx = \frac{L}{2\pi m} = \frac{L}{2\pi} \frac{1}{N^{1/s}} \quad \text{and} \quad \frac{1}{m^s} \sum_{k_1,\dots,k_s=0}^{m-1} g(x_k) = 0.$$

Hence $e(g, \Gamma_{m,s}) = \frac{L}{2\pi} \frac{1}{N^{1/s}}$. This means that

$$\frac{L}{2\pi} \frac{1}{N^{1/s}} \le \sup_{f \in \mathcal{F}_{\mathrm{Lip}}} |e(f, \Gamma_{m,s})| \le \frac{L}{2} \frac{1}{N^{1/s}}.$$

For example, for $m = 2$, the node set consists of $N = 2^s$ points, which is a huge number already for moderate dimensions s, say $s = 50$, but nevertheless we only have $L/(4\pi) \le \sup_{f \in \mathcal{F}_{\mathrm{Lip}}} |e(f, \Gamma_{2,s})| \le L/4$.

The question arises whether there are algorithms for multivariate integration for which the error convergence does not show such a poor dependence on the dimension s. This question can be answered in the affirmative, which can be seen by the following considerations.

1.2 Monte Carlo Integration

We aim at approximating the integral of a function $f : [0, 1]^s \to \mathbb{R}$ by an equal weight quadrature rule of the form

$$\frac{1}{N} \sum_{n=0}^{N-1} f(\boldsymbol{x}_n),$$

where the quadrature points $\mathcal{P} = \{\boldsymbol{x}_0, \dots, \boldsymbol{x}_{N-1}\}$ are from $[0, 1)^s$. We are interested in the integration error

$$e(f, \mathcal{P}) := \int_{[0,1]^s} f(\boldsymbol{x}) \, \mathrm{d}\boldsymbol{x} - \frac{1}{N} \sum_{n=0}^{N-1} f(\boldsymbol{x}_n).$$

But how should we choose the quadrature points? One idea is to choose realizations of N independent and uniformly distributed random variables $X_0, \dots, X_{N-1}$ in $[0, 1]^s$ and to check what we can expect for the resulting error. This means that we use

$$Q_{N,s}(f) := \frac{1}{N} \sum_{n=0}^{N-1} f(X_n) \tag{1.3}$$

as an estimator for the integral. Note that a measurable function $f : [0, 1]^s \to \mathbb{R}$ can be considered as a random variable on the probability space $([0, 1]^s, \mathcal{B}, \lambda_s)$ where $\mathcal{B}$ is the Borel σ-algebra on $[0, 1]^s$ and λ_s the Lebesgue measure. Then the expectation of this random variable equals the integral we want to compute, i.e., $\mathbb{E}[f] = \int_{[0,1]^s} f(\boldsymbol{x}) \, \mathrm{d}\boldsymbol{x}$. Using the linearity of the expected value we have

$$\mathbb{E}[Q_{N,s}(f)] = \frac{1}{N} \sum_{n=0}^{N-1} \mathbb{E}[f] = \mathbb{E}[f] = \int_{[0,1]^s} f(\boldsymbol{x}) \, \mathrm{d}\boldsymbol{x}$$

and hence $Q_{N,s}(f)$ is an unbiased estimator for the integral $\int_{[0,1]^s} f(x)\,dx$. The strong law of large numbers guarantees that

$$\mathbb{P}\left[\lim_{N\to\infty} Q_{N,s}(f) = \int_{[0,1]^s} f(x)\,dx\right] = 1,$$

where $\mathbb{P}[\cdot]$ is the probability on an arbitrary probability space supporting an independent sequence $(X_n)_{n\in\mathbb{N}_0}$ of random variables uniformly distributed on $[0,1]^s$.

The *variance* of f is given by $\mathrm{Var}[f] := \int_{[0,1]^s}\left(f(x) - \int_{[0,1]^s} f(y)\,dy\right)^2 dx$. Since $X_0,\ldots,X_{N-1}$ are independent, we obtain from the Bienaymé formula the following result for the variance of the estimator $Q_{N,s}(f)$.

Theorem 1.1 *Let $f \in L_2([0,1]^s)$. Then for any $N \in \mathbb{N}$ we have*

$$\mathrm{Var}[Q_{N,s}(f)] = \frac{\mathrm{Var}[f]}{N}.$$

Note that

$$\mathrm{Var}[Q_{N,s}(f)] = \mathbb{E}[(Q_{N,s}(f) - \mathbb{E}[f])^2] = \mathbb{E}[e^2(f,\cdot)],$$

where $e(f,\cdot)$ is the error estimator

$$e(f,\cdot) := \int_{[0,1]^s} f(x)\,dx - \frac{1}{N}\sum_{n=0}^{N-1} f(X_n).$$

Hence it follows from Theorem 1.1 that

$$\mathbb{E}[|e(f,\cdot)|] \leq \sqrt{\mathbb{E}[e^2(f,\cdot)]} = \frac{\sigma[f]}{\sqrt{N}},$$

where $\sigma[f] = (\mathrm{Var}[f])^{1/2}$ denotes the *standard deviation* of f. This means that the absolute value of the integration error is, on average, bounded by $\sigma[f]/\sqrt{N}$. It is remarkable that the convergence rate of the expected integration error does not depend on the dimension s. We have $N^{-1/2} < N^{-1/s}$ for $s > 2$. Hence, roughly speaking, for $s > 2$ it is on average better to use random points for the approximation of the integral of f rather than using classical product rules. Here f does not even have to be continuous if one chooses random samples.

The method of using random sample points is called *(plain) Monte Carlo (MC) method* and the integration rule (1.3) is called a *Monte Carlo rule (or algorithm)*. The phrase "Monte Carlo" goes back to J. von Neumann who chose it as a code name for this method when he worked on a secret project together with S. Ulam in 1946 at Los Alamos Scientific Laboratory.

To summarize, the advantages of the MC method are as follows:

- It suffices that the integrands are quadratic integrable.
- The convergence rate of order $O(N^{-1/2})$ is independent of the dimension s. This is a surprising fact although it does not mean that MC breaks the curse of dimensionality because the standard deviation $\sigma[f]$ is in general not independent of s.

On the other hand, the MC method also has some disadvantages, which are:

- The error bound is only "probabilistic," that is, in any one instance one cannot be sure of the integration error. However, further probabilistic information is obtained from the central limit theorem, which states that, if $0 < \sigma[f] < \infty$, then

$$\lim_{N \to \infty} \mathbb{P}\left[|e(f, \{X_0, X_1, \ldots, X_{N-1}\})| \le \frac{c\,\sigma[f]}{\sqrt{N}} \right] = 2\Phi(c) - 1,$$

for any $c > 0$, where $\Phi(x) := \frac{1}{\sqrt{2\pi}} \int_{-\infty}^{x} \exp(-t^2/2)\, dt$ is the cumulative distribution function of the standard normal distribution.

If we would like to have

$$\mathbb{P}\left[|e(f, \{X_0 X_1, \ldots, X_{N-1}\})| > \varepsilon \right] < \alpha$$

for small $\varepsilon, \alpha > 0$, then we need to require

$$1 - \alpha < \mathbb{P}\left[|e(f, \{X_0, X_1, \ldots, X_{N-1}\})| \le \varepsilon \right]$$
$$= \mathbb{P}\left[-\frac{\varepsilon\sqrt{N}}{\sigma[f]} \le \sqrt{N}\frac{e(f, \{X_0, \ldots, X_{N-1}\})}{\sigma[f]} \le \frac{\varepsilon\sqrt{N}}{\sigma[f]} \right]$$
$$\approx 2\Phi\left(\frac{\varepsilon\sqrt{N}}{\sigma[f]} \right) - 1.$$

This means that $1 - \frac{\alpha}{2} < \Phi\left(\frac{\varepsilon\sqrt{N}}{\sigma[f]} \right)$ or

$$N > \frac{\mathrm{Var}[f]}{\varepsilon^2} \left(\Phi^{-1}\left(1 - \frac{\alpha}{2} \right) \right)^2.$$

Note that $\mathrm{Var}[f]$ is in general not explicitly known. However it can be estimated per MC by using the same function evaluations as for the MC integration of f. This will be explained in more detail shortly.
- A second problem is that the generation of random samples is difficult. This problem is a topic on its own which cannot be further discussed in this book. References for readings in this direction are provided at the end of this chapter.

Fig. 1.1 2000 random points
in $[0, 1]^2$

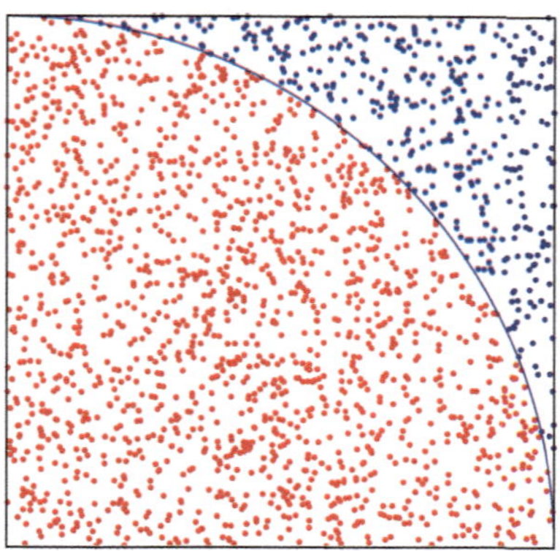

- For some applications the convergence rate of $O(N^{-1/2})$ is too slow.
- The convergence rate of $O(N^{-1/2})$ does not reflect some possible regularity of the integrand.

Example 1.2 A popular first example for the MC method is the approximation of the number π. The idea is to distribute a certain number of points independently and uniformly in the unit square and to count the number of them which have distance of at most 1 from the origin; see Fig. 1.1. This number, normalized by the total number of distributed points, is an approximation for the number $\pi/4$ which is the area of a quadrant with radius 1.

Let $f : [0, 1]^2 \to \mathbb{R}$, $f(x_1, x_2) = 1$ if $x_1^2 + x_2^2 \leq 1$ and 0 otherwise. Obviously we have $\int_0^1 \int_0^1 f(x_1, x_2)\, dx_1\, dx_2 = \frac{\pi}{4}$. Then the method described above for the approximation of $\pi/4$ amounts to calculating the MC mean $\frac{1}{N} \sum_{n=0}^{N-1} f(X_n)$ with N independent and uniformly distributed random variables $X_0, \ldots, X_{N-1}$ in $[0, 1]^2$. Performing for fixed N one hundred random experiments and plotting the mean MC error for this random experiments leads to a picture as shown in Fig. 1.2.

A practical error estimate for the MC method can be obtained from the estimator

$$V_N := \frac{1}{N-1} \sum_{n=0}^{N-1} \left(f(X_n) - Q_{N,s}(f) \right)^2,$$

for the variance $\mathrm{Var}[f]$ whenever $f \in L_2([0, 1]^s)$. This estimator is unbiased since

$$V_N = \frac{1}{N-1} \sum_{n=0}^{N-1} (f(X_n) - \mathbb{E}[f])^2 - \frac{N}{N-1} (\mathbb{E}[f] - Q_{N,s}(f))^2,$$

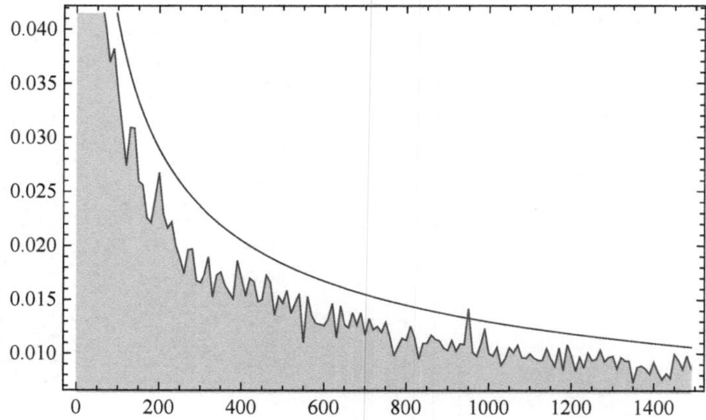

Fig. 1.2 Average MC error (blue) for $N = 1, \ldots, 1500$ and 100 random experiments per given N compared to $\sigma[f]/\sqrt{N}$ (red)

and hence, by using Theorem 1.1,

$$
\mathbb{E}[V_N] = \frac{1}{N-1} \sum_{n=0}^{N-1} \mathbb{E}[(f(X_n) - \mathbb{E}[f])^2] - \frac{N}{N-1} \mathbb{E}[(\mathbb{E}[f] - Q_{N,s}(f))^2]
$$

$$
= \frac{1}{N-1} \sum_{n=0}^{N-1} \mathrm{Var}[f] - \frac{N}{N-1} \frac{\mathrm{Var}[f]}{N}
$$

$$
= \mathrm{Var}[f].
$$

Hence

$$
\frac{V_N}{N} = \frac{1}{N(N-1)} \left(\sum_{n=0}^{N-1} f^2(X_n) - N(Q_{N,s}(f))^2 \right)
$$

provides an unbiased estimator for $\mathrm{Var}[Q_{N,s}(f)] = \mathbb{E}[e^2(f, \cdot)]$. Note that this estimator requires only $O(N)$ function evaluations and hence it is very efficient for practical implementation.

As already mentioned, one of the disadvantages of MC is the slow convergence rate of order $O(N^{-1/2})$. Although in practice one can use variance reduction techniques, such as importance sampling, stratified sampling, or correlated sampling, these techniques usually do not improve the rate of error convergence. One very simple variance reduction technique for dimension $s = 1$ is discussed in Exercise 1.2. This is the main motivation for the following strategy.

The aim is to find *deterministic* constructions of quadrature points which are at least as good as the average. This method is then called *quasi-Monte Carlo (QMC) method* as opposed to MC, where one uses randomly chosen quadrature points, and

$$Q_{N,s}(f) := \frac{1}{N} \sum_{n=0}^{N-1} f(\boldsymbol{x}_n)$$

with *deterministic* quadrature points $\boldsymbol{x}_0, \ldots, \boldsymbol{x}_{N-1}$ in $[0, 1)^s$ is called a *quasi-Monte Carlo (QMC) rule (or algorithm)*. Hence in the deterministic case we need quadrature points which are in some sense "well" distributed in $[0, 1)^s$. To this end we need to clarify what we mean by "well" distributed. This question is subject to the theory of Uniform Distribution Modulo One, which is the topic of our next chapter.

Further Reading

A coherent theory underlying numerical quadrature is provided in the book of Brass and Petras [14]. Early classical results about the complexity of numerical integration are well summarized in [115, Sect. 1]. An excellent introduction to the Monte Carlo method offers the book of Müller-Gronbach, Novak and Ritter [107]. We also recommend the books of Niederreiter [110], of Lemieux [97], of Kalos and Whitlock [78], and of Glasserman [47], where the latter deals with the application of Monte Carlo for financial problems. Information concerning the generation of (pseudo) random samples can be found in the books of Knuth [81, Chap. 3], of Niederreiter [110, Chap. 7–10], of Tezuka [141] and in the survey articles of L'Ecuyer and Hellekalek [95] and of Topuzoğlu and Winterhof [142].

Exercises

1.1 Let $f \in C^1([0, 1])$. For $m \in \mathbb{N}$ the trapezoidal rule is given by

$$T_m f := \sum_{n=0}^{m-1} \frac{f(n/m) + f((n+1)/m)}{2m} = \sum_{n=0}^{m} q_n f\left(\frac{n}{m}\right)$$

where $q_0 = q_m = 1/(2m)$ and $q_n = 1/m$ for $n = 1, \ldots, m-1$. Show that

$$\int_0^1 f(x)\, dx - T_m f = \int_0^1 K(t) f'(t)\, dt \tag{1.4}$$

where $K(t) = \frac{n}{m} + \frac{1}{2m} - t$ for $\frac{n}{m} < t \le \frac{n+1}{m}$ and $n = 0, \ldots, m-1$. Deduce from (1.4) that the integration error is of order $O(m^{-1})$. *Hint*: Consider first the term

$$\int_{n/m}^{(n+1)/m} f(x)\, dx - \frac{f(n/m) + f((n+1)/m)}{2m}$$

and use that $f(x) = f(n/m) + \int_{n/m}^x f'(t)\,dt$ for $n/m \leq x \leq (n+1)/m$.

1.2 Assume that even $f \in C^2([0, 1])$. Show that then the integration error of the trapezoidal rule is of order $O(m^{-2})$. *Hint*: Use (1.4) and integration by parts on $[\frac{n}{m}, \frac{n+1}{m}]$.

1.3 Let $f : [0, 1]^s \to \mathbb{R}$ with continuous second-order partial derivatives $\partial^2 f/\partial x_i^2$ for all $i = 1, \ldots, s$. Consider the product trapezoidal rule of the form

$$T_m^{(s)} f := \sum_{n_1=0}^{m} \cdots \sum_{n_s=0}^{m} w_{n_1} \cdots w_{n_s} f\left(\frac{n_1}{m}, \ldots, \frac{n_s}{m}\right).$$

Show that if f depends only on one variable and is constant with respect to the remaining $s - 1$ variables, then the integration error is of order $O(m^{-2})$. Is this result satisfactory if we consider the number of required integration nodes?

1.4 Let $f : [0, 1]^2 \to \mathbb{R}$, $f(x_1, x_2) = 1$ if $x_1^2 + x_2^2 \leq 1$ and $f(x_1, x_2) = 0$, otherwise, be the function from Example 1.2. Write a computer program for MC integration and run several experiments. Compare the MC error with $1/\sqrt{N}$ where N denotes the number of nodes involved.

1.5 Let $f : [0, 1]^3 \to \mathbb{R}$, $f(x, y, z) = x^2 y - \exp(2y + z)$. Compute

$$\int_{[0,1]^3} f(x, y, z)\,dx\,dy\,dz$$

as well as $\mathrm{Var}[f]$ and $\mathbb{E}[e^2(f, \mathcal{P})]$.

1.6 Compute the integral from Exercise 1.2 with the MC method. Run for fixed N one hundred experiments and compare the average MC error with $\mathbb{E}[e^2(f, \mathcal{P})]$.

1.7 Let $f, g : [0, 1] \to \mathbb{R}$, $f(x) = x^{-1/4}$ and $g(x) = x^{-3/4}$. Use the MC method to estimate the integrals $\int_0^1 f(x)\,dx$ and $\int_0^1 g(x)\,dx$. Repeat the random experiment for fixed N one hundred times and plot the average error for $N \in \{1, \ldots, 5000\}$ against N (for example, in steps of 50). Why is there such a big difference between the results for f and g? *Remark*: This example is taken from [107] where one can find a very detailed discussion of it.

1.8 Let $f \in L_2([0, 1]^s)$ and let $X_0, \ldots X_{N-1}$ be independent and uniformly distributed random variables in $[0, 1]^s$.

(a) Show that

$$\mathbb{P}\left[\left|Q_{N,s}(f) - \mathbb{E}[f]\right| \geq \varepsilon\right] \leq \frac{\mathrm{Var}[f]}{N\varepsilon^2}.$$

(b) Let $\delta > 0$ and $N \in \mathbb{N}$. Determine confidence intervals of the form $I_{N,\delta} = \left[Q_{N,s}(f) - L_N, Q_{N,s}(f) + L_N\right]$ such that

$$\mathbb{P}\left[\mathbb{E}[f] \in I_{N,\delta}\right] \geq 1 - \delta.$$

1.9 Show that the estimator $\frac{1}{N} \sum_{n=0}^{N-1} \left(f(X_n) - Q_{N,s} \right)^2$ for $\mathrm{Var}[f]$ is *not* unbiased.

1.10 Let $f : [0, 1] \to \mathbb{R}$ be continuous and monotone. Put $g(x) = (f(x) + f(1 - x))/2$. Show that $\int_0^1 f(x)\,\mathrm{d}x = \int_0^1 g(x)\,\mathrm{d}x$ and

$$\mathrm{Var}[g] \leq \frac{1}{2}\mathrm{Var}[f].$$

Hint: A proof of this simple variance reduction technique can be found in [110, Proposition 1.3].

Uniform Distribution Modulo One

2

The theory of *Uniform Distribution Modulo One* is a branch of Number Theory which goes back to the seminal work of H. Weyl from 1916. For us the main motivation to study this topic lies in its application for numerical integration based on QMC rules.

2.1 Definition and Basic Properties

On first sight, it is not immediately clear what it should mean that a finite point set $\mathcal{P}$ is uniformly distributed in the unit cube. Intuitively one may suppose that every region contains a proper portion of elements from $\mathcal{P}$, that there are no accumulations of points and that there are no big gaps between the points from $\mathcal{P}$. In other words, every region should contain a number of elements from $\mathcal{P}$ that is proportional to the size of the region. To put this intuition into exact mathematical terms we consider infinite sequences instead of finite point sets. Since a sequence is a discrete object it is clear that we need some restriction on the demand "*every* region". In Uniform Distribution Theory, these regions are usually restricted to intervals of the form $[\boldsymbol{a}, \boldsymbol{b})$, where $\boldsymbol{a} = (a_1, \ldots, a_s)$, $\boldsymbol{b} = (b_1, \ldots, b_s)$ are elements of $[0, 1]^s$ and $[\boldsymbol{a}, \boldsymbol{b}) := [a_1, b_1) \times \ldots \times [a_s, b_s)$. See Fig. 2.1 for a memorable picture.

For a sequence $\mathcal{S} = (\boldsymbol{x}_n)_{n \in \mathbb{N}_0}$ in $[0, 1)^s$ and an interval $[\boldsymbol{a}, \boldsymbol{b}) \subseteq [0, 1)^s$ the number of indices $n \in \{0, \ldots, N - 1\}$ for which $\boldsymbol{x}_n \in [\boldsymbol{a}, \boldsymbol{b})$ is denoted by $A([\boldsymbol{a}, \boldsymbol{b}), \mathcal{S}, N)$ (or in short by $A([\boldsymbol{a}, \boldsymbol{b}), N)$, if there is no risk of confusion). In mathematical terms

$$A([\boldsymbol{a}, \boldsymbol{b}), \mathcal{S}, N) := \#\{n \in \mathbb{N}_0 : 0 \le n \le N - 1 \text{ and } \boldsymbol{x}_n \in [\boldsymbol{a}, \boldsymbol{b})\}.$$

© The Author(s), under exclusive license to Springer Nature Switzerland AG 2026
G. Leobacher and F. Pillichshammer, *Introduction to Quasi-Monte Carlo Integration and Applications*, Compact Textbooks in Mathematics,
https://doi.org/10.1007/978-3-032-05446-3_2

Fig. 2.1 Every interval
should contain the proper
portion of points

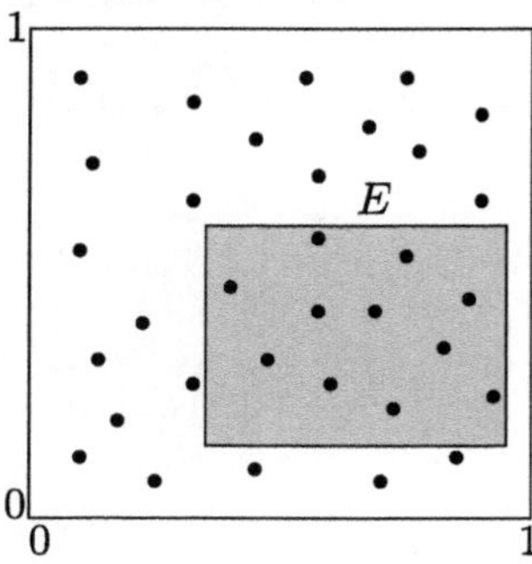

We use this notion also for finite point sets $\mathcal{P} = \{x_0, \ldots, x_{N-1}\}$, which we always interpret in the sense of the combinatorial notion of "multiset," i.e., a set where the multiplicity of elements matters. I.e.,

$$A([a, b), \mathcal{P}, N) := \#\{n \in \mathbb{N}_0 : 0 \le n \le N - 1 \text{ and } x_n \in [a, b)\}.$$

Let λ_s denote the s-dimensional Lebesgue measure. For intervals of the form $[a, b)$ this reduces to $\lambda_s([a, b)) = \prod_{j=1}^{s} (b_j - a_j)$.

Definition 2.1 An infinite sequence $\mathcal{S}$ in $[0, 1)^s$ is said to be *uniformly distributed modulo one* (or equidistributed), if for every interval of the form $[a, b) \subseteq [0, 1)^s$ we have

$$\lim_{N \to \infty} \frac{A([a, b), \mathcal{S}, N)}{N} = \lambda_s([a, b)). \tag{2.1}$$

We will encounter sequences which enjoy this property later on.

Let χ_J be the *characteristic function* of a set $J \subseteq \mathbb{R}^s$, i.e., $\chi_J(x) = 1$, if $x \in J$ and 0, otherwise. Then we can write

$$A([a, b), \mathcal{S}, N) = \sum_{n=0}^{N-1} \chi_{[a,b)}(x_n),$$

and hence Eq. (2.1) is equivalent to

$$\lim_{N \to \infty} \frac{1}{N} \sum_{n=0}^{N-1} \chi_{[a,b)}(x_n) = \int_{[0,1]^s} \chi_{[a,b)}(x) \, dx. \tag{2.2}$$

Equation (2.2) leads to the following characterization of uniform distribution modulo one:

Theorem 2.2 *A sequence $(\boldsymbol{x}_n)_{n\in\mathbb{N}_0}$ in $[0, 1)^s$ is uniformly distributed modulo one, if and only if for every Riemann integrable function $f : [0, 1]^s \to \mathbb{R}$ we have*

$$\lim_{N\to\infty} \frac{1}{N} \sum_{n=0}^{N-1} f(\boldsymbol{x}_n) = \int_{[0,1]^s} f(\boldsymbol{x})\,\mathrm{d}\boldsymbol{x}. \tag{2.3}$$

Observe that Theorem 2.2 gives the link to QMC integration. It is now clear that, in order to obtain a QMC rule which converges to the actual value of the integral, the underlying nodes must come from a uniformly distributed sequence.

Remark 2.3 There is no sequence which satisfies (2.3) for all Lebesgue integrable functions. Moreover, it is known that for every Lebesgue integrable function which is not integrable in the Riemann sense there exists a uniformly distributed sequence for which (2.3) does not hold true (see [21]).

Proof of Theorem 2.2. Assume that (2.3) holds for any Riemann integrable function $f : [0, 1]^s \to \mathbb{R}$. But then (2.2) holds true as well since for any sub-interval J of $[0, 1]^s$ the characteristic function χ_J is Riemann integrable. Thus the sequence $(\boldsymbol{x}_n)_{n\in\mathbb{N}_0}$ is uniformly distributed modulo one.

On the other hand, assume that the sequence $(\boldsymbol{x}_n)_{n\in\mathbb{N}_0}$ is uniformly distributed modulo one. We will show in a first step, that then Eq. (2.3) is satisfied for every step function on $[0, 1)^s$ (i.e., every finite linear combination of characteristic functions of intervals of the form $[\boldsymbol{a}, \boldsymbol{b})$). Let the intervals $E_1, \ldots, E_m$ form a partition of $[0, 1]^s$ and put

$$g(\boldsymbol{x}) = \sum_{i=1}^{m} d_i \chi_{E_i}(\boldsymbol{x}),$$

where $d_i \in \mathbb{R}$ for $i = 1, 2, \ldots, m$. Then we have

$$\lim_{N\to\infty} \frac{1}{N} \sum_{n=0}^{N-1} g(\boldsymbol{x}_n) = \lim_{N\to\infty} \frac{1}{N} \sum_{n=0}^{N-1} \sum_{i=1}^{m} d_i \chi_{E_i}(\boldsymbol{x}_n)$$

$$= \sum_{i=1}^{m} d_i \lim_{N\to\infty} \frac{1}{N} \sum_{n=0}^{N-1} \chi_{E_i}(\boldsymbol{x}_n) = \sum_{i=1}^{m} d_i \lambda(E_i) = \int_{[0,1]^s} g(\boldsymbol{x})\,\mathrm{d}\boldsymbol{x}.$$

Hence Eq. (2.3) is true for all step functions on $[0, 1]^s$.

Now we consider a Riemann integrable function $f : [0, 1]^s \to \mathbb{R}$. According to the definition of the Riemann integral, for every $\varepsilon > 0$ there exist step functions g_1

and g_2 defined on $[0, 1]^s$ such that $g_1 \leq f \leq g_2$ and $\int_{[0,1]^s} (g_2(x) - g_1(x)) \, \mathrm{d}x < \varepsilon$. Hence we have

$$\int_{[0,1]^s} f(x) \, \mathrm{d}x - \varepsilon \leq \int_{[0,1]^s} g_1(x) \, \mathrm{d}x = \lim_{N \to \infty} \frac{1}{N} \sum_{n=0}^{N-1} g_1(x_n)$$

$$\leq \liminf_{N \to \infty} \frac{1}{N} \sum_{n=0}^{N-1} f(x_n) \leq \limsup_{N \to \infty} \frac{1}{N} \sum_{n=0}^{N-1} f(x_n)$$

$$\leq \lim_{N \to \infty} \frac{1}{N} \sum_{n=0}^{N-1} g_2(x_n) = \int_{[0,1]^s} g_2(x) \, \mathrm{d}x \leq \int_{[0,1]^s} f(x) \, \mathrm{d}x + \varepsilon.$$

Since ε can be chosen arbitrarily close to zero, the result follows. $\qquad\square$

In the same manner one can prove the following result:

Theorem 2.4 *A sequence $(x_n)_{n \in \mathbb{N}_0}$ in $[0, 1)^s$ is uniformly distributed modulo one, if and only if (2.3) holds for every continuous one-periodic and complex-valued function $f : [0, 1]^s \to \mathbb{C}$.*

Proof The necessity follows from applying Theorem 2.2 to the real and imaginary parts of f. The sufficiency of (2.3) for every continuous one-periodic and complex-valued function $f : [0, 1]^s \to \mathbb{C}$ follows from the elementary fact that the characteristic function of every interval $[a, b) \subseteq [0, 1)^s$ can be approximated from above and below by continuous one-periodic functions, i.e., for all $\varepsilon > 0$ there exist continuous one-periodic f_1, f_2 on $[0, 1]^s$ such that $f_1 \leq \chi_{[a,b)} \leq f_2$ and $0 \leq \int_{[0,1]^s} (f_2(x) - f_1(x)) \, \mathrm{d}x < \varepsilon$. We leave the details as exercise. $\qquad\square$

In the already mentioned paper entitled *Über die Gleichverteilung von Zahlen mod. Eins* H. Weyl proved the following criterion for uniform distribution modulo one, which can be considered as a simplification of Theorem 2.4 in the sense that it is sufficient to check condition (2.3) for trigonometric functions of the form $\exp(2\pi \mathrm{i} h \cdot x)$ for integer vectors $h \neq 0$, rather than for all continuous one-periodic and complex-valued functions.

Theorem 2.5 (Weyl's criterion) *A sequence $(x_n)_{n \in \mathbb{N}_0}$ in $[0, 1)^s$ is uniformly distributed modulo one, if and only if*

$$\lim_{N \to \infty} \frac{1}{N} \sum_{n=0}^{N-1} \exp(2\pi \mathrm{i} h \cdot x_n) = 0 \tag{2.4}$$

for all vectors $h \in \mathbb{Z}^s \setminus \{0\}$. Here $x \cdot y$ denotes the usual inner product of two elements $x, y \in \mathbb{R}^s$.

Proof If the sequence $(x_n)_{n \in \mathbb{N}_0}$ is uniformly distributed modulo one, then Theorem 2.4 with the special choice $f(x) = \exp(2\pi \mathrm{i} h \cdot x)$ implies (2.4).

Now let $f : [0, 1]^s \to \mathbb{C}$ be a continuous one-periodic and complex-valued function. According to the trigonometric version of Weierstrass' approximation theorem this function can be uniformly approximated as closely as desired by a finite linear combination of the functions $\exp(2\pi \mathrm{i} h \cdot x)$, $h \in \mathbb{Z}^s$, with complex coefficients. In other words, for any $\varepsilon > 0$ there exists a trigonometric polynomial of the form

$$P(x) = \sum_{\substack{h \in \mathbb{Z}^s \\ \|h\|_1 < R}} a_h \exp(2\pi \mathrm{i} h \cdot x)$$

with complex coefficients a_h such that

$$\|f - P\|_{L_\infty} < \frac{\varepsilon}{3},$$

where $R = R(\varepsilon)$. Here for $h = (h_1, \ldots, h_s)$ we put $\|h\|_1 = |h_1| + \cdots + |h_s|$, and for a function $g : [0, 1]^s \to \mathbb{C}$ we put $\|g\|_{L_\infty} := \sup_{x \in [0,1]^s} |g(x)|$. Then we have

$$\left| \int_{[0,1]^s} f(x)\,\mathrm{d}x - \frac{1}{N} \sum_{n=0}^{N-1} f(x_n) \right| \leq \left| \int_{[0,1]^s} f(x)\,\mathrm{d}x - \int_{[0,1]^s} P(x)\,\mathrm{d}x \right|$$

$$+ \left| \int_{[0,1]^s} P(x)\,\mathrm{d}x - \frac{1}{N} \sum_{n=0}^{N-1} P(x_n) \right|$$

$$+ \left| \frac{1}{N} \sum_{n=0}^{N-1} P(x_n) - \frac{1}{N} \sum_{n=0}^{N-1} f(x_n) \right|.$$

We have

$$\left| \int_{[0,1]^s} f(x)\,\mathrm{d}x - \int_{[0,1]^s} P(x)\,\mathrm{d}x \right| \leq \|f - P\|_{L_\infty} < \frac{\varepsilon}{3}$$

and

$$\left| \frac{1}{N} \sum_{n=0}^{N-1} P(x_n) - \frac{1}{N} \sum_{n=0}^{N-1} f(x_n) \right| \leq \|f - P\|_{L_\infty} < \frac{\varepsilon}{3}.$$

Furthermore, it follows from (2.4), that

$$\lim_{N \to \infty} \frac{1}{N} \sum_{n=0}^{N-1} P(x_n) = \sum_{\substack{h \in \mathbb{Z}^s \\ \|h\|_1 < R}} a_h \lim_{N \to \infty} \frac{1}{N} \sum_{n=0}^{N-1} \exp(2\pi \mathrm{i} h \cdot x_n) = a_0$$

and

$$\int_{[0,1]^s} P(x)\,\mathrm{d}x = \sum_{\substack{h \in \mathbb{Z}^s \\ \|h\|_1 < R}} a_h \int_{[0,1]^s} \exp(2\pi \mathrm{i} h \cdot x)\,\mathrm{d}x = a_0.$$

Hence, for large enough N, it follows that

$$\left| \int_{[0,1]^s} P(x)\,dx - \frac{1}{N} \sum_{n=0}^{N-1} P(x_n) \right| < \frac{\varepsilon}{3}.$$

Altogether, for large enough N, we obtain

$$\left| \int_{[0,1]^s} f(x)\,dx - \frac{1}{N} \sum_{n=0}^{N-1} f(x_n) \right| < \varepsilon$$

and the result follows. $\qquad\square$

Based on Weyl's criterion we can give a first example of a uniformly distributed sequence.

Proposition 2.6 *Let $\alpha = (\alpha_1, \ldots, \alpha_s) \in \mathbb{R}^s$. The sequence $(\{n\alpha\})_{n\in\mathbb{N}_0}$, where the fractional part $\{\cdot\}$ is applied component-wise, is uniformly distributed modulo one, if and only if the numbers $1, \alpha_1, \ldots, \alpha_s$ are linearly independent over the rationals. In particular, the one-dimensional sequence $(\{n\alpha\})_{n\in\mathbb{N}_0}$ is uniformly distributed modulo one, if and only if $\alpha \in \mathbb{R} \setminus \mathbb{Q}$.*

Proof Let $h \in \mathbb{Z}^s \setminus \{0\}$. Since the function $x \mapsto \exp(2\pi i h \cdot x)$ is one periodic we have

$$\frac{1}{N} \sum_{n=0}^{N-1} \exp(2\pi i h \cdot \{n\alpha\}) = \frac{1}{N} \sum_{n=0}^{N-1} \exp(2\pi i h \cdot \alpha)^n.$$

If the numbers $1, \alpha_1, \ldots, \alpha_s$ are linearly independent over the rationals, then $h \cdot \alpha \notin \mathbb{Q}$ and therefore $\exp(2\pi i h \cdot \alpha) \neq 1$. Using the formula for geometric sums we obtain

$$\left| \frac{1}{N} \sum_{n=0}^{N-1} \exp(2\pi i h \cdot \alpha)^n \right| = \frac{1}{N} \left| \frac{\exp(2\pi i N h \cdot \alpha) - 1}{\exp(2\pi i h \cdot \alpha) - 1} \right|$$

$$\leq \frac{1}{N} \frac{2}{|\exp(2\pi i h \cdot \alpha) - 1|}.$$

An application of Weyl's criterion shows that the sequence is uniformly distributed modulo one. The other direction is left as an exercise. $\qquad\square$

Remark 2.7 The sequences considered in Proposition 2.6 are sometimes called *Kronecker sequences* or *$n\alpha$-sequences*.

We provide another example of a uniformly distributed sequence. To this end we need the following definition:

Definition 2.8 Let $b \in \mathbb{N}$, $b \geq 2$.

- The *b-adic radical inverse function* is defined as $\phi_b : \mathbb{N}_0 \to [0, 1)$,

$$\phi_b(n) := \frac{n_0}{b} + \frac{n_1}{b^2} + \frac{n_2}{b^3} \cdots ,$$

  for $n \in \mathbb{N}_0$ with b-adic digit expansion $n = n_0 + n_1 b + n_2 b^2 + \cdots$, where $n_i \in \{0, 1, \ldots, b - 1\}$. In other words, $\phi_b(n)$ is the reflection of the b-adic digit expansion of n at the comma.
- The *van der Corput sequence in base b* is defined as $(x_n)_{n \in \mathbb{N}_0}$ with $x_n := \phi_b(n)$.

Example 2.9 For example, for $b = 2$, we have

n	0	1	2	3	4	5	6	7	8
$(n)_2$	0.	1.	10.	11.	100.	101.	110.	111.	1000.
$(\phi_2(n))_2$	.0	.1	.01	.11	.001	.101	.011	.111	.0001
$\phi_2(n)$	0	$\frac{1}{2}$	$\frac{1}{4}$	$\frac{3}{4}$	$\frac{1}{8}$	$\frac{5}{8}$	$\frac{3}{8}$	$\frac{7}{8}$	$\frac{1}{16}$

Proposition 2.10 *The van der Corput sequence in base b is uniformly distributed modulo one.*

Proof Fix $m \in \mathbb{N}$. For every $a \in \{0, 1, \ldots, b^m - 1\}$ with b-adic digit expansion $a = a_0 b^{m-1} + a_1 b^{m-2} + \cdots + a_{m-2} b + a_{m-1}$ we consider the so-called elementary interval in base b of the form $J_a = \left[\frac{a}{b^m}, \frac{a+1}{b^m} \right)$. For $n \in \mathbb{N}_0$ with b-adic digit expansion $n = n_0 + n_1 b + n_2 b^2 + \cdots$ the element $x_n = \phi_b(n)$ belongs to J_a, if and only if

$$\frac{a}{b^m} \leq \frac{n_0}{b} + \frac{n_1}{b^2} + \frac{n_2}{b^3} + \cdots < \frac{a+1}{b^m}.$$

Multiplying with b^m gives

$$a \leq \underbrace{n_0 b^{m-1} + n_1 b^{m-2} + \cdots + n_{m-1}}_{\in \mathbb{N}_0} + \underbrace{\frac{n_m}{b} + \cdots}_{\in [0,1)} < a + 1.$$

This is equivalent to $a = n_0 b^{m-1} + \cdots + n_{m-1}$ or, in terms of b-adic digits of a, to

$$n_0 = a_0, n_1 = a_1, \ldots, n_{m-1} = a_{m-1}.$$

This, in turn, is equivalent to

$$n \equiv a' \pmod{b^m} \quad \text{where} \quad a' = a_0 + a_1 b + \cdots + a_{m-1} b^{m-1}.$$

Since the congruence $x \equiv a' \pmod{b^m}$ has a unique solution modulo b^m it follows that exactly one of b^m consecutive elements of the van der Corput sequence belongs to J_a. Hence, for $N \in \mathbb{N}$, it follows that

$$A(J_a, N) = \left\lfloor \frac{N}{b^m} \right\rfloor + \theta$$

with $\theta \in \{0, 1\}$. Therefore and since $\lim_{N \to \infty} \frac{1}{N} \left\lfloor \frac{N}{b^m} \right\rfloor = \frac{1}{b^m}$ we obtain

$$\lim_{N \to \infty} \frac{A(J_a, N)}{N} = \frac{1}{b^m} = \lambda_1(J_a)$$

where λ_1 denotes the one-dimensional Lebesgue measure.

Now let $J = [\alpha, \beta) \subseteq [0, 1)$ and let

$$\underline{J} = \bigcup_{a:\, J_a \subseteq J} J_a \quad \text{and} \quad \overline{J} = \bigcup_{a:\, J_a \cap J \neq \emptyset} J_a.$$

Then we have $\underline{J} \subseteq J \subseteq \overline{J}$ and $\lambda_1(\overline{J}) - \lambda_1(\underline{J}) \leq 2/b^m$ and

$$\frac{A(\underline{J}, N)}{N} - \lambda_1(\overline{J}) \leq \frac{A(J, N)}{N} - \lambda_1(J) \leq \frac{A(\overline{J}, N)}{N} - \lambda_1(\underline{J}). \qquad (2.5)$$

Note that for arbitrary $x, y, z \in \mathbb{R}$ and $\delta > 0$ we have

$$y - \delta \leq x \leq z + \delta \quad \Rightarrow \quad |x| \leq \delta + \max(|y|, |z|). \qquad (2.6)$$

Therefore it follows from (2.5) that

$$\left| \frac{A(J, N)}{N} - \lambda_1(J) \right|$$

$$\leq \lambda_1(\overline{J}) - \lambda_1(\underline{J}) + \max \left(\left| \frac{A(\underline{J}, N)}{N} - \lambda_1(\underline{J}) \right|, \left| \frac{A(\overline{J}, N)}{N} - \lambda_1(\overline{J}) \right| \right)$$

$$\leq \frac{2}{b^m} + \max \left(\left| \frac{A(\underline{J}, N)}{N} - \lambda_1(\underline{J}) \right|, \left| \frac{A(\overline{J}, N)}{N} - \lambda_1(\overline{J}) \right| \right).$$

Since

$$\lim_{N \to \infty} \left| \frac{A(\underline{J}, N)}{N} - \lambda_1(\underline{J}) \right| \leq \sum_{a:\, J_a \subseteq J} \lim_{N \to \infty} \left| \frac{A(J_a, N)}{N} - \lambda_1(J_a) \right| = 0$$

and, in the same manner, $\lim_{N \to \infty} \left| \frac{A(\overline{J}, N)}{N} - \lambda_1(\overline{J}) \right| = 0$, it follows that

$$\lim_{N \to \infty} \left| \frac{A(J, N)}{N} - \lambda_1(J) \right| \leq \frac{2}{b^m}.$$

Since $m \in \mathbb{N}$ was arbitrary we obtain

$$\lim_{N \to \infty} \left| \frac{A(J, N)}{N} - \lambda_1(J) \right| = 0.$$

This means that the van der Corput sequence in base b is uniformly distributed modulo one. $\qquad\square$

Remark 2.11 It is not difficult to show that every uniformly distributed sequence in $[0, 1)^s$ is dense in $[0, 1]^s$. The converse is not true. A proof of these facts is left as an exercise. In particular, a Kronecker sequence $(\{n\alpha\})_{n\in\mathbb{N}_0}$ is dense in $[0, 1]^s$ if the numbers $1, \alpha_1, \ldots, \alpha_s$ are linearly independent over the rationals. This is a well-known result from Number Theory known as *Kronecker's approximation theorem*.

2.2 Discrepancy

In this section we introduce quantitative measures for the assessment of the quality of uniform distribution. For $a, b \in [0, 1]^s$ with $a = (a_1, \ldots, a_s)$ and $b = (b_1, \ldots, b_s)$ the inequality $a \le b$ means that $a_j \le b_j$ for all $j = 1, \ldots, s$.

Definition 2.12 Let $\mathcal{P}$ be an N-element point set in $[0, 1)^s$. The *extreme discrepancy* D_N of this point set is defined as

$$D_N(\mathcal{P}) := \sup_{\substack{a, b \in [0, 1]^s \\ a \le b}} \left| \frac{A([a, b), \mathcal{P}, N)}{N} - \lambda_s([a, b)) \right|.$$

For an infinite sequence $\mathcal{S}$ the discrepancy $D_N(\mathcal{S})$ is the discrepancy of the first N elements of $\mathcal{S}$.

Often it is enough to consider a slightly simplified notion of discrepancy:

Definition 2.13 Let $\mathcal{P}$ be an N-element point set in $[0, 1)^s$. The *star discrepancy* D_N^* of this point set is defined as

$$D_N^*(\mathcal{P}) := \sup_{a \in [0, 1]^s} \left| \frac{A([\mathbf{0}, a), \mathcal{P}, N)}{N} - \lambda_s([\mathbf{0}, a)) \right|.$$

For an infinite sequence $\mathcal{S}$ the star discrepancy $D_N^*(\mathcal{S})$ is the star discrepancy of the first N elements of $\mathcal{S}$.

We have the following relation between extreme- and star discrepancy:

Proposition 2.14 *For every N-element point set $\mathcal{P}$ in $[0, 1)^s$ we have*

$$D_N^*(\mathcal{P}) \le D_N(\mathcal{P}) \le 2^s D_N^*(\mathcal{P}).$$

Proof The left inequality is clear from the definitions of extreme- and star discrepancy. We prove the right-hand inequality for dimension $s = 1$ and $s = 2$ only. The

Fig. 2.2 Inclusion-exclusion
principle

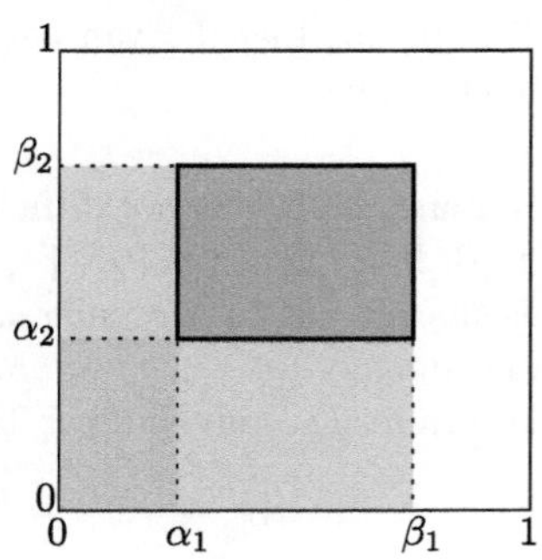

case $s \geq 3$ can be shown in the same manner. Let first $s = 1$ and $[\alpha, \beta) \subseteq [0, 1)$.
Then we have $[\alpha, \beta) = [0, \beta) \setminus [0, \alpha)$,

$$A([\alpha, \beta), N) = A([0, \beta), N) - A([0, \alpha), N)$$

and

$$\lambda_1([\alpha, \beta)) = \lambda_1([0, \beta)) - \lambda_1([0, \alpha)).$$

Hence we have

$$\frac{A([\alpha, \beta), N)}{N} - \lambda_1([\alpha, \beta)) = \frac{A([0, \beta), N)}{N} - \lambda_1([0, \beta)) - \left(\frac{A([0, \alpha), N)}{N} - \lambda_1([0, \alpha)) \right).$$

Taking the absolute value and applying the triangle inequality leads to

$$\left| \frac{A([\alpha, \beta), N)}{N} - \lambda_1([\alpha, \beta)) \right| \leq 2\, D_N^*(\mathcal{P})$$

and therefore we get

$$D_N(\mathcal{P}) \leq 2\, D_N^*(\mathcal{P}),$$

as desired.

In the case $s = 2$ for $J = [\alpha_1, \beta_1) \times [\alpha_2, \beta_2)$ we have the following inclusion-exclusion principle (see also Fig. 2.2)

$$J = \left([0, \beta_1) \times [0, \beta_2) \setminus [0, \alpha_1) \times [0, \beta_2) \right) \setminus \left([0, \beta_1) \times [0, \alpha_2) \setminus [0, \alpha_1) \times [0, \alpha_2) \right)$$

and hence

$$\begin{aligned}
A(J, N) =\, & A([0, \beta_1) \times [0, \beta_2), N) \\
& - A([0, \alpha_1) \times [0, \beta_2), N) - A([0, \beta_1) \times [0, \alpha_2), N) \\
& + A([0, \alpha_1) \times [0, \alpha_2), N).
\end{aligned}$$

A similar equality is true for λ_2. Now the result follows as for the case $s = 1$. $\square$

Based on the notions of discrepancy we can give the following criterion for uniform distribution modulo one:

Theorem 2.15 *Let $\mathcal{S}$ be a sequence in $[0, 1)^s$. The following assertions are equivalent:*

(a) $\mathcal{S}$ is uniformly distributed modulo one;
(b) $\lim_{N\to\infty} D_N^*(\mathcal{S}) = 0$;
(c) $\lim_{N\to\infty} D_N(\mathcal{S}) = 0$.

For the proof of this result we need the following lemma:

Lemma 2.16 *For $j = 1, \ldots, s$ let $u_j, v_j \in [0, 1]$ with $|u_j - v_j| \leq \delta$. Then we have*

$$\left| \prod_{j=1}^{s} u_j - \prod_{j=1}^{s} v_j \right| \leq 1 - (1 - \delta)^s \leq s\,\delta.$$

Proof We prove the result using induction on s. Trivially, the assertion holds true for $s = 1$. Assume that the result holds true for $s \in \mathbb{N}$. Without loss of generality we may assume that $u_{s+1} \geq v_{s+1}$. Then we have

$$\left| \prod_{j=1}^{s+1} u_j - \prod_{j=1}^{s+1} v_j \right| = \left| (u_{s+1} - v_{s+1}) \prod_{j=1}^{s} u_j + v_{s+1}\left(\prod_{j=1}^{s} u_j - \prod_{j=1}^{s} v_j \right) \right|$$

$$\leq |u_{s+1} - v_{s+1}| \cdot 1 + v_{s+1}(1 - (1 - \delta)^s)$$

$$= u_{s+1} - v_{s+1}(1 - \delta)^s$$

$$= u_{s+1}(1 - (1 - \delta)^s) + (u_{s+1} - v_{s+1})(1 - \delta)^s$$

$$\leq 1 - (1 - \delta)^s + \delta(1 - \delta)^s$$

$$= 1 - (1 - \delta)^{s+1}.$$

Hence the left-hand inequality is shown.

According to the Mean Value Theorem we have for all $y, z \in \mathbb{R}$ with $z \leq y$ that $y^s - z^s = s\xi^{s-1}(y - z)$ for some $\xi \in (z, y)$. Choosing $y = 1$ and $z = 1 - \delta$ yields the right-hand inequality of the lemma. $\qquad\square$

Proof of Theorem 2.15. The equivalence of *(b)* and *(c)* follows from Proposition 2.14.

(c) $\Rightarrow$ *(a)*: Assume that $\lim_{N\to\infty} D_N(\mathcal{S}) = 0$, i.e.,

$$\lim_{N\to\infty} \sup_{\substack{a,b\in[0,1]^s \\ a\leq b}} \left| \frac{A([a,b),\mathcal{S},N)}{N} - \lambda_s([a,b)) \right| = 0.$$

Then obviously

$$\lim_{N\to\infty} \frac{A([a,b),\mathcal{S},N)}{N} = \lambda_s([a,b))$$

for all intervals of the form $[a,b) \subseteq [0,1)^s$, and so the sequence $\mathcal{S}$ is uniformly distributed modulo one.

(a) $\Rightarrow$ *(b)*: Assume that the sequence $\mathcal{S}$ is uniformly distributed modulo one. Let J be a sub-interval of $[0,1)^s$ of the form $[0,a)$ and choose $m \in \mathbb{N}$, $m \geq 2$. Now we proceed in a similar way to what we did in the proof of Proposition 2.10. For $k = (k_1,\ldots,k_s) \in \{0,1,\ldots,m-1\}^s$ put $Q_k = \prod_{j=1}^{s} \left[\frac{k_j}{m}, \frac{k_j+1}{m} \right)$ and define

$$\underline{J} = \bigcup_{k:\, Q_k \subseteq J} Q_k \qquad \text{and} \qquad \overline{J} = \bigcup_{k:\, Q_k \cap J \neq \emptyset} Q_k.$$

Obviously, $\underline{J}$ and $\overline{J}$ are intervals anchored at the origin ($\underline{J}$ may be empty) which satisfy $\underline{J} \subseteq J \subseteq \overline{J}$ and corresponding edge-lengths of $\underline{J}$ and $\overline{J}$ differ at most $1/m$. Hence we obtain from Lemma 2.16 that

$$\lambda_s(\overline{J}) - \lambda_s(\underline{J}) \leq \frac{s}{m}.$$

Now we have

$$\frac{A(\underline{J},\mathcal{S},N)}{N} - \lambda_s(\overline{J}) \leq \frac{A(J,\mathcal{S},N)}{N} - \lambda_s(J) \leq \frac{A(\overline{J},\mathcal{S},N)}{N} - \lambda_s(\underline{J}),$$

whence

$$\frac{A(J,\mathcal{S},N)}{N} - \lambda_s(J) \begin{cases} \geq \frac{A(\underline{J},\mathcal{S},N)}{N} - \lambda_s(\underline{J}) - (\lambda_s(\overline{J}) - \lambda_s(\underline{J})), \\ \leq \frac{A(\overline{J},\mathcal{S},N)}{N} - \lambda_s(\overline{J}) + (\lambda_s(\overline{J}) - \lambda_s(\underline{J})). \end{cases}$$

Using (2.6) we obtain

$$\left| \frac{A(J,\mathcal{S},N)}{N} - \lambda_s(J) \right| \leq \lambda_s(\overline{J}) - \lambda_s(\underline{J})$$

$$+ \max \left\{ \left| \frac{A(\underline{J},\mathcal{S},N)}{N} - \lambda_s(\underline{J}) \right|, \left| \frac{A(\overline{J},\mathcal{S},N)}{N} - \lambda_s(\overline{J}) \right| \right\}$$

$$\leq \frac{s}{m} + \max_{k\in\{0,1,\ldots,m\}^s} \left| \frac{A([0,\frac{1}{m}k),\mathcal{S},N)}{N} - \lambda_s\left(\left[0,\frac{1}{m}k\right) \right) \right|.$$

Note that the final term does not depend on J.

Since the sequence $\mathcal{S}$ is uniformly distributed modulo one we have, for every $\boldsymbol{k} \in \{0, 1, \ldots, m\}^s$,

$$\lim_{N \to \infty} \left| \frac{A([\boldsymbol{0}, \frac{1}{m}\boldsymbol{k}), \mathcal{S}, N)}{N} - \lambda_s \left(\left[\boldsymbol{0}, \frac{1}{m}\boldsymbol{k} \right) \right) \right| = 0,$$

and thus also

$$\lim_{N \to \infty} \max_{\boldsymbol{k} \in \{0,1,\ldots,m\}^s} \left| \frac{A([\boldsymbol{0}, \frac{1}{m}\boldsymbol{k}), \mathcal{S}, N)}{N} - \lambda_s \left(\left[\boldsymbol{0}, \frac{1}{m}\boldsymbol{k} \right) \right) \right| = 0.$$

Consequently,

$$0 \le \limsup_{N \to \infty} D_N^*(\mathcal{S}) \le \frac{s}{m},$$

and since m can be chosen arbitrarily large, the result follows. $\qquad\square$

Remark 2.17 We learn from the proof of Theorem 2.15 that for the star discrepancy of every N-element point set in $[0, 1)^s$ and for every $m \in \mathbb{N}$, $m \ge 2$, we have

$$D_N^*(\mathcal{P}) \le \frac{s}{m} + \max_{\boldsymbol{k} \in \{0,1,\ldots,m\}^s} \left| \frac{A([\boldsymbol{0}, \frac{1}{m}\boldsymbol{k}), \mathcal{P}, N)}{N} - \lambda_s \left(\left[\boldsymbol{0}, \frac{1}{m}\boldsymbol{k} \right) \right) \right|.$$

This means that the supremum in the definition of star discrepancy can be replaced by a maximum over a finite set consisting of $(m + 1)^s$ elements with an error of at most s/m.

Theorem 2.15 shows that a sequence is uniformly distributed modulo one, if and only if its (star) discrepancy converges to zero as N approaches infinity. We now show that the order of convergence to zero cannot be faster than $1/N$.

Proposition 2.18 *For every N-element point set $\mathcal{P}$ in $[0, 1)^s$ we have*

$$D_N(\mathcal{P}) \ge \frac{1}{N} \qquad and \qquad D_N^*(\mathcal{P}) \ge \frac{1}{2^s N}.$$

Proof Choose $0 < \varepsilon \le 1/N$ and let $\boldsymbol{x} = (x_1, \ldots, x_s)$ be an element of the point set $\mathcal{P}$. Put

$$J = \left([x_1, x_1 + \varepsilon^{1/s}) \times \ldots \times [x_s, x_s + \varepsilon^{1/s}) \right) \cap [0, 1]^s.$$

Since $\boldsymbol{x} \in J$, it follows that

$$D_N(\mathcal{P}) \ge \frac{A(J, N)}{N} - \lambda_s(J) \ge \frac{1}{N} - \varepsilon.$$

Since ε can be chosen arbitrarily close to zero, it follows that $D_N(\mathcal{P}) \ge \frac{1}{N}$. The result for the star discrepancy can be obtained from Proposition 2.14. $\qquad\square$

If one thinks of a point set whose elements are evenly distributed in the unit cube on first sight one might have a regular lattice in mind. By a *regular lattice* in $[0, 1)^s$ consisting of m^s elements, where $m \in \mathbb{N}$, $m \geq 2$, we mean the point set

$$\Gamma_{m,s} := \left\{ \left(\frac{2n_1 + 1}{2m}, \ldots, \frac{2n_s + 1}{2m} \right) : n_1, \ldots, n_s \in \{0, \ldots, m - 1\} \right\}. \qquad (2.7)$$

Sometimes one also speaks of a *centered* regular lattice.

However, for dimension $s \geq 2$ it turns out that this is not a good choice.

Theorem 2.19 *Let $s, m \in \mathbb{N}$, $m \geq 2$. For the star discrepancy of the regular lattice $\Gamma_{m,s}$ with $N = m^s$ elements in $[0, 1)^s$ it holds that*

$$D_N^*(\Gamma_{m,s}) = 1 - \left(1 - \frac{1}{2m} \right)^s.$$

Proof Since $\Gamma_{m,s} \subseteq [0, 1 - 1/(2m)]^s =: B_{m,s}$, we find that

$$D_N^*(\Gamma_{m,s}) \geq \left| \frac{A(B_{m,s}, \Gamma_{m,s}, N)}{N} - \lambda_s(B_{m,s}) \right| = 1 - \left(1 - \frac{1}{2m} \right)^s.$$

To prove an upper bound, let $J = [0, \alpha_1) \times \ldots \times [0, \alpha_s) \subseteq [0, 1)^s$. If $\min_{j=1,\ldots,s} \alpha_j < \frac{1}{2m}$, then

$$A(J, \Gamma_{m,s}, N) = 0 \quad \text{and} \quad \lambda_s(J) \leq \frac{1}{2m} \leq 1 - \left(1 - \frac{1}{2m} \right)^s.$$

Otherwise, let $a_j \in \{0, \ldots, m - 1\}$ for $j = 1, \ldots, s$ be such that $\frac{2a_j+1}{2m} < \alpha_j \leq \frac{2a_j+3}{2m}$. Then we have $A(J, \Gamma_{m,s}, N) = \prod_{j=1}^{s} (a_j + 1)$ and

$$0 \leq \frac{A(J, \Gamma_{m,s}, N)}{N} - \lambda_s(J) \leq \prod_{j=1}^{s} \frac{a_j + 1}{m} - \prod_{j=1}^{s} \frac{2a_j + 1}{2m}.$$

Therefore and by Lemma 2.16 we obtain

$$\left| \frac{A(J, \Gamma_{m,s}, N)}{N} - \lambda_s(J) \right| \leq \left| \prod_{j=1}^{s} \frac{a_j + 1}{m} - \prod_{j=1}^{s} \frac{2a_j + 1}{2m} \right| \leq 1 - \left(1 - \frac{1}{2m} \right)^s.$$

Since this upper bound is independent of the specific choice of the interval J, we obtain that

$$D_N^*(\Gamma_{m,s}) \leq 1 - \left(1 - \frac{1}{2m} \right)^s.$$

Hence the result follows. $\qquad\qquad\qquad\qquad\qquad\qquad\qquad\qquad\qquad\qquad\qquad\qquad\qquad\square$

Remark 2.20 For $s = 1$ we obtain $D_N^*(\Gamma_{m,1}) = 1/(2N)$ which is best possible, as we can see from Proposition 2.18. Hence in dimension $s = 1$ we have found, for any $N \in \mathbb{N}$, the N-element point set with lowest star discrepancy. Unfortunately, this is not true for dimensions $s \geq 2$. We have

$$\frac{1}{2\,N^{1/s}} = \frac{1}{2\,m} \leq 1 - \left(1 - \frac{1}{2\,m}\right)^s \leq \frac{s}{2\,m} = \frac{s}{2\,N^{1/s}}$$

and so

$$D_N^*(\Gamma_{m,s}) \asymp_s \frac{1}{N^{1/s}}.$$

For large dimensions s this convergence rate is rather weak, as we shall see in Sect. 2.4 where we will encounter point sets with much lower discrepancy.

The star discrepancy of a point set $\mathcal{P}$ may also be defined as the supremum norm of the function $\Delta_{\mathcal{P},N} : [0, 1]^s \to \mathbb{R}$,

$$\Delta_{\mathcal{P},N}(\mathbf{y}) := \frac{A([\mathbf{0}, \mathbf{y}), \mathcal{P}, N)}{N} - \lambda_s([\mathbf{0}, \mathbf{y})),$$

i.e.,

$$D_N^*(\mathcal{P}) = \|\Delta_{\mathcal{P},N}\|_{L_\infty}.$$

The function $\Delta_{\mathcal{P},N}$ is sometimes called *local discrepancy* or *discrepancy function* of the point set $\mathcal{P}$. From this point of view it is near at hand to also consider other norms of the local discrepancy, e.g., the L_p-norm for $p \geq 1$.

Definition 2.21 Let $\mathcal{P}$ be an N-element point set in $[0, 1)^s$ and let $p \in [1, \infty)$. The L_p *discrepancy* (sometimes also referred to as L_p star discrepancy) of $\mathcal{P}$ is defined as

$$L_{p,N}(\mathcal{P}) := \|\Delta_{\mathcal{P},N}\|_{L_p} = \left(\int_{[0,1]^s} \left|\frac{A([\mathbf{0}, \mathbf{y}), \mathcal{P}, N)}{N} - \lambda_s([\mathbf{0}, \mathbf{y}))\right|^p \, d\mathbf{y}\right)^{1/p}.$$

For an infinite sequence $\mathcal{S}$ the L_p discrepancy $L_{p,N}(\mathcal{S})$ is the L_p discrepancy of the first N elements of $\mathcal{S}$.

From the monotonicity of the L_p-norm it follows immediately that $L_{p_1,N}(\mathcal{P}) \leq L_{p_2,N}(\mathcal{P})$ whenever $p_1 \leq p_2$. Also there is a relation between the star- and the L_p discrepancy which we state here without proof.

Proposition 2.22 *For every N-element point set $\mathcal{P}$ in $[0, 1)^s$ we have*

$$L_{p,N}(\mathcal{P}) \leq D_N^*(\mathcal{P}) \leq c(s, p) L_{p,N}^{\frac{p}{p+s}}(\mathcal{P}),$$

where the positive quantity $c(s, p)$ only depends on s and p, but not on N.

As a consequence we obtain a further criterion for uniform distribution modulo one of infinite sequences.

Corollary 2.23 *Let $p \in [1, \infty)$. A sequence $\mathcal{S}$ in $[0, 1)^s$ is uniformly distributed modulo one, if and only if $\lim_{N \to \infty} L_{p,N}(\mathcal{S}) = 0$.*

2.3 Bounds on the Discrepancy

The lower bound on the star discrepancy from Proposition 2.18 is rather weak, at least for large dimensions. Only for dimension one it is best possible. Now we present a much stronger result which was first shown by K. F. Roth in the year 1954.

Theorem 2.24 (Roth) *For every dimension $s \in \mathbb{N}$ there exists a quantity $c_s > 0$ with the following property: for every N-element point set $\mathcal{P}$ in $[0, 1)^s$ we have*

$$D_N(\mathcal{P}) \geq D_N^*(\mathcal{P}) \geq L_{2,N}(\mathcal{P}) \geq c_s \frac{(\log N)^{\frac{s-1}{2}}}{N}.$$

For the proof of these results we use the method of Roth but we follow the exposition in [105]. For the sake of simplicity, we restrict ourselves to the case of dimension $s = 2$. The general result can be shown in the same manner but with more technical effort. The proof is technical and can be skipped by beginners.

Proof of Theorem 2.24 For $x \in [0, 1]^2$ let

$$D(x) := N\lambda_2([\mathbf{0}, x)) - A([\mathbf{0}, x), \mathcal{P}, N).$$

By the Cauchy-Schwarz inequality, for any function $F : [0, 1]^2 \to \mathbb{R}$

$$\int_{[0,1]^2} F(x) D(x)\, \mathrm{d}x \leq \left(\int_{[0,1]^2} F^2(x)\, \mathrm{d}x \right)^{1/2} \left(\int_{[0,1]^2} D^2(x)\, \mathrm{d}x \right)^{1/2},$$

and hence, provided that $\int_{[0,1]^2} F^2(x)\, \mathrm{d}x > 0$,

$$N L_{2,N}(\mathcal{P}) = \left(\int_{[0,1]^2} D^2(x)\, \mathrm{d}x \right)^{1/2} \geq \frac{\int_{[0,1]^2} F(x) D(x)\, \mathrm{d}x}{\left(\int_{[0,1]^2} F^2(x)\, \mathrm{d}x \right)^{1/2}}.$$

Now it is our aim to chose the function F such that $\int_{[0,1]^2} F^2(x)\, \mathrm{d}x = O(\log N)$, but $\int_{[0,1]^2} F(x) D(x)\, \mathrm{d}x$ is at least of the order of magnitude $\log N$.

The function F will depend on the given point set $\mathcal{P}$. For $x \in \mathbb{R}$ define

$$\psi(x) := \begin{cases} 1 & \text{if } x \in [k, k+1/2) \text{ for some } k \in \mathbb{Z}, \\ -1 & \text{if } x \in [k+1/2, k+1) \text{ for some } k \in \mathbb{Z}. \end{cases}$$

Note that for $i \in \mathbb{N}_0$ and $a \in \{0, \ldots, 2^i - 1\}$ we have

$$\int_{a/2^i}^{(a+1)/2^i} \psi(2^i y)\,\mathrm{d}y = \frac{1}{2^i} \int_a^{a+1} \psi(y)\,\mathrm{d}y = 0.$$

Choose $m \in \mathbb{N}$ such that $2N \le 2^m < 4N$. For $j \in \{0, 1, \ldots, m\}$ define functions $f_j : [0, 1]^2 \to \{-1, 0, 1\}$ in the following way: let

$$R_{a,b}^{(j)} = \left[\frac{a}{2^{m-j}}, \frac{a+1}{2^{m-j}}\right) \times \left[\frac{b}{2^j}, \frac{b+1}{2^j}\right),$$

where $a \in \{0, 1, \ldots, 2^{m-j} - 1\}$ and $b \in \{0, 1, \ldots, 2^j - 1\}$. Then, for $\boldsymbol{y} = (y_1, y_2) \in R_{a,b}^{(j)}$, we set

$$f_j(\boldsymbol{y}) := \begin{cases} 0 & \text{if } R_{a,b}^{(j)} \cap \mathcal{P} \neq \emptyset, \\ \psi(2^{m-j} y_1)\psi(2^j y_2) & \text{if } R_{a,b}^{(j)} \cap \mathcal{P} = \emptyset. \end{cases} \tag{2.8}$$

Now we show that the functions f_j are mutually orthogonal. For $i < j$ there exists some $b \in \{0, \ldots, 2^i - 1\}$ such that either

$$\left[\frac{a}{2^j}, \frac{a+1}{2^j}\right) \subseteq \left[\frac{b}{2^i}, \frac{b}{2^i} + \frac{1}{2^{i+1}}\right), \quad \text{or} \quad \left[\frac{a}{2^j}, \frac{a+1}{2^j}\right) \subseteq \left[\frac{b}{2^i} + \frac{1}{2^{i+1}}, \frac{b+1}{2^i}\right).$$

Hence $\psi(2^i y) = c \in \{-1, 0, 1\}$ is constant on the intervals $[a/2^j, (a+1)/2^j)$ and therefore

$$\int_{a/2^j}^{(a+1)/2^j} \psi(2^j y)\psi(2^i y)\,\mathrm{d}y = c \int_{a/2^j}^{(a+1)/2^j} \psi(2^j y)\,\mathrm{d}y = 0.$$

For $a \in \{0, \ldots, 2^{m-i} - 1\}$ and $b \in \{0, \ldots, 2^j - 1\}$ let

$$R_{a,b}^{(i,j)} = \left[\frac{a}{2^{m-i}}, \frac{a+1}{2^{m-i}}\right) \times \left[\frac{b}{2^j}, \frac{b+1}{2^j}\right).$$

Note that $i < j$ implies that $R_{a,b}^{(i,j)} \subseteq R_{a,\lfloor b2^{i-j}\rfloor}^{(i)}$. Hence if $R_{a,b}^{(i,j)} \cap \mathcal{P} \neq \emptyset$, then also $R_{a,\lfloor b2^{i-j}\rfloor}^{(i)} \cap \mathcal{P} \neq \emptyset$. This implies that $f_i(\boldsymbol{y}) = 0$ for all $\boldsymbol{y} \in R_{a,\lfloor b2^{i-j}\rfloor}^{(i)}$ which in turn yields $f_i(\boldsymbol{y}) = 0$ for all $\boldsymbol{y} \in R_{a,b}^{(i,j)}$. Consequently,

$$\int_{[0,1]^2} f_i(\boldsymbol{y})f_j(\boldsymbol{y})\,\mathrm{d}\boldsymbol{y} = \sum_{a=0}^{2^{m-i}-1}\sum_{b=0}^{2^j-1} \int_{R_{a,b}^{(i,j)}} f_i(\boldsymbol{y})f_j(\boldsymbol{y})\,\mathrm{d}\boldsymbol{y}$$

$$= \underbrace{\sum_{a=0}^{2^{m-i}-1}\sum_{b=0}^{2^j-1} \int_{R_{a,b}^{(i,j)}} \psi(2^{m-i}y_1)\psi(2^i y_2)\psi(2^{m-j}y_1)\psi(2^j y_2)\,\mathrm{d}y_1\,\mathrm{d}y_2}_{R_{a,b}^{(i,j)} \cap \mathcal{P} = \emptyset}$$

$$= \underbrace{\sum_{a=0}^{2^{m-i}-1} \sum_{b=0}^{2^j-1} \int_{a/2^{m-i}}^{(a+1)/2^{m-i}} \psi(2^{m-i} y_1)\psi(2^{m-j} y_1)\,\mathrm{d}y_1}_{R_{a,b}^{(i,j)} \cap \mathcal{P} = \emptyset}$$

$$\times \int_{b/2^j}^{(b+1)/2^j} \psi(2^i y_2)\psi(2^j y_2)\,\mathrm{d}y_2$$

$$= 0,$$

i.e., the f_j are mutually orthogonal.

Now we put

$$F(x) := f_0(x) + f_1(x) + \cdots + f_m(x). \tag{2.9}$$

Using the orthogonality of the f_j and the fact that $2^m < 4N$ it follows that

$$\int_{[0,1]^2} F^2(x)\,\mathrm{d}x = \sum_{i,j=0}^{m} \int_{[0,1]^2} f_i(x)f_j(x)\,\mathrm{d}x = \sum_{i=0}^{m} \int_{[0,1]^2} f_i^2(x)\,\mathrm{d}x$$

$$\leq \sum_{i=0}^{m} 1 = m + 1 < \log_2(4N) + 1 = \log_2 N + 3.$$

Next we show that for all $j \in \{0, 1, \ldots, m\}$ we have

$$\int_{[0,1]^2} f_j(x)D(x)\,\mathrm{d}x \geq \frac{1}{2^8}. \tag{2.10}$$

To this end we consider $f_j D$ on the rectangles $R_{a,b}^{(j)}$ from the definition of the function f_j. We only have to consider empty $R_{a,b}^{(j)}$, since otherwise $f_j(x) = 0$ for every $x \in R_{a,b}^{(j)}$. Here and in the following we speak of an *empty* rectangle $R_{a,b}^{(j)}$ whenever $R_{a,b}^{(j)} \cap \mathcal{P} = \emptyset$.

In total, there are $2^m \geq 2N$ rectangles $R_{a,b}^{(j)}$, of which at most N contain a point from $\mathcal{P}$. Hence, there are at least N empty rectangles $R_{a,b}^{(j)}$. Thus it suffices to show that for every empty rectangle $R_{a,b}^{(j)}$ we have

$$\int_{R_{a,b}^{(j)}} f_j(x)D(x)\,\mathrm{d}x \geq \frac{1}{2^8 N}. \tag{2.11}$$

Let $R_{a,b}^{(j)}$ be an empty rectangle and let R_{ll} the lower left quadrant of this rectangle and accordingly R_{lr}, R_{ul}, R_{ur} (see Fig. 2.3). Let a, b be given as in Fig. 2.3. Then

Fig. 2.3 An empty rectangle $R_{a,b}^{(j)}$ with quadrants $R_{ll}, R_{lr}, R_{ul}, R_{ur}$

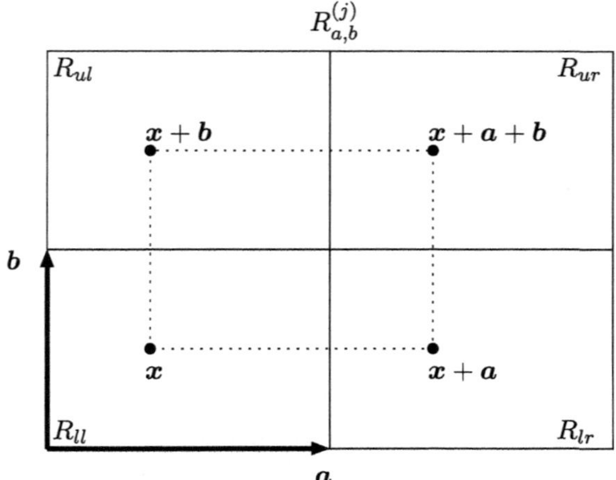

$$\int_{R_{a,b}^{(j)}} f_j(x) D(x)\, dx$$

$$= \int_{R_{ll}} D(x)\, dx - \int_{R_{lr}} D(x)\, dx - \int_{R_{ul}} D(x)\, dx + \int_{R_{ur}} D(x)\, dx$$

$$= \int_{R_{ll}} [D(x) - D(x + a) - D(x + b) + D(x + a + b)]\, dx.$$

Since

$$\lambda_2(R_{ll}) = \lambda_2([x, x + a + b))$$
$$= \lambda_2([0, x)) - \lambda_2([0, x + a)) - \lambda_2([0, x + b)) + \lambda_2([0, x + a + b))$$

and since

$$0 = A([x, x + a + b), \mathcal{P}, N)$$
$$= A([0, x), \mathcal{P}, N) - A([0, x + a), \mathcal{P}, N)$$
$$\quad - A([0, x + b), \mathcal{P}, N) + A([0, x + a + b), \mathcal{P}, N)$$

it follows that

$$D(x) - D(x + a) - D(x + b) + D(x + a + b) = N \lambda_2(R_{ll}).$$

Hence

$$\int_{R_{a,b}^{(j)}} f_j(x) D(x)\, dx = \int_{R_{ll}} N \lambda_2(R_{ll})\, dx = N \lambda_2(R_{ll})^2$$

$$= \frac{N}{2^{2m+4}} > \frac{1}{2^{m+6}} > \frac{1}{2^8\, N}.$$

Hence (2.11) and (2.10) are shown.

From (2.10) we obtain

$$\int_{[0,1]^2} F(x)D(x)\,\mathrm{d}x \geq \frac{m+1}{2^8} \geq \frac{\log_2 N}{2^8}.$$

Altogether it follows that

$$NL_{2,N}(\mathcal{P}) \geq \frac{\log_2 N}{2^8\sqrt{\log_2 N + 3}} \geq \frac{\sqrt{\log_2 N}}{2^9}$$

as claimed. $\qquad\square$

Roth's lower bound also holds for the L_p discrepancy for every $p \in (1,\infty)$ in the sense that there exists a quantity $c_{s,p} > 0$, which depends only on s and p, such that for every N-element point set $\mathcal{P} \in [0,1)^s$ we have

$$L_{p,N}(\mathcal{P}) \geq c_{s,p}\frac{(\log N)^{\frac{s-1}{2}}}{N}.$$

This follows immediately from Theorem 2.24 for every $p \in [2,\infty)$. The general result has been shown by Schmidt in 1977.

It is known that Roth's and Schmidt's lower bound is best possible for the L_p discrepancy in the sense that for every s, $N \in \mathbb{N}$, $N \geq 2$, and every $p \in (1,\infty)$ there exists an N-element point set $\mathcal{P}$ in $[0,1)^s$ with

$$L_{p,N}(\mathcal{P}) \leq \widetilde{c}_{s,p}\frac{(\log N)^{\frac{s-1}{2}}}{N},$$

where the positive $\widetilde{c}_{s,p}$ depends on the dimension s and on p, but not on N. For $p = 2$ a first explicit construction of such point sets in arbitrary dimension s has been provided by W.W.L. Chen and M.M. Skriganov in the year 2002. An alternative construction has been presented by J. Dick and F. Pillichshammer in 2014. For general $p \in (1,\infty)$ there is a construction according to M.M. Skriganov in 2006.

On the other hand, Roth's lower bound is not best possible for the star discrepancy. In 2008, D. Bilyk, M.T. Lacey, and A. Vagharshakyan proved in a joint work that, for every dimension $s \in \mathbb{N}$, $s \geq 2$, there exist quantities $c_s > 0$ and $\eta_s \in (0,\frac{1}{2})$ with the property that for every N-element point set $\mathcal{P}$ in $[0,1)^s$ we have

$$D_N^*(\mathcal{P}) \geq c_s\frac{(\log N)^{(s-1)/2+\eta_s}}{N}.$$

This result is the currently best lower estimate for the star discrepancy of finite point sets in dimension $s \geq 3$. For dimension $s = 2$ W.M. Schmidt proved the following result in 1972:

Theorem 2.25 (Schmidt) *There exists a constant $c > 0$ such that for the star discrepancy of any N-element point set $\mathcal{P}$ in $[0, 1)^2$ we have*

$$D_N^*(\mathcal{P}) \geq c \frac{\log N}{N}.$$

We shall prove this result in Exercise 2.4. Schmidt's lower bound in dimension two is best possible in the order of magnitude in N. For $s \geq 3$ the exact asymptotic order of star discrepancy is still unknown. Although it is conjectured that the lower bound on the star discrepancy is significantly larger than Roth's lower bound there is presently no consensus among the experts about its sharp form. There are many people who conjecture that the sharp exponent of the logarithm for $s \geq 3$ should be $s - 1$. But there are also other opinions such as, for example, $s/2$. Both conjectures are consistent with Schmidt's lower bound in dimension $s = 2$. From above we know constructions of N-element point sets in dimension s with star discrepancy of order $(\log N)^{s-1}/N$. We will encounter such constructions later on. However, the exact determination of the sharp lower bound on star discrepancy seems to be a very difficult problem.

For infinite sequences the situation is slightly different. The following lower bound can be deduced from Roth's result for finite point sets:

Theorem 2.26 *For every $s \in \mathbb{N}$ there exists a quantity $c_s' > 0$ such that for every infinite sequence $\mathcal{S}$ in $[0, 1)^s$ we have*

$$D_N^*(\mathcal{S}) \geq c_s' \frac{(\log N)^{s/2}}{N} \quad \text{for infinitely many } N \in \mathbb{N}.$$

Proof Let $\mathcal{S} = (x_n)_{n \in \mathbb{N}_0}$ where $x_n = (x_{n,1}, \ldots, x_{n,s})$. For fixed $N \in \mathbb{N}$ we consider the N-element point set $\mathcal{P} = \{y_0, \ldots, y_{N-1}\}$ in $[0, 1)^{s+1}$ given by

$$y_n := (x_{n,1}, \ldots, x_{n,s}, n/N), \quad \text{for } n = 0, \ldots, N - 1.$$

According to Theorem 2.24, there exist $a_1, \ldots, a_{s+1} \in [0, 1]$ such that

$$\left| \frac{A(\prod_{j=1}^{s+1}[0, a_j), \mathcal{P}, N)}{N} - \prod_{j=1}^{s+1} a_j \right| \geq c_{s+1} \frac{(\log N)^{s/2}}{N}.$$

Choose $m \in \mathbb{N}$ with the property $\frac{m-1}{N} < a_{s+1} \leq \frac{m}{N}$ or, equivalently, $-1 < Na_{s+1} - m \leq 0$.

By definition $A(\prod_{j=1}^{s+1}[0, a_j), \mathcal{P}, N)$ is the number of all $n \in \{0, \ldots, N-1\}$ for which we have $x_{n,j} \in [0, a_j)$ for all $j = 1, \ldots, s$ and $\frac{n}{N} \in [0, a_{s+1})$. Here the latter condition is equivalent to $n \in \{0, \ldots, m-1\}$. Therefore,

$$A\left(\prod_{j=1}^{s+1}[0,a_j),\mathcal{P},N\right) = A\left(\prod_{j=1}^{s}[0,a_j),\mathcal{S},m\right).$$

From this and the triangle inequality we obtain

$$\left|A\left(\prod_{j=1}^{s}[0,a_j),\mathcal{S},m\right) - m\,a_1\cdots a_s\right| \geq \left|A\left(\prod_{j=1}^{s+1}[0,a_j),\mathcal{P},N\right) - N\,a_1\cdots a_{s+1}\right|$$
$$- |N a_1\cdots a_{s+1} - m\,a_1\cdots a_s|$$
$$\geq c_{s+1}\,(\log N)^{s/2} - a_1\cdots a_s\,|N a_{s+1} - m|$$
$$\geq c_{s+1}\,(\log N)^{s/2} - 1.$$

It follows that for $c_s' < c_{s+1}$ and $N \in \mathbb{N}$ large enough there exists some $m \in \{1,\ldots,N\}$ such that

$$m D_m^*(\mathcal{S}) \geq \left|A\left(\prod_{j=1}^{s}[0,a_j),\mathcal{S},m\right) - m\,a_1\cdots a_s\right| > c_s'\,(\log N)^{s/2} \geq c_s'(\log m)^{s/2}.$$

Now we show that the inequality

$$m D_m^*(\mathcal{S}) > c_s'(\log m)^{s/2} \tag{2.12}$$

holds true for infinitely many $m \in \mathbb{N}$. Assume, on the contrary, that (2.12) only holds for finitely many $m \in \mathbb{N}$ and let m^* be the maximal integer with this property. Then choose $N \in \mathbb{N}$ such that

$$c_s'(\log N)^{s/2} > \max_{1 \leq k \leq m^*} k D_k^*(\mathcal{S}).$$

For this N it is possible, as shown above, to find an $m \in \{1,\ldots,N\}$ with the property that

$$m D_m^*(\mathcal{S}) > c_s'(\log N)^{s/2} \geq c_s'(\log m)^{s/2}.$$

If $m \leq m^*$, then we would have a contradiction, since

$$m D_m^*(\mathcal{S}) > c_s'(\log N)^{s/2} > \max_{1 \leq k \leq m^*} k D_k^*(\mathcal{S}).$$

Hence $m > m^*$. But since $m D_m^*(\mathcal{S}) \geq c_s'(\log m)^{s/2}$, this contradicts the maximality of m^*. Therefore, (2.12) is true for infinitely many $m \in \mathbb{N}$. $\qquad\square$

For dimension $s = 1$ there is an improvement due to W.M. Schmidt from the year 1972 whose proof we leave as exercise (cf. Exercise 2.4).

Theorem 2.27 (Schmidt) *There exists a constant $c > 0$ such that for every infinite sequence $\mathcal{S}$ in $[0, 1)$ we have*

$$D_N^*(\mathcal{S}) \geq c\frac{\log N}{N} \quad \text{for infinitely many } N \in \mathbb{N}.$$

Schmidt's lower bound in dimension $s = 1$ is best possible in the order of magnitude in N. For dimensions $s \geq 2$ it is conjectured that the sharp form of the lower bound for the star discrepancy is significantly larger than the result from Theorem 2.26. Corresponding to the conjectures for finite point sets there are some opinions that s is the correct exponent of the $\log N$ term and others that $\frac{s+1}{2}$ is the correct exponent. Again, both conjectures are consistent with Schmidt's result for dimension $s = 1$.

One can extend the method of proof of Theorem 2.26 to show that for every $s \in \mathbb{N}$ there exists a quantity $c_s'' > 0$ such that for every infinite sequence $\mathcal{S}$ in $[0, 1)^s$ we have

$$L_{2,N}(\mathcal{S}) \geq c_s'' \frac{(\log N)^{s/2}}{N} \quad \text{for infinitely many } N \in \mathbb{N}.$$

This was first shown by P.D. Proinov in 1985. In 2014 J. Dick and F. Pillichshammer presented the first explicit construction of an infinite sequence whose L_2 discrepancy is of order of magnitude $(\log N)^{s/2}/N$ which proves that Proinov's lower bound is asymptotically best possible in N. Corresponding results hold for the L_p discrepancy for $p \in (1, \infty)$.

Now we turn our attention to inequalities which are often useful to prove upper bounds on the star discrepancy of point sets and sequences. The first result that we mention, though without proof, is the so-called inequality of Erdős-Turán-Koksma. This fundamental inequality connects the behavior of discrepancy with that of exponential sums of the form as they appear in the Weyl criterion. This way, the inequality of Erdős-Turán-Koksma can be viewed as quantitative version of the Weyl criterion.

Theorem 2.28 (Inequality of Erdős-Turán-Koksma) *For every $s \in \mathbb{N}$ there exists a $c_s > 0$ such that for every N-element point set $\mathcal{P} = \{x_0, \ldots, x_{N-1}\}$ in $[0, 1)^s$ we have*

$$D_N(\mathcal{P}) \leq c_s \left(\frac{1}{m+1} + \sum_{0 < \|\boldsymbol{h}\|_\infty \leq m} \frac{1}{r_1(\boldsymbol{h})} \left| \frac{1}{N} \sum_{n=0}^{N-1} \exp(2\pi i \boldsymbol{h} \cdot \boldsymbol{x}_n) \right| \right),$$

where $m \in \mathbb{N}$ and where $r_1(\boldsymbol{h}) := \prod_{j=1}^{s} \max(1, |h_j|)$ and $\|\boldsymbol{h}\|_\infty := \max_{j=1,\ldots,s} |h_j|$ for $\boldsymbol{h} = (h_1, \ldots, h_s) \in \mathbb{Z}^s$.

References for the proof of this result can be found at the end of this section.

Example 2.29 For $N \in \mathbb{N}$ let $\mathcal{P} = \{0, 1/N, 2/N, \ldots, (N-1)/N\}$. For all $h = 1, \ldots, N-1$ we have $\sum_{n=0}^{N-1} \exp(2\pi \mathrm{i} hn/N) = 0$. Hence, with the choice $m = N-1$, from Theorem 2.28 we obtain $D_N(\mathcal{P}) \leq c_1/N$.

Sometimes one has to estimate the star discrepancy of point sets whose components are rational numbers. In 1977 H. Niederreiter proved a discrepancy estimate for such point sets which is related to the inequality of Erdős-Turán-Koksma. To state Niederreiter's result we have to introduce some notation.

For $M \in \mathbb{N}$, $M \geq 2$, put $C(M) := (-M/2, M/2] \cap \mathbb{Z}$ and $C_s(M) := C(M)^s$ the s-fold Cartesian product of $C(M)$. Further we write $C_s^*(M) := C_s(M) \setminus \{0\}$. For $h \in C(M)$ put

$$r(h, M) := \begin{cases} M \sin(\pi |h|/M) & \text{if } h \neq 0, \\ 1 & \text{if } h = 0, \end{cases}$$

and for $\boldsymbol{h} = (h_1, \ldots, h_s) \in C_s(M)$ put $r(\boldsymbol{h}, M) := \prod_{j=1}^{s} r(h_j, M)$.

Theorem 2.30 (Niederreiter) *For an integer $M \geq 2$ and $\boldsymbol{y}_0, \ldots, \boldsymbol{y}_{N-1} \in \mathbb{Z}^s$, let $\mathcal{P} = \{\boldsymbol{x}_0, \ldots, \boldsymbol{x}_{N-1}\}$ be the point set consisting of the fractional parts $\boldsymbol{x}_n = \{\boldsymbol{y}_n/M\}$ for $n = 0, \ldots, N-1$. Then*

$$D_N(\mathcal{P}) \leq 1 - \left(1 - \frac{1}{M}\right)^s + \sum_{\boldsymbol{h} \in C_s^*(M)} \frac{1}{r(\boldsymbol{h}, M)} \left| \frac{1}{N} \sum_{n=0}^{N-1} \exp\left(2\pi \mathrm{i} \frac{\boldsymbol{h} \cdot \boldsymbol{y}_n}{M}\right) \right|.$$

Proof For $\boldsymbol{k} = (k_1, \ldots, k_s) \in \mathbb{Z}^s$ let

$$A(\boldsymbol{k}) := \#\{n \in \mathbb{N}_0 : 0 \leq n < N \text{ and } \boldsymbol{y}_n \equiv \boldsymbol{k} \pmod{M}\},$$

where a congruence between vectors is meant to be component-wise. We have

$$\frac{1}{M} \sum_{h \in C(M)} \exp\left(2\pi \mathrm{i} \frac{ha}{M}\right) = \begin{cases} 1 & \text{if } a \equiv 0 \pmod{M}, \\ 0 & \text{if } a \not\equiv 0 \pmod{M}, \end{cases}$$

and therefore

$$\sum_{n=0}^{N-1} \frac{1}{M^s} \sum_{\boldsymbol{h} \in C_s(M)} \exp\left(2\pi \mathrm{i} \frac{\boldsymbol{h} \cdot (\boldsymbol{y}_n - \boldsymbol{k})}{M}\right)$$

$$= \sum_{n=0}^{N-1} \prod_{j=1}^{s} \underbrace{\left(\frac{1}{M} \sum_{h_j \in C(M)} \exp\left(2\pi \mathrm{i} \frac{h_j(y_{n,j} - k_j)}{M}\right)\right)}_{= \begin{cases} 1 & \text{if } y_{n,j} \equiv k_j \pmod{M}, \\ 0 & \text{otherwise,} \end{cases}} = A(\boldsymbol{k}).$$

Consequently,

$$A(\boldsymbol{k}) - \frac{N}{M^s} = \frac{1}{M^s} \sum_{\boldsymbol{h} \in C_s^*(M)} \exp\left(-2\pi\mathrm{i}\frac{\boldsymbol{h}\cdot\boldsymbol{k}}{M}\right) \sum_{n=0}^{N-1} \exp\left(2\pi\mathrm{i}\frac{\boldsymbol{h}\cdot\boldsymbol{y}_n}{M}\right).$$

Now let $J = \prod_{j=1}^s [u_j, v_j) \subseteq [0,1)^s$. For $j = 1, \ldots, s$ let $a_j \in \mathbb{Z}$ be minimal such that $u_j \le a_j/M$ and $b_j \in \mathbb{Z}$ maximal such that $b_j/M < v_j$. In particular, we have $[a_j/M, b_j/M] \subseteq [u_j, v_j)$.

If $[a_j/M, b_j/M] = \emptyset$ for some $j \in \{1, \ldots, s\}$, then $A(J, \mathcal{P}, N) = 0$ and $v_j - u_j < 1/M$. Therefore we have

$$\left| \frac{A(J, \mathcal{P}, N)}{N} - \lambda_s(J) \right| = \lambda_s(J) < \frac{1}{M} \le 1 - \left(1 - \frac{1}{M}\right)^s.$$

Now assume that $[a_j/M, b_j/M] \neq \emptyset$ for all $j = 1, \ldots, s$. We have

$$A(J, \mathcal{P}, N) = \sum_{\boldsymbol{a} \le \boldsymbol{k} \le \boldsymbol{b}} A(\boldsymbol{k}) \quad \text{and} \quad \sum_{\boldsymbol{a} \le \boldsymbol{k} \le \boldsymbol{b}} \frac{1}{M^s} = \frac{1}{M^s} \prod_{j=1}^s (b_j - a_j + 1),$$

where $\boldsymbol{a} \le \boldsymbol{k} \le \boldsymbol{b}$ means summation over all $\boldsymbol{k} \in \mathbb{Z}^s$ for which $a_j \le k_j \le b_j$ for all $j = 1, 2, \ldots, s$, and hence

$$\frac{A(J, \mathcal{P}, N)}{N} - \lambda_s(J)$$

$$= \sum_{\boldsymbol{a} \le \boldsymbol{k} \le \boldsymbol{b}} \left(\frac{A(\boldsymbol{k})}{N} - \frac{1}{M^s} \right) + \frac{1}{M^s} \prod_{j=1}^s (b_j - a_j + 1) - \lambda_s(J)$$

$$= \frac{1}{M^s} \sum_{\boldsymbol{h} \in C_s^*(M)} \left(\sum_{\boldsymbol{a} \le \boldsymbol{k} \le \boldsymbol{b}} \exp\left(-2\pi\mathrm{i}\frac{\boldsymbol{h}\cdot\boldsymbol{k}}{M}\right) \right) \frac{1}{N} \sum_{n=0}^{N-1} \exp\left(2\pi\mathrm{i}\frac{\boldsymbol{h}\cdot\boldsymbol{y}_n}{M}\right)$$

$$+ \prod_{j=1}^s \frac{b_j - a_j + 1}{M} - \prod_{j=1}^s (v_j - u_j).$$

For all $j \in \{1, \ldots, s\}$ we have

$$\left| \frac{b_j - a_j + 1}{M} - (v_j - u_j) \right| < \frac{1}{M},$$

and so it follows from Lemma 2.16 that

$$\left| \prod_{j=1}^s \frac{b_j - a_j + 1}{M} - \prod_{j=1}^s (v_j - u_j) \right| \le 1 - \left(1 - \frac{1}{M}\right)^s.$$

Consequently,

$$\left| \frac{A(J, \mathcal{P}, N)}{N} - \lambda_s(J) \right| \le 1 - \left(1 - \frac{1}{M}\right)^s$$

$$+ \frac{1}{M^s} \sum_{\boldsymbol{h} \in C_s^*(M)} \underbrace{\left| \sum_{\boldsymbol{a} \le \boldsymbol{k} \le \boldsymbol{b}} \exp\left(-2\pi \mathrm{i} \frac{\boldsymbol{h} \cdot \boldsymbol{k}}{M}\right) \right|}_{=: r^*(\boldsymbol{h}, M)} \cdot \left| \frac{1}{N} \sum_{n=0}^{N-1} \exp\left(2\pi \mathrm{i} \frac{\boldsymbol{h} \cdot \boldsymbol{y}_n}{M}\right) \right|.$$

Since $|z| = |\bar{z}|$ for any complex number z and $|\exp(2\pi \mathrm{i} t)| = 1$ for any real number t, we obtain

$$r^*(\boldsymbol{h}, M) = \prod_{j=1}^{s} \left| \sum_{k_j = a_j}^{b_j} \exp\left(2\pi \mathrm{i} \frac{h_j k_j}{M}\right) \right|$$

$$= \prod_{j=1}^{s} \left| \sum_{k_j = 0}^{b_j - a_j} \exp\left(2\pi \mathrm{i} \frac{h_j k_j}{M}\right) \exp\left(2\pi \mathrm{i} \frac{h_j a_j}{M}\right) \right|$$

$$= \prod_{j=1}^{s} \left| \sum_{k_j = 0}^{b_j - a_j} \exp\left(2\pi \mathrm{i} \frac{h_j k_j}{M}\right) \right|.$$

If $h_j = 0$, then

$$\left| \sum_{k_j = 0}^{b_j - a_j} \exp\left(2\pi \mathrm{i} \frac{h_j k_j}{M}\right) \right| = b_j - a_j + 1 \le M = \frac{M}{r(h_j, M)}.$$

If now $h_j \in C^*(M)$, then

$$\left| \sum_{k_j = 0}^{b_j - a_j} \exp\left(2\pi \mathrm{i} \frac{h_j k_j}{M}\right) \right| = \left| \frac{\exp(2\pi \mathrm{i} h_j (b_j - a_j + 1)/M) - 1}{\exp(2\pi \mathrm{i} h_j/M) - 1} \right|$$

$$= \left| \frac{\sin(\pi h_j (b_j - a_j + 1)/M)}{\sin(\pi h_j/M)} \right|$$

$$\le \frac{1}{\sin(\pi |h_j|/M)}$$

$$= \frac{M}{r(h_j, M)}.$$

In any case,

$$r^*(\boldsymbol{h}, M) \le \prod_{j=1}^{s} \frac{M}{r(h_j, M)} = \frac{M^s}{r(\boldsymbol{h}, M)},$$

and therefore

$$\left| \frac{A(J, \mathcal{P}, N)}{N} - \lambda_s(J) \right| \leq 1 - \left(1 - \frac{1}{M}\right)^s$$

$$+ \sum_{\mathbf{h} \in C_s^*(M)} \frac{1}{r(\mathbf{h}, M)} \left| \frac{1}{N} \sum_{n=0}^{N-1} \exp\left(2\pi \mathrm{i} \frac{\mathbf{h} \cdot \mathbf{y}_n}{M}\right) \right|.$$

The right-hand side of this inequality is independent of the specific choice of the interval J and hence the result follows. $\square$

We close this section with a formula for the star discrepancy of finite one-dimensional point sets. This formula was first proved in 1972 by H. Niederreiter.

Proposition 2.31 (Niederreiter) *Let* $\mathcal{P} = \{x_1, \ldots, x_N\}$ *be a point set in the unit interval* $[0, 1)$ *satisfying* $x_1 \leq x_2 \leq \ldots \leq x_N$. *Then we have*

$$D_N^*(\mathcal{P}) = \frac{1}{2N} + \max_{n=1,2,\ldots,N} \left| x_n - \frac{2n-1}{2N} \right|.$$

(We remark that the requirement $x_1 \leq x_2 \leq \ldots \leq x_N$ *does not pose any restriction on the point set.)*

Remark 2.32 Applying Proposition 2.31 to the regular lattice $\Gamma_{N,1} : \{\frac{2n-1}{2N} : n = 1, \ldots, N\}$ we find again that $D_N^*(\Gamma_{N,1}) = \frac{1}{2N}$ and this is best possible for any N-element point set in dimension $s = 1$ (cf. Remark 2.20).

Proof of Proposition 2.31. Put $x_0 := 0$ and $x_{N+1} := 1$. We have

$$D_N^*(\mathcal{P}) = \sup_{0 < \alpha \leq 1} \left| \frac{A([0, \alpha), N)}{N} - \alpha \right| = \max_{\substack{n=0,\ldots,N \\ x_n < x_{n+1}}} \sup_{x_n < \alpha \leq x_{n+1}} \left| \frac{A([0, \alpha), N)}{N} - \alpha \right|$$

$$= \max_{\substack{n=0,\ldots,N \\ x_n < x_{n+1}}} \sup_{x_n < \alpha \leq x_{n+1}} \left| \frac{n}{N} - \alpha \right|. \tag{2.13}$$

Assume that $x_n < x_{n+1}$. The function $g_n(x) = \left| \frac{n}{N} - x \right|$ is convex and hence it achieves its maximum on the compact interval $[x_n, x_{n+1}]$ in one of the end-points of this interval. Therefore,

$$D_N^*(\mathcal{P}) = \max_{\substack{n=0,\ldots,N \\ x_n < x_{n+1}}} \max \left(\left| \frac{n}{N} - x_n \right|, \left| \frac{n}{N} - x_{n+1} \right| \right). \tag{2.14}$$

Now we show that we may omit the condition $x_n < x_{n+1}$ in the first of the above maxima. Assume that

$$x_n < x_{n+1} = x_{n+2} = \ldots = x_{n+r} < x_{n+r+1},$$

with some $r \geq 2$. Hence the indices $n + j$ for $j = 1, \ldots, r - 1$ do not appear in the above maximum. We show that

$$\left| \frac{n + j}{N} - x_{n+j} \right| \quad \text{and} \quad \left| \frac{n + j}{N} - x_{n+j+1} \right|$$

for $1 \leq j \leq r - 1$ are always less or equal to some number which already appears in the first maximum in (2.14). Indeed, let $1 \leq j \leq r - 1$. Then

$$\left| \frac{n + j}{N} - x_{n+j} \right| = \left| \frac{n + j}{N} - x_{n+1} \right| \leq \max \left(\left| \frac{n}{N} - x_{n+1} \right|, \left| \frac{n + r}{N} - x_{n+1} \right| \right)$$

$$= \max \left(\left| \frac{n}{N} - x_{n+1} \right|, \left| \frac{n + r}{N} - x_{n+r} \right| \right).$$

However, both terms in the latter maximum already appear in the first maximum in (2.14). The same can be shown for the term $\left| \frac{n+j}{N} - x_{n+j+1} \right|$. Hence,

$$D_N^*(\mathcal{P}) = \max_{n=0,\ldots,N} \max \left(\left| \frac{n}{N} - x_n \right|, \left| \frac{n}{N} - x_{n+1} \right| \right)$$

$$= \max_{n=1,\ldots,N} \max \left(\left| \frac{n}{N} - x_n \right|, \left| \frac{n - 1}{N} - x_n \right| \right)$$

$$= \max_{n=1,\ldots,N} \max \left(\left| \frac{2n - 1}{2N} - x_n + \frac{1}{2N} \right|, \left| \frac{2n - 1}{2N} - x_n - \frac{1}{2N} \right| \right)$$

$$= \frac{1}{2N} + \max_{n=1,\ldots,N} \left| \frac{2n - 1}{2N} - x_n \right|,$$

since $\max(|a - b|, |a + b|) = |a| + |b|$. $\square$

2.4 A Classical Construction: The Halton Sequence

In Sect. 2.1 we introduced the one-dimensional van der Corput sequence in base b and showed its uniform distribution modulo one. This sequence is in fact the prototype of many uniformly distributed sequences and point sets with low star discrepancy, even in higher dimensions. In this section we introduce two generalizations of the van der Corput sequence which are already classical. The first one is the infinite Halton sequence and the second one its finite version, the so-called Hammersley point set.

The Halton Sequence

We construct an infinite sequence by component-wise concatenation of corresponding elements of van der Corput sequences in different bases. Recall that $\phi_b(n)$ denotes the b-adic radical inverse function as defined in Definition 2.8.

Fig. 2.4 The first 1000 elements of the Halton sequence in bases $b_1 = 2$ and $b_2 = 3$

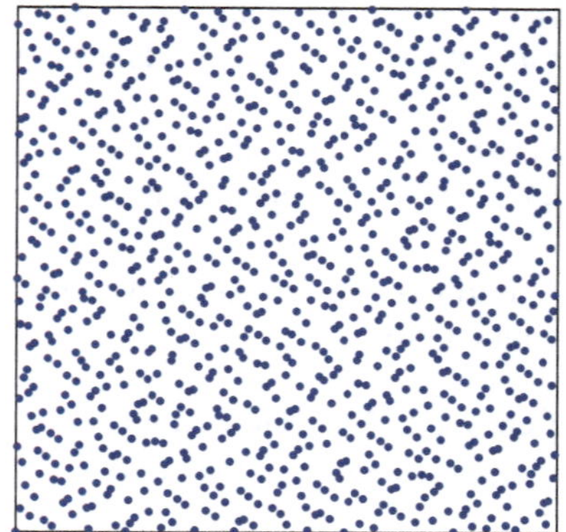

Definition 2.33 Let $s \in \mathbb{N}$ and let $b_1, \ldots, b_s \geq 2$ be integers. The *Halton sequence* in bases $b_1, \ldots, b_s$ is the sequence $\mathcal{S}_{b_1,\ldots,b_s} = (\boldsymbol{x}_n)_{n \in \mathbb{N}_0}$ whose nth element is given by

$$\boldsymbol{x}_n := (\phi_{b_1}(n), \phi_{b_2}(n), \ldots, \phi_{b_s}(n)) \quad \text{for} \quad n \in \mathbb{N}_0.$$

For $s = 1$ we recover the van der Corput sequence $(\phi_b(n))_{n \in \mathbb{N}_0}$ in base $b = b_1$.

Example 2.34 For example in dimension $s = 2$ with bases $b_1 = 2$ and $b_2 = 3$ we have $\boldsymbol{x}_0 = (0, 0)$, $\boldsymbol{x}_1 = (1/2, 1/3)$, $\boldsymbol{x}_2 = (1/4, 2/3)$, $\boldsymbol{x}_3 = (3/4, 1/9)$, $\boldsymbol{x}_4 = (1/8, 4/9)$, ... Figure 2.4 shows the first 1000 elements of this sequence.

Theorem 2.35 *Let $s \in \mathbb{N}$, and let $b_1, \ldots, b_s \geq 2$ be pairwise coprime integers. Then for the star discrepancy of the Halton sequence $\mathcal{S}_{b_1,\ldots,b_s}$ in bases $b_1, \ldots, b_s$ we have*

$$D_N^*(\mathcal{S}_{b_1,\ldots,b_s}) \leq \left(\prod_{j=1}^{s} \frac{b_j \log(b_j N)}{\log b_j} \right) \frac{1}{N} \quad \text{for} \quad N \in \mathbb{N}.$$

Hence, asymptotically for $N \to \infty$ we have $D_N^*(\mathcal{S}_{b_1,\ldots,b_s}) = O((\log N)^s / N)$.

In dimension one the Halton sequence coincides with the van der Corput sequence, as given in Definition 2.8. Thus Theorem 2.35 shows that the star discrepancy of the van der Corput sequence $\mathcal{S}_b = (\phi_b(n))_{n \in \mathbb{N}_0}$ satisfies

$$D_N^*(\mathcal{S}_b) \leq \frac{b}{\log b} \frac{\log(Nb)}{N}.$$

Here the asymptotic order of magnitude $(\log N)/N$ is best possible according to Schmidt's lower bound from Theorem 2.25.

For the proof of Theorem 2.35 we require the following lemmas:

Lemma 2.36 *Let $a, n, m \in \mathbb{Z}$. The congruence $x \equiv a \pmod{m}$ has exactly $\lfloor \frac{n}{m} \rfloor + \theta$ solutions $x \in \{0, \ldots, n-1\}$, where θ is either 0 or 1.*

Proof The result follows from the elementary fact that among any m consecutive integers the congruence has exactly one solution. $\qquad\qquad\square$

Furthermore, we need the Chinese Remainder Theorem which we recall here without proof.

Lemma 2.37 (Chinese Remainder Theorem) *Let $m_1, \ldots, m_s$ be pairwise coprime integers. Then the system of congruences*

$$x \equiv a_1 \pmod{m_1}$$
$$x \equiv a_2 \pmod{m_2}$$
$$\vdots$$
$$x \equiv a_s \pmod{m_s}$$

has exactly one solution modulo $m_1 m_2 \cdots m_s$.

Now we prove Theorem 2.35 for dimension $s = 1$ and $s = 2$. The general case follows by the same arguments. The proof presented here follows [71].

Proof of Theorem 2.35 For $N < \max_{j=1,2,\ldots,s} b_j$ the bound on the right-hand side is always larger than 1 and therefore the assertion trivially holds true. Hence we may assume in the following that $N \geq \max_{j=1,2,\ldots,s} b_j$.

The case $s = 1$:   Let $b = b_1$. Let $\alpha \in (0, 1]$ with infinite b-adic expansion

$$\alpha = \frac{\alpha_0}{b} + \frac{\alpha_1}{b^2} + \frac{\alpha_2}{b^3} + \cdots.$$

If the b-adic expansion of α is finite, say $\alpha = (0, \alpha_0 \alpha_1 \ldots \alpha_M)_b$ with $\alpha_M \neq 0$, then we switch to the infinite expansion by writing

$$\alpha = (0, \alpha_0 \ldots \alpha_{M-1} \alpha'_M \alpha'_{M+1} \ldots)_b$$

with $\alpha'_M = \alpha_M - 1$ and $\alpha'_i = b - 1$ for $i \geq M + 1$.

Let $k \in \{0, \ldots, N-1\}$ with b-adic expansion

$$k = k_0 + k_1 b + k_2 b^2 + \cdots + k_m b^m$$

where $k_m \neq 0$. Then we have $b^m \leq k < b^{m+1}$ and hence $m = \lfloor \frac{\log k}{\log b} \rfloor$. This means that k has exactly $m + 1 = \lfloor \frac{\log k}{\log b} \rfloor + 1 \leq M + 1$ digits in its b-adic expansion, where $M = \lfloor \frac{\log N}{\log b} \rfloor$. Hence for $k \in \{0, \ldots, N - 1\}$ we have

$$k = k_0 + k_1 b + \cdots + k_M b^M$$

with b-adic digits $k_0, k_1, \ldots, k_M \in \{0, \ldots, b - 1\}$, and the corresponding value of the b-adic radical inverse function equals

$$\phi_b(k) = \frac{k_0}{b} + \frac{k_1}{b^2} + \cdots + \frac{k_M}{b^{M+1}}.$$

Let us determine the number

$$A([0, \alpha), \mathcal{S}_b, N) = \#\{k \in \{0, \ldots, N - 1\} \ : \ \phi_b(k) \in [0, \alpha)\}$$

of indices $k \in \{0, \ldots, N - 1\}$ for which $x_k = \phi_b(k)$ belongs to the interval $[0, \alpha)$. According to the definition of the radical inverse function $\phi_b(k)$, we have $\phi_b(k) < \alpha$, if and only if one of the following conditions is satisfied:

$$
\begin{aligned}
&(1)\ k_0 < \alpha_0 \\
&(2)\ k_0 = \alpha_0,\, k_1 < \alpha_1 \\
&(3)\ k_0 = \alpha_0,\, k_1 = \alpha_1,\, k_2 < \alpha_2 \\
&\qquad\vdots \\
&(M + 1)\ k_0 = \alpha_0, \ldots, k_{M-1} = \alpha_{M-1},\, k_M < \alpha_M \\
&(M + 2)\ k_0 = \alpha_0, \ldots, k_{M-1} = \alpha_{M-1},\, k_M = \alpha_M.
\end{aligned}
$$

Note that conditions $(1), \ldots, (M+2)$ are mutually exclusive. Now recast these $M + 2$ conditions as congruences in the following way:

$$
\begin{aligned}
&(1)\ k \equiv k_0 \quad (\mathrm{mod}\ b),\ 0 \leq k_0 < \alpha_0 \\
&(2)\ k \equiv \alpha_0 + k_1 b \quad (\mathrm{mod}\ b^2),\ 0 \leq k_1 < \alpha_1 \\
&(3)\ k \equiv \alpha_0 + \alpha_1 b + k_2 b^2 \quad (\mathrm{mod}\ b^3),\ 0 \leq k_2 < \alpha_2 \\
&\qquad\vdots \\
&(M + 1)\ k \equiv \alpha_0 + \cdots + \alpha_{M-1} b^{M-1} + k_M b^M \quad (\mathrm{mod}\ b^{M+1}),\ 0 \leq k_M < \alpha_M \\
&(M + 2)\ k \equiv \alpha_0 + \cdots + \alpha_M b^M \quad (\mathrm{mod}\ b^{M+2}).
\end{aligned}
$$

By Lemma 2.36, the number of solutions of the respective congruences in $\{0, \ldots, N - 1\}$ equals

$$(1)\ \alpha_0 \left(\lfloor N/b \rfloor + \epsilon_0 \right)$$
$$(2)\ \alpha_1 \left(\lfloor N/b^2 \rfloor + \epsilon_1 \right)$$
$$\vdots$$
$$(M+1)\ \alpha_M \left(\lfloor N/b^{M+1} \rfloor + \epsilon_M \right)$$
$$(M+2)\ \lfloor N/b^{M+2} \rfloor + \epsilon_{M+1} = \epsilon_{M+1}$$

with appropriate $\epsilon_j \in [0, 1]$ for $j = 0, \ldots, M+1$ (note that Lemma 2.36 is applied in every line (j) exactly α_j times with possibly different $\theta_{j,0}, \ldots, \theta_{j,\alpha_j-1} \in \{0, 1\}$, depending on $k_j \in \{0, 1, \ldots, \alpha_j - 1\}$, such that $\epsilon_j = (\theta_{j,0} + \cdots + \theta_{j,\alpha_j-1})/\alpha_j \in [0, 1]$). For the last line we mention that $b^M \le N < b^{M+1}$ implies $\lfloor N/b^{M+2} \rfloor = 0$. Now we have

$$A([0, \alpha), \mathcal{S}_b, N) = \sum_{i=0}^{M} \alpha_i \left(\left\lfloor \frac{N}{b^{i+1}} \right\rfloor + \epsilon_i \right) + \epsilon_{M+1}$$

and

$$\left| A([0, \alpha), \mathcal{S}_b, N) - N \sum_{i=0}^{M} \frac{\alpha_i}{b^{i+1}} \right| = \left| \sum_{i=0}^{M} \alpha_i \left(\left\lfloor \frac{N}{b^{i+1}} \right\rfloor + \epsilon_i - \frac{N}{b^{i+1}} \right) + \epsilon_{M+1} \right|$$

$$\le \sum_{i=0}^{M} \alpha_i \left| \frac{N}{b^{i+1}} - \left\lfloor \frac{N}{b^{i+1}} \right\rfloor - \epsilon_i \right| + 1$$

$$\le \sum_{i=0}^{M} \alpha_i + 1$$

$$\le (M+1)(b-1) + 1,$$

since $\alpha_i \le b - 1$. Therefore,

$$|A([0, \alpha), \mathcal{S}_b, N) - \alpha N| \le (M+1)(b-1) + 1 + N \sum_{i=M+1}^{\infty} \frac{\alpha_i}{b^{i+1}}.$$

Using again the fact that $\alpha_i \le b - 1$, we obtain

$$N \sum_{i=M+1}^{\infty} \frac{\alpha_i}{b^{i+1}} \le N \frac{b-1}{b^{M+2}} \sum_{i=0}^{\infty} \frac{1}{b^i} = \frac{N}{b^{M+1}} < 1 \tag{2.15}$$

and hence

$$|A([0, \alpha), \mathcal{S}_b, N) - \alpha N| \le (M+1)(b-1) + 2.$$

For the final term we have

$$(M + 1)(b - 1) + 2 \le \left(\frac{\log N}{\log b} + 1 \right) (b - 1) + 2$$

$$= b \frac{\log(Nb)}{\log b} + 2 - \frac{\log(Nb)}{\log b}$$

$$\le b \frac{\log(Nb)}{\log b}, \tag{2.16}$$

since

$$2 - \frac{\log(Nb)}{\log b} = \frac{\log b^2 - \log(Nb)}{\log b} < 0$$

for $N \ge b$. Thus we have shown that for all $\alpha \in (0, 1]$ and all $N \ge b$ we have

$$\left| \frac{A([0, \alpha), \mathcal{S}_b, N)}{N} - \alpha \right| \le \frac{b}{N} \frac{\log(Nb)}{\log b}.$$

Since this bound is independent of the specific choice of α the result follows for $s = 1$.

The case $s = 2$:   We write $b = b_1$ and $c = b_2$. Let $\alpha, \beta \in (0, 1]$ with infinite digit expansions

$$\alpha = (0, \alpha_0 \alpha_1 \ldots)_b \quad \text{and} \quad \beta = (0, \beta_0 \beta_1 \ldots)_c,$$

in bases b and c, respectively. Put

$$M := \left\lfloor \frac{\log N}{\log b} \right\rfloor \quad \text{and} \quad L := \left\lfloor \frac{\log N}{\log c} \right\rfloor.$$

The number $A([0, \alpha) \times [0, \beta), \mathcal{S}_{b,c}, N)$ is the number of indices $k \in \{0, \ldots, N - 1\}$ with

$$\phi_b(k) < \alpha \text{ and } \phi_c(k) < \beta.$$

Let

$$k = l_0 + l_1 c + \cdots + l_L c^L,$$

with $0 \le l_i \le c - 1$, be the c-adic expansion of k. Then we have $\phi_b(k) < \alpha$ and $\phi_c(k) < \beta$ if and only if, in addition to the conditions (1), (2), ..., (M+2) for the first component from the case $s = 1$ also one of the following conditions for the second component is satisfied:

$$\overline{(1)} \ k \equiv l_0 \quad (\mathrm{mod}\ c),\ 0 \le l_0 < \beta_0$$
$$\overline{(2)} \ k \equiv \beta_0 + l_1 c \quad (\mathrm{mod}\ c^2),\ 0 \le l_1 < \beta_1$$
$$\overline{(3)} \ k \equiv \beta_0 + \beta_1 c + l_2 c^2 \quad (\mathrm{mod}\ c^3),\ 0 \le l_2 < \beta_2$$
$$\vdots$$
$$\overline{(L+1)} \ k \equiv \beta_0 + \cdots + \beta_{L-1} c^{L-1} + l_L c^L \quad (\mathrm{mod}\ c^{L+1}),\ 0 \le l_L < \beta_L$$
$$\overline{(L+2)} \ k \equiv \beta_0 + \cdots + \beta_L c^L \quad (\mathrm{mod}\ c^{L+2}).$$

Recall that by our global assumption we have $\gcd(b, c) = 1$ and thus also $\gcd(b^m, c^l) = 1$ for $m, l \in \mathbb{N}$. According to Lemmas 2.36 and 2.37, the number of $k \in \{0, \ldots, N-1\}$ which satisfy the congruences (m) and (l) for

- $m \in \{1, \ldots, M+1\}$ and $l \in \{1, \ldots, L+1\}$ equals

$$\alpha_{m-1} \beta_{l-1} \left(\left\lfloor \frac{N}{b^m c^l} \right\rfloor + \theta_{m-1,l-1} \right),$$

  where $\theta_{m-1,l-1} \in [0, 1]$;
- for $m \in \{1, \ldots, M+1\}$ and $l = L+2$ equals

$$\alpha_{m-1} \theta_{m-1},$$

  where $\theta_{m-1} \in [0, 1]$;
- for $m = M+2$ and $l \in \{1, \ldots, L+1\}$ equals

$$\beta_{l-1} \theta_{l-1},$$

  where $\theta_{l-1} \in [0, 1]$; and
- for $m = M+2$ and $l = L+2$ equals

$$\theta_{M+1,L+1} \in [0, 1].$$

Therefore,

$$A([0, \alpha) \times [0, \beta), \mathcal{S}_{b,c}, N) = \sum_{m=1}^{M+1} \sum_{l=1}^{L+1} \alpha_{m-1} \beta_{l-1} \left(\left\lfloor \frac{N}{b^m c^l} \right\rfloor + \theta_{m-1,l-1} \right) +$$
$$+ \sum_{m=1}^{M+1} \alpha_{m-1} \theta_{m-1} + \sum_{l=1}^{L+1} \beta_{l-1} \theta_{l-1} + \theta_{M+1,L+1}.$$

We obtain

$$\left| A([0, \alpha) \times [0, \beta), \mathcal{S}_{b,c}, N) - \sum_{m=1}^{M+1} \sum_{l=1}^{L+1} \alpha_{m-1} \beta_{l-1} \frac{N}{b^m c^l} \right|$$

$$\leq \sum_{m=1}^{M+1} \sum_{l=1}^{L+1} \alpha_{m-1} \beta_{l-1} + \sum_{m=1}^{M+1} \alpha_{m-1} + \sum_{l=1}^{L+1} \beta_{l-1} + 1$$

$$= \left(\sum_{m=1}^{M+1} \alpha_{m-1} + 1 \right) \left(\sum_{l=1}^{L+1} \beta_{l-1} + 1 \right),$$

and further, using (2.15) and (2.16),

$$|A([0, \alpha) \times [0, \beta), \mathcal{S}_{b,c}, N) - \alpha \beta N|$$

$$\leq \left(\sum_{m=1}^{M+1} \alpha_{m-1} + 1 \right) \left(\sum_{l=1}^{L+1} \beta_{l-1} + 1 \right)$$

$$+ N \sum_{m=M+2}^{\infty} \frac{\alpha_{m-1}}{b^m} + N \sum_{l=L+2}^{\infty} \frac{\beta_{l-1}}{c^l} + N \sum_{m=M+2}^{\infty} \sum_{l=L+2}^{\infty} \frac{\alpha_{m-1}}{b^m} \frac{\beta_{l-1}}{c^l}$$

$$\leq ((b - 1)(M + 1) + 1)((c - 1)(L + 1) + 1) + 3$$

$$\leq ((b - 1)(M + 1) + 2)((c - 1)(L + 1) + 2)$$

$$\leq \frac{b \log(bN)}{\log b} \frac{c \log(cN)}{\log c},$$

since $N \geq b$ and $N \geq c$. This proves also the case $s = 2$ and indicates how the proof works for arbitrary dimensions $s \geq 3$. $\qquad\square$

Theorem 2.35 shows that the star discrepancy of the Halton sequence is asymptotically of the order of magnitude $(\log N)^s / N$ for $N \to \infty$ and this is the best we can show. In the course of time it has become customary to speak of *low-discrepancy sequences* in case where their star discrepancy is of order of magnitude $(\log N)^s / N$. Later, we will be acquainted with other examples of low-discrepancy sequences. Recall from Sect. 2.3 that many people conjecture, though without proof, that $(\log N)^s / N$ is the best that one achieve at all for the star discrepancy. In dimension $s = 1$ the Halton result shows that Schmidt's lower bound from Theorem 2.27 is best possible in the order of magnitude in N.

On the other hand, the bound from Theorem 2.35 has a slightly defective appearance. To explain this, we consider the quantity

$$d^*(\mathcal{S}) := \limsup_{N \to \infty} \frac{N D_N^*(\mathcal{S})}{(\log N)^s} \tag{2.17}$$

for Halton sequences $\mathcal{S}_{b_1,\ldots,b_s}$. From Theorem 2.35 it follows that

$$d^*(\mathcal{S}_{b_1,\ldots,b_s}) \leq \prod_{j=1}^{s} \frac{b_j}{\log b_j} =: c_s.$$

This bound c_s is large and grows very fast to infinity for growing dimensions s. For example, assume that $b_1,\ldots,b_s$ are the first s prime numbers, which is the "smallest" choice in order to keep the bases pairwise coprime. The prime number theorem implies that $b_j \approx j \log j$ for large j and hence

$$c_s = \prod_{j=1}^{s} \frac{b_j}{\log b_j} \approx \prod_{j=1}^{s} \frac{j \log j}{\log(j \log j)} \approx \prod_{j=1}^{s} \frac{j \log j}{\log j} = s!.$$

In 2004, Atanassov [6] was able to overcome this particular disadvantage of the bound in Theorem 2.35. We state his result without proof here.

Theorem 2.38 (Atanassov) *Let $s \in \mathbb{N}$, and let $b_1,\ldots,b_s \geq 2$ be pairwise coprime integers. Then for the star discrepancy of the Halton sequence $\mathcal{S}_{b_1,\ldots,b_s}$ in bases $b_1,\ldots,b_s$ for $N \geq 2$ we have*

$$D_N^*(\mathcal{S}_{b_1,\ldots,b_s}) \leq \left[\frac{1}{s!} \prod_{j=1}^{s} \left(\frac{\lfloor b_j/2 \rfloor \log N}{\log b_j} + s \right) + \sum_{k=0}^{s-1} \frac{b_{k+1}}{k!} \prod_{j=1}^{k} \left(\frac{\lfloor b_j/2 \rfloor \log N}{\log b_j} + k \right) \right] \frac{1}{N}.$$

This result implies that

$$d^*(\mathcal{S}_{b_1,\ldots,b_s}) \leq \frac{1}{s!} \prod_{j=1}^{s} \frac{\lfloor b_j/2 \rfloor}{\log b_j}.$$

If again $b_1,\ldots,b_s$ are the first s prime numbers, then it can be shown that $\frac{1}{s!} \prod_{j=1}^{s} \frac{\lfloor b_j/2 \rfloor}{\log b_j} \leq \frac{7}{2^s s}$, and hence

$$\limsup_{s \to \infty} \frac{\log d^*(\mathcal{S}_{b_1,\ldots,b_s})}{s} \leq -\log 2. \tag{2.18}$$

This shows that the quantity $d^*(\mathcal{S}_{b_1,\ldots,b_s})$ tends to zero at an exponential rate as $s \to \infty$.

The discrepancy bounds for the Halton sequence suggest to choose the bases $b_1,\ldots,b_s$ as small as possible such that they are still pairwise coprime. Hence the best we can do is to choose them to be the first s prime numbers. However, in higher dimensions already this "minimal" choice leads to the problem that the initial segments of the Halton sequence, or, more exact, projections to lower dimensional faces thereof, have very poor distribution properties. For many practical applications such a behavior is not desirable. Just as an example, consider the projection of a Halton sequence in dimension $s \geq 36$ with prime bases $b_1, b_2, \ldots, b_s$ to the components

Fig. 2.5 The projection of
the first 1000 elements of the
Halton sequence to the
coordinates 35 and 36

with numbers 35 and 36, that is with $b_{35} = 149$ and $b_{36} = 151$. Then most points
of the initial segment are concentrated around the main diagonal and very large
areas of the unit square contain no point; see Fig. 2.5 where we plotted the first
1000 elements. Similar pictures appear when we consider other two-dimensional
projections to "high" coordinates. There are two ways to overcome this problem, at
least to a certain extent: the first one is the concept of *generalized Halton sequences*
which uses s sequences of permutations $(\pi_{j,k})_{k \in \mathbb{N}_0}$ for $j = 1, 2, \ldots, s$, where $\pi_{j,k}$
is a permutation of $\{0, 1, \ldots, b_j - 1\}$ for each $k \in \mathbb{N}_0$. Then the jth component of
the nth element of a generalized Halton sequence in bases $b_1, \ldots, b_s$ is given by

$$x_{n,j} = \sum_{k=0}^{\infty} \frac{\pi_{j,k}(n_{j,k})}{b_j^{k+1}} \quad \text{whenever} \quad n = n_{j,0} + n_{j,1}b_j + n_{j,2}b_j^2 + \cdots \quad (2.19)$$

for $j = 1, 2, \ldots, s$ and $n \in \mathbb{N}_0$. A second possibility is a generalization of the Halton
sequence to *digital (t, s)-sequences* (see Chap. 5) which use for all s components the
same fixed base b in the digital construction. However, also in this case there might
appear some defects in lower dimensional projections.

The Hammersley Point Set

We use the infinite Halton sequence to construct for every $N \in \mathbb{N}$, $N \geq 2$, an
N-element point set with low star discrepancy.

Definition 2.39 Let $s \in \mathbb{N}$, and let $b_1, \ldots, b_{s-1} \geq 2$ be pairwise coprime integers.
The N-element *Hammersley point set* $\mathcal{H}_{N,b_1,\ldots,b_{s-1}}$ in bases $b_1, \ldots, b_{s-1}$ is the finite
point set $\{x_0, \ldots, x_{N-1}\}$ in $[0, 1)^s$ where

$$x_n := \left(\frac{n}{N}, \phi_{b_1}(n), \ldots, \phi_{b_{s-1}}(n) \right) \quad \text{for} \quad n = 0, \ldots, N - 1.$$

Fig. 2.6 The 128-element
Hammersley point set $\mathcal{H}_{128,2}$
in base 2

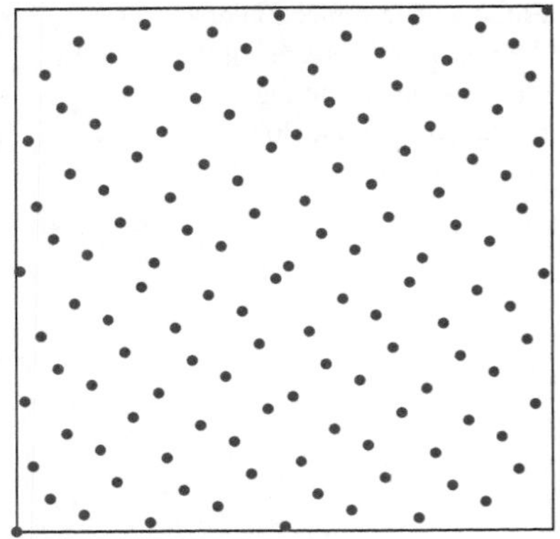

Example 2.40 For example, in dimension $s = 2$ with $N = 8$ and $b = 2$ the Hammersley point set $\mathcal{H}_{8,2}$ consists of the elements $x_0 = (0,0)$, $x_1 = (1/8, 1/2)$, $x_2 = (2/8, 1/4)$, $x_3 = (3/8, 3/4)$, $x_4 = (4/8, 1/8)$, $x_5 = (5/8, 5/8)$, $x_6 = (6/8, 3/8)$ and $x_7 = (7/8, 7/8)$. A further example is illustrated in Fig. 2.6.

Theorem 2.41 *For the star discrepancy of the N-element Hammersley point set* $\mathcal{H}_{N,b_1,\ldots,b_{s-1}}$ *in* $[0, 1)^s$ *with pairwise coprime bases* $b_1, \ldots, b_{s-1}$ *we have*

$$D_N^*(\mathcal{H}_{N,b_1,\ldots,b_{s-1}}) \leq \frac{1}{N}\left(\prod_{j=1}^{s-1} \frac{b_j \log(b_j N)}{\log b_j} + 1\right).$$

Hence the Hammersley point set in pairwise coprime bases satisfies

$$D_N^*(\mathcal{H}_{N,b_1,\ldots,b_{s-1}}) = O\left(\frac{(\log N)^{s-1}}{N}\right).$$

Note that this is a substantial improvement compared to the discrepancy of the regular lattice which is of the order of magnitude $N^{-1/s}$ only. It has become customary to speak of *low-discrepancy point sets* in the case where their star discrepancy is of order of magnitude $(\log N)^{s-1}/N$ in dimension s, with $N \geq 2$ being the cardinality of the considered point set.

For the proof of Theorem 2.41 we use the following general principle which is usually attributed to Roth.

Lemma 2.42 *Let* $s \in \mathbb{N}$, $s \geq 2$, *and let* $\mathcal{S} = (y_n)_{n \in \mathbb{N}_0}$ *be an infinite sequence in* $[0, 1)^{s-1}$ *with star discrepancy* $D_M^*(\mathcal{S})$ *for* $M \in \mathbb{N}$. *Let* $N \in \mathbb{N}$ *and let* $\mathcal{P} =$

$\{\boldsymbol{x}_0, \ldots, \boldsymbol{x}_{N-1}\}$ where $\boldsymbol{x}_n := (\frac{n}{N}, \boldsymbol{y}_n)$ for $n = 0, 1, \ldots, N - 1$. Then

$$D_N^*(\mathcal{P}) \le \frac{1}{N} \left(\max_{M=1,2,\ldots,N} M D_M^*(\mathcal{S}) + 1 \right).$$

Proof Consider an interval of the form $J := \prod_{j=1}^{s}[0, \alpha_j) \subseteq [0, 1)^s$. An element $\boldsymbol{x}_n$ from $\mathcal{P}$ belongs to J, if and only if

$$0 \le \frac{n}{N} < \alpha_1 \quad \text{and} \quad \boldsymbol{y}_n \in \prod_{j=2}^{s}[0, \alpha_j) =: \widetilde{J}.$$

Now we have $A(J, \mathcal{P}, N) = A(\widetilde{J}, \mathcal{S}, M)$, where $M = \lceil N\alpha_1 \rceil$. Consequently,

$$|A(J, \mathcal{P}, N) - N\lambda_s(J)| \le |A(\widetilde{J}, \mathcal{S}, M) - M\lambda_{s-1}(\widetilde{J})| + |M\lambda_{s-1}(\widetilde{J}) - N\lambda_s(J)|.$$

Since $|M\lambda_{s-1}(\widetilde{J}) - N\lambda_s(J)| = |(\lceil N\alpha_1 \rceil - N\alpha_1) \prod_{j=2}^{s} \alpha_j| \le 1$, it follows that

$$\begin{aligned}
|A(J, \mathcal{P}, N) - N\lambda_s(J)| &\le |A(\widetilde{J}, \mathcal{S}, M) - M\lambda_{s-1}(\widetilde{J})| + 1 \\
&\le M D_M^*(\mathcal{S}) + 1 \\
&\le \max_{M=1,2,\ldots,N} M D_M^*(\mathcal{S}) + 1
\end{aligned}$$

and the result follows. $\qquad\square$

Proof of Theorem 2.41. Let $\mathcal{S}_{b_1,\ldots,b_{s-1}}$ be an $(s - 1)$-dimensional Halton sequence in pairwise coprime bases $b_1, \ldots, b_{s-1}$. Then we obtain from Theorem 2.35

$$M D_M^*(\mathcal{S}_{b_1,\ldots,b_{s-1}}) \le \prod_{j=1}^{s-1} \frac{b_j \log(b_j M)}{\log b_j},$$

whence

$$\max_{M=1,2,\ldots,N} M D_M^*(\mathcal{S}_{b_1,\ldots,b_{s-1}}) \le \prod_{j=1}^{s-1} \frac{b_j \log(b_j N)}{\log b_j}.$$

Now the result follows from Lemma 2.42. $\qquad\square$

Further Reading

The theory of Uniform Distribution Modulo One was initiated with the seminal work of Weyl [145]. A standard reference is the book of Kuipers and Niederreiter [85]. Furthermore, the book of Matoušek [105] offers a readable introduction into discrepancy theory. Also the books of Beck and Chen [10], of Dick and Pillichshammer [33], of Drmota and Tichy [40], and of Hlawka [68] can be warmly recommended.

The lower bound for the L_p discrepancy for $p \in (1, \infty)$ is due to Schmidt [131]. A convenient way to prove these lower bounds is by means of Haar functions. An overview on this regard can be found in [24]. Explicit constructions of point sets with L_2 discrepancy of optimal order of magnitude can be found in [18,36] and for the general L_p discrepancy where $p \in (1, \infty)$ in [134]. For infinite sequences see [23,36]. The survey article [12] of Bilyk and Lacey gives an overview on results concerning the sharp order of magnitude for the smallest possible value of star discrepancy and the survey [35] discusses constructions of point sets with best order of L_2 discrepancy.

A proof of Proposition 2.22 can be found in [40] and of the inequality of Erdős-Turán-Koksma (Theorem 2.28) in [40] or in [85] (for $s = 1$). The current best value for c_s follows from Cochrane [19]. The proof of Theorem 2.30 follows the exposition in [110].

The van der Corput sequence is the prototype of a uniformly distributed sequence of low discrepancy. It is a well-studied object in literature. A comprehensive overview can be found in the survey article [44]. More information on the Halton sequence can be found in the books of Hua and Wang [71], of Niederreiter [110], and of Dick and Pillichshammer [33]. The currently best asymptotic estimate for the star discrepancy of the Halton sequence was shown by Atanassov [6]. The presented version of this result (Theorem 2.38) is due to [33, Theorem 3.36]. A proof of the Chinese Remainder Theorem can be found, e.g., in [50]. A lower bound on the star discrepancy of the Halton sequence of order of magnitude $(\log N)^s / N$ has been shown by Levin [101]. This bound matches the upper bounds presented in this chapter and supports the long-standing conjecture that the best possible order of the star discrepancy should be $(\log N)^s / N$. An upper bound on the L_2 discrepancy of the Halton sequence in pairwise coprime bases of the optimal order of magnitude $O((\log N)^{s/2} / N)$ has been proven by Levin [102]. The problem of defects in lower dimensional projections of Halton sequences and other sequences based on digit expansions is discussed, e.g., by Lemieux [97] and Schmid [130].

Known approaches and algorithms to compute discrepancy measures of point sets are surveyed in the book chapter [39] by Doerr, Gnewuch and Wahlström. Information concerning applications of discrepancy theory to a wide variety of topics in computer science can be found in the book by Chazelle [17].

Exercises

2.1 Let $\mathcal{S}$ be a sequence in $[0, 1)^s$. Provide a Lebesgue integrable function $f :$ $[0, 1] \to \mathbb{R}$ for which (2.3) does not hold.

2.2 Show that an infinite sequence $\mathcal{S}$ in $[0, 1)^s$ is uniformly distributed modulo one, if and only if for every closed interval of the form $[\boldsymbol{a}, \boldsymbol{b}] \subseteq [0, 1)^s$ we have

$$\lim_{N \to \infty} \frac{A([\boldsymbol{a}, \boldsymbol{b}], \mathcal{S}, N)}{N} = \lambda_s([\boldsymbol{a}, \boldsymbol{b}]),$$

where for $\boldsymbol{a} = (a_1, \ldots, a_s)$, $\boldsymbol{b} = (b_1, \ldots, b_s)$ we write $[\boldsymbol{a}, \boldsymbol{b}] = [a_1, b_1] \times \ldots \times [a_s, b_s]$. An analogous assertion can be shown for open intervals $(\boldsymbol{a}, \boldsymbol{b})$.

2.3 Give a detailed proof of Theorem 2.4.

2.4 Show that a necessary condition for the uniform distribution of $(\{\boldsymbol{\alpha} n\})_{n \in \mathbb{N}_0}$, where $\boldsymbol{\alpha} = (\alpha_1, \ldots, \alpha_s) \in \mathbb{R}^s$, is that $1, \alpha_1, \ldots, \alpha_s$ are linearly independent over the rationals.

2.5 Let $b, c \geq 2$ be positive integers satisfying $\gcd(b, c) = 1$. Show directly, i.e., without the knowledge of the Halton result, that the two-dimensional sequence given by $\boldsymbol{x}_n = (\phi_b(n), \phi_c(n))$ for $n \in \mathbb{N}_0$ is uniformly distributed modulo one. *Hint*: First consider intervals of the form $[a_1/b^{m_1}, (a_1+1)/b^{m_1}) \times [a_2/c^{m_2}, (a_2 + 1)/c^{m_2})$ where $m_j \in \mathbb{N}_0$ for $j \in \{1, 2\}$, $a_1 \in \{0, \ldots, b^{m_1} - 1\}$ and $a_2 \in \{0, \ldots, c^{m_2} - 1\}$.

2.6 Determine both the L_1 discrepancy and the star discrepancy of the following point sets:

(a) $\mathcal{P} = \{n/N : n = 0, \ldots, N - 1\}$.

(b) $\mathcal{P} = \{n/(2N) : n = 0, \ldots, N - 1\}$.

2.7 Show the so-called *Warnock formula* (historically more correct would be the term *Formula of Koksma*; see [82])

$$(L_{2,N}(\mathcal{P}))^2 = \frac{1}{3^s} - \frac{2}{N} \sum_{n=0}^{N-1} \prod_{j=1}^{s} \frac{1 - x_{n,j}^2}{2} + \frac{1}{N^2} \sum_{n,m=0}^{N-1} \prod_{j=1}^{s} \min(1 - x_{m,j}, 1 - x_{n,j})$$

for the squared L_2 discrepancy of a point set $\mathcal{P} = \{\boldsymbol{x}_0, \ldots, \boldsymbol{x}_{N-1}\}$ in $[0, 1)^s$, where $x_{n,j}$ denotes the jth component of the point $\boldsymbol{x}_n$. *Hint*: Start with

$$(L_{2,N}(\mathcal{P}))^2 = \int_0^1 \cdots \int_0^1 \left(\frac{1}{N} \sum_{n=0}^{N-1} \prod_{j=1}^{s} \chi_{[0,t_j)}(x_{n,j}) - t_1 \cdots t_s \right)^2 dt_1 \ldots dt_s.$$

2.8 Determine the L_2 discrepancy of the point sets from Exercise 2.4.

2.9 Determine the L_1 discrepancy and the L_2 discrepancy of the one-dimensional centered regular lattice $\Gamma_{N,1} : \{(2n + 1)/(2N) : n = 0, \ldots, N - 1\}$.

2.10 Compute $\int_{[0,1]^{sN}} [L_{2,N}(\{\boldsymbol{t}_1, \ldots, \boldsymbol{t}_N\})]^2 d\boldsymbol{t}_1 \ldots d\boldsymbol{t}_N$ and interpret the result.

2.11 Let $\mathcal{T}_{s,N}(\alpha)$ be the set of all tuples $(t_1, \ldots, t_N) \in [0,1)^{sN}$ for which

$$L_{2,N}(\{t_1, \ldots, t_N\}) \leq \frac{\alpha}{\sqrt{N}} \left(\frac{1}{2^s} - \frac{1}{3^s} \right)^{1/2}.$$

Show that for all $\alpha \geq 1$ we have $\lambda_{sN}(\mathcal{T}_{s,N}(\alpha)) > 1 - \alpha^{-2}$.

2.12 For $s, m \in \mathbb{N}$, $m \geq 2$, let

$$\Gamma_{m,s}^* := \left\{ \left(\frac{n_1}{m}, \ldots, \frac{n_s}{m} \right) \ : \ n_1, \ldots, n_s \in \{0, \ldots, m-1\} \right\}.$$

(a) Show that $D_N^*(\Gamma_{m,s}^*) = 1 - (1 - 1/m)^s$ and deduce $D_N^*(\Gamma_{m,s}^*) \asymp_s N^{-1/s}$.

(b) Determine the L_2 discrepancy $L_{2,N}(\Gamma_{m,s}^*)$ and deduce from the obtained formula that $L_{2,N}(\Gamma_{m,s}^*) \asymp_s N^{-1/s}$.

2.13 Let $m_1, \ldots, m_s \geq 2$ be integers and let

$$\Gamma_{m_1, \ldots, m_s} := \left\{ \left(\frac{k_1}{m_1}, \ldots, \frac{k_s}{m_s} \right) \ : \ k_j \in \{0, \ldots, m_j - 1\} \ \text{for} \ j = 1, \ldots, s \right\}$$

be a regular lattice consisting of $N = m_1 \cdots m_s$ elements in $[0,1)^s$. Show that

$$D_N^*(\Gamma_{m_1, \ldots, m_s}) = 1 - \prod_{j=1}^{s} \left(1 - \frac{1}{m_j} \right).$$

2.14 For $i = 1, \ldots, k$ let $\mathcal{P}_i$ be an N_i-element point set with discrepancy $D_{N_i}(\mathcal{P}_i)$. Let $\mathcal{P}$ be a superposition of the point sets $\mathcal{P}_i$, $i = 1, \ldots, k$ and put $N := N_1 + \cdots + N_k$. Show the so-called *triangle inequality for the discrepancy*

$$D_N(\mathcal{P}) \leq \sum_{i=1}^{k} \frac{N_i}{N} D_{N_i}(\mathcal{P}_i),$$

and similar for the star discrepancy.

2.15 We prove Theorem 2.25: The proof is based on the proof of Theorem 2.24. Choose $n \in \mathbb{N}$ such that $2^{n-1} < 2N \leq 2^n$. For $i = 0, \ldots, n$ define the functions $f_i : [0,1)^s \to \{-1, 0, 1\}$ as in (2.8). Furthermore, define the auxiliary function $H : [0,1)^2 \to \mathbb{R}$ by

$$H(\boldsymbol{x}) := \prod_{i=0}^{n} (1 + \alpha f_i(\boldsymbol{x})) - 1,$$

where $\alpha \in (0, 1/2)$ which will be fixed in a later stage of the proof. Put $D(\boldsymbol{x}) := N x_1 x_2 - A([\boldsymbol{0}, \boldsymbol{x}), \mathcal{P}, N)$, where $\boldsymbol{x} = (x_1, x_2)$. We split the proof into several steps:

(a) Show the following lemma:

Lemma 2.43 *For any set of integers $\{i_1, \ldots, i_k\}$ satisfying $0 \le i_1 < \ldots < i_k \le n$ we have*

$$\int_{[0,1]^2} f_{i_1}(\boldsymbol{x}) \cdots f_{i_k}(\boldsymbol{x}) \, d\boldsymbol{x} = 0.$$

Hint: Partition the square $[0, 1)^2$ into $2^{n-i_1+i_k}$ rectangles by dividing $[0, 1)$ on the x-axes into 2^{i_k} and on the y-axes into 2^{n-i_1} intervals of the same length. Then argue as in the case $k = 2$.

(b) Show that

$$\prod_{i=0}^{n}(1 + \alpha f_i(\boldsymbol{x})) = 1 + \alpha F(\boldsymbol{x}) + \sum_{k=2}^{n+1} \alpha^k F_k(\boldsymbol{x}),$$

where the function F is defined as in (2.9) and

$$F_k(\boldsymbol{x}) := \sum_{0 \le i_1 < \ldots < i_k \le n} f_{i_1}(\boldsymbol{x}) \cdots f_{i_k}(\boldsymbol{x})$$

for $k = 2, \ldots, n + 1$.

(c) Prove that $\int_{[0,1]^2} |H(\boldsymbol{x})| \, d\boldsymbol{x} \le 2$. *Hint*: From $\alpha < 1/2$ we obtain $|H(\boldsymbol{x})| \le \prod_{i=0}^{n}(1 + \alpha f_i(\boldsymbol{x})) + 1$. Now use Lemma 2.43.

(d) Show the following lemma:

Lemma 2.44 *For any set of integers $\{i_1, \ldots, i_k\}$ satisfying $0 \le i_1 < \ldots < i_k \le n$ we have*

$$\left| \int_{[0,1]^2} f_{i_1}(\boldsymbol{x}) \cdots f_{i_k}(\boldsymbol{x}) D(\boldsymbol{x}) \, d\boldsymbol{x} \right| \le N 2^{-n+i_1-i_k-4}.$$

Hint: Let R be one of the $2^{n-i_1+i_k}$ rectangles from the proof of Lemma 2.43. Show that $\int f_{i_1} \cdots f_{i_k} D = \pm(\lambda_2(R)/4)^2$. Proceed as in the special case $k = 1$ in the proof of Theorem 2.24.

(e) Show the following lemma:

Lemma 2.45 *For every $k = 2, \ldots, n + 1$ we have*

$$\left| \int_{[0,1]^2} F_k(\boldsymbol{x}) D(\boldsymbol{x}) \, d\boldsymbol{x} \right| \le \sum_{i=0}^{n-k+1} \sum_{h=1}^{n-i} \frac{N}{2^{n+h+4}} \binom{h-1}{k-2}.$$

Hint: Use Lemma 2.44 and put $i_1 = i \in \{0, \dots, n - k + 1\}$ and $i_k = i + h$ where $h \in \{1, \dots, n - i\}$.

(f) Show that

$$\left| \sum_{k=2}^{n+1} \alpha^k \int_{[0,1]^2} F_k(\boldsymbol{x}) D(\boldsymbol{x}) \, \mathrm{d}\boldsymbol{x} \right| \leq \alpha^2 N \frac{n}{2^{n+2}}.$$

(g) Show that for sufficiently small $\alpha \in (0, 1/2)$ there exists a constant $C > 0$ such that

$$\left| \int_{[0,1]^2} H(\boldsymbol{x}) D(\boldsymbol{x}) \, \mathrm{d}\boldsymbol{x} \right| \geq C \log N.$$

Hint: From the triangle inequality it follows that $\left| \int HD \right| \geq \alpha \left| \int FD \right| - \left| \sum_{k=2}^{n+1} \alpha^k \int F_k D \right|$.

(h) Deduce from Exercises 2.15c and 2.15g that

$$\sup_{\boldsymbol{x} \in [0,1]^2} |D(\boldsymbol{x})| > c \log N$$

for some positive constant c.

(i) Use 2.12a–2.12h to finish the proof of Theorem 2.25.

2.16 Deduce Theorem 2.27 from Theorem 2.25. *Hint*: Use Lemma 2.42.

2.17 Let $\mathcal{S}_{b_1,\dots,b_s}^{\Pi_1,\dots,\Pi_s}$ be the generalized Halton sequence in pairwise coprime bases $b_1, \dots, b_s$ and with s sequences of permutations $\Pi_j = (\pi_{j,k})_{k \in \mathbb{N}_0}$ of $\{0, 1, \dots, b_j - 1\}$ with $\pi_{j,k}(0) = 0$ for $j = 1, 2, \dots, s$ and $k \in \mathbb{N}_0$, as defined in (2.19). Show that the star discrepancy of $\mathcal{S}_{b_1,\dots,b_s}^{\Pi_1,\dots,\Pi_s}$ satisfies the bound from Theorem 2.35, i.e.,

$$D_N^*(\mathcal{S}_{b_1,\dots,b_s}^{\Pi_1,\dots,\Pi_s}) \leq \left(\prod_{j=1}^{s} \frac{b_j \log(b_j N)}{\log b_j} \right) \frac{1}{N} \quad \text{for all} \ \ N \in \mathbb{N}.$$

QMC Integration in Reproducing Kernel Hilbert Spaces

3

We return to the problem of numerical integration of multivariate functions. As already mentioned in Sect. 1.1 we normalize the integration domain to be the compact unit cube $[0, 1]^s$, and hence the integrals considered are of the form (1.1). We aim at approximating such integrals by QMC rules of the form $Q_{N,s}(f) := \frac{1}{N} \sum_{n=0}^{N-1} f(\boldsymbol{x}_n)$ with fixed integration nodes $\boldsymbol{x}_0, \ldots, \boldsymbol{x}_{N-1}$ taken from $[0, 1)^s$, i.e.:

$$\int_{[0,1]^s} f(\boldsymbol{x})\, \mathrm{d}\boldsymbol{x} \approx Q_{N,s}(f).$$

On first sight this approach looks quite simple but the crux of this method is the choice of underlying nodes. On the other hand, as already mentioned, the knowledge of the integration nodes is insufficient for solving the integration problem in full generality. It is easy to construct two functions $f, g : [0, 1]^s \to \mathbb{R}$ for which we have $f(\boldsymbol{x}_n) = g(\boldsymbol{x}_n)$ for all $n = 0, 1, \ldots, N - 1$, but $\int_{[0,1]^s} f(\boldsymbol{x})\, \mathrm{d}\boldsymbol{x} - \int_{[0,1]^s} g(\boldsymbol{x})\, \mathrm{d}\boldsymbol{x}$ can be any number. This means that we require some global information on the functions to be integrated. To tackle this problem we consider function classes with certain smoothness properties and study the worst-case integration error of QMC rules therein. Based on this error analysis, we find criteria for point sets to be used as underlying nodes of QMC rules for the specific function class.

3.1 Univariate QMC Integration

To begin with, we consider QMC integration of univariate real-valued functions $f : [0, 1] \to \mathbb{R}$ with continuous first derivative on $[0, 1]$. For such functions we obtain from the fundamental theorem of calculus that for any $x \in [0, 1]$ we have

$$f(x) = f(1) - \int_x^1 f'(y)\,\mathrm{d}y.$$

For the error

$$e(f, \mathcal{P}) := \int_0^1 f(x)\,\mathrm{d}x - \frac{1}{N}\sum_{n=0}^{N-1} f(x_n)$$

of a QMC rule based on sample nodes $\mathcal{P} = \{x_0, \ldots, x_{N-1}\}$ from $[0, 1)$ we now have

$$e(f, \mathcal{P}) = \frac{1}{N}\sum_{n=0}^{N-1}\int_{x_n}^1 f'(y)\,\mathrm{d}y - \int_0^1\int_x^1 f'(y)\,\mathrm{d}y\,\mathrm{d}x$$

$$= \int_0^1 \frac{1}{N}\sum_{n=0}^{N-1}\chi_{(x_n,1]}(y)f'(y)\,\mathrm{d}y - \int_0^1\int_0^y f'(y)\,\mathrm{d}x\,\mathrm{d}y$$

$$= \int_0^1 f'(y)\left[\frac{1}{N}\sum_{n=0}^{N-1}\chi_{(x_n,1]}(y) - y\right]\mathrm{d}y.$$

Since

$$\sum_{n=0}^{N-1}\chi_{(x_n,1]}(y) = \sum_{n=0}^{N-1}\chi_{[0,y)}(x_n) = A([0, y), \mathcal{P}, N)$$

is the number of indices $n \in \{0, \ldots, N - 1\}$ for which x_n belongs to $[0, y)$, we find that

$$\frac{1}{N}\sum_{n=0}^{N-1}\chi_{(x_n,1]}(y) - y = \Delta_{\mathcal{P},N}(y),$$

i.e., the local discrepancy of $\mathcal{P}$ in y. Hence we have

$$e(f, \mathcal{P}) = \int_0^1 f'(y)\Delta_{\mathcal{P},N}(y)\,\mathrm{d}y. \tag{3.1}$$

Taking the absolute value and applying first the triangle inequality for integrals and then the Hölder inequality, we get

$$|e(f, \mathcal{P})| \le \int_0^1 |f'(y)||\Delta_{\mathcal{P},N}(y)|\,\mathrm{d}y$$

$$\le \left(\int_0^1 |f'(y)|^q\,\mathrm{d}y\right)^{1/q}\left(\int_0^1 |\Delta_{\mathcal{P},N}(y)|^p\,\mathrm{d}y\right)^{1/p}$$

$$= \|f'\|_{L_q}\|\Delta_{\mathcal{P},N}\|_{L_p}, \tag{3.2}$$

where, as usual, $p, q \in [1, \infty]$ with $\frac{1}{p} + \frac{1}{q} = 1$.

The estimate in (3.2) separates the influence of the integrand f and of the underlying point set $\mathcal{P}$ on the absolute integration error. Hence, for QMC integration of functions f for which $\|f'\|_{L_q} < \infty$, one should choose sample nodes $\mathcal{P}$ with low L_p discrepancy $L_{p,N}(\mathcal{P}) = \|\Delta_{\mathcal{P},N}\|_{L_p}$.

Note that $\|f'\|_{L_q}$ is just a semi-norm of the function f, while

$$\|f\|_{1,q} := \left(|f(1)|^q + \int_0^1 |f'(y)|^q \, \mathrm{d}y \right)^{1/q} \tag{3.3}$$

is a proper norm. For this reason, (3.2) is often stated in the form:

$$|e(f, \mathcal{P})| \leq \|f\|_{1,q} \, L_{p,N}(\mathcal{P}) \quad \text{for} \ \ p, q \geq 1 \text{ and } \frac{1}{p} + \frac{1}{q} = 1.$$

For $p = \infty$ and $q = 1$ this is a simplified version of the inequality of Koksma.

Theorem 3.1 (Koksma inequality) *For any $f \in C^1([0, 1])$ and any N-element point set $\mathcal{P}$ in $[0, 1)$ we have*

$$|e(f, \mathcal{P})| \leq \|f\|_{1,1} D_N^*(\mathcal{P}).$$

Now we aim at developing a similar theory for multivariate functions. This can be done in a very elegant way by using the notion of reproducing kernel Hilbert spaces.

3.2 Reproducing Kernel Hilbert Spaces

For an integrable function f defined on $[0, 1]^s$ and a point set $\mathcal{P} = \{\boldsymbol{x}_0, \ldots, \boldsymbol{x}_{N-1}\}$ in $[0, 1)^s$ we use the notation

$$e(f, \mathcal{P}) := \int_{[0,1]^s} f(\boldsymbol{x}) \, \mathrm{d}\boldsymbol{x} - \frac{1}{N} \sum_{n=0}^{N-1} f(\boldsymbol{x}_n),$$

in full analogy to the univariate case.

We want to introduce the notion of the worst-case error of a QMC rule in a Hilbert space of functions. We denote the inner product in a Hilbert space $\mathcal{H}$ by $\langle \cdot, \cdot \rangle$ and the corresponding norm by $\|\cdot\| = \langle \cdot, \cdot \rangle^{1/2}$.

Definition 3.2 The *worst-case error* of a QMC rule based on a point set $\mathcal{P} = \{\boldsymbol{x}_0, \ldots, \boldsymbol{x}_{N-1}\}$ in $[0, 1)^s$ in a Hilbert space $\mathcal{H}$ of integrable functions on $[0, 1]^s$ is defined as

$$e(\mathcal{H}, \mathcal{P}) := \sup_{f \in \mathcal{H}, \|f\| \leq 1} |e(f, \mathcal{P})|.$$

One can develop a particularly nice error theory by using a reproducing kernel for $\mathcal{H}$.

Definition 3.3 A Hilbert space $\mathcal{H}$ of functions on $[0, 1]^s$ is called a *reproducing kernel Hilbert space* on $[0, 1]^s$ if there exists a function $K : [0, 1]^s \times [0, 1]^s \to \mathbb{C}$ such that

K1: $K(\cdot, y) \in \mathcal{H}$ for all $y \in [0, 1]^s$ and

K2: $\langle f, K(\cdot, y) \rangle = f(y)$ for all $y \in [0, 1]^s$ and for all $f \in \mathcal{H}$.

The function K is called the *reproducing kernel* of $\mathcal{H}$.

Note that in **K1** and **K2** we consider the reproducing kernel K as a function of the first variable, denoted by $\cdot$, and in $\langle f, K(\cdot, y) \rangle$ we apply the inner product with respect to the first variable of K. The second property **K2** is the so-called *reproducing property*, i.e., the evaluation of the function f can be expressed as the inner product of the function and the kernel function.

A function which satisfies the properties **K1** and **K2** is automatically symmetric, uniquely defined and positive semi-definite:

K3 (symmetry): $K(x, y) = \overline{K(y, x)}$ for all $x, y \in [0, 1]^s$.

K4 (uniqueness):   for any function $\widetilde{K}$ satisfying **K1** and **K2** we have $\widetilde{K} = K$.

K5 (positive semi-definiteness):   for any choice of $a_0, \ldots, a_{N-1} \in \mathbb{C}$ and $x_0, \ldots, x_{N-1} \in [0, 1]^s$ we have $\sum_{m,n=0}^{N-1} \overline{a}_m a_n K(x_m, x_n) \geq 0$.

Proof **K3:** $K(x, y) = \langle K(\cdot, y), K(\cdot, x) \rangle = \overline{\langle K(\cdot, x), K(\cdot, y) \rangle} = \overline{K(y, x)}$.

K4: $\widetilde{K}(x, y) = \langle \widetilde{K}(\cdot, y), K(\cdot, x) \rangle = \overline{\langle K(\cdot, x), \widetilde{K}(\cdot, y) \rangle} = \overline{K(y, x)} = K(x, y)$.

K5:

$$\sum_{m,n=0}^{N-1} \overline{a}_m a_n K(x_m, x_n) = \sum_{m,n=0}^{N-1} \overline{a}_m a_n \langle K(\cdot, x_n), K(\cdot, x_m) \rangle$$

$$= \left\langle \sum_{n=0}^{N-1} a_n K(\cdot, x_n), \sum_{m=0}^{N-1} a_m K(\cdot, x_m) \right\rangle$$

$$= \left\| \sum_{m=0}^{N-1} a_m K(\cdot, x_m) \right\|^2 \geq 0.$$

$\square$

Conversely, one can show that a function K satisfying properties **K3** and **K5** uniquely determines a Hilbert space of functions for which it satisfies **K1** and **K2** (and hence also **K4**). Therefore it makes sense to speak of a reproducing kernel without explicitly specifying the corresponding Hilbert space of functions.

Now we revisit Sect. 3.1 and consider the one-dimensional integration problem in the light of reproducing kernel Hilbert spaces.

For $f, g \in C^1([0, 1])$ define an inner product by

$$\langle f, g \rangle_1 := f(1)g(1) + \int_0^1 f'(x)g'(x)\,dx. \tag{3.4}$$

The corresponding norm $\|f\|_{1,2} := \sqrt{\langle f, f \rangle_1}$ is exactly the norm considered in (3.3) for $q = 2$. We have $\|f\|_{1,2} < \infty$ whenever the first derivative f' is in $L_2([0, 1])$. Based on this norm we define a Hilbert space $\mathcal{H}_1$ by

$$\mathcal{H}_1 := \{f : [0, 1] \to \mathbb{R} : f \text{ absolutely continuous and } \|f\|_{1,2} < \infty\}.$$

Recall that a function $f : [0, 1] \to \mathbb{R}$ is called *absolutely continuous*, if for all $\varepsilon > 0$ there exists a $\delta > 0$ such that for all $n \in \mathbb{N}$ and any choice of pairwise disjoint intervals (x_k, y_k), $k = 1, \ldots, n$, in $[0, 1]$ with $\sum_{k=1}^n (y_k - x_k) < \delta$ we have $\sum_{k=1}^n |f(x_k) - f(y_k)| < \varepsilon$. For example, any Lipschitz continuous function is absolutely continuous. In particular, if $f : [0, 1] \to \mathbb{R}$ is differentiable on $[0, 1]$ with bounded first derivative f', then f is absolutely continuous.

The definition of an absolutely continuous function is a bit unwieldy, but there is a well-known equivalent condition that is more suited to our purpose: one can show that a function $f : [0, 1] \to \mathbb{R}$ is absolutely continuous, if and only if there exists a Lebesgue integrable function $g : [0, 1] \to \mathbb{R}$ such that f can be written in the form:

$$f(x) = f(0) + \int_0^x g(t)\,dt = f(1) - \int_x^1 g(t)\,dt \quad \text{for all} \ \ x \in [0, 1].$$

In this case f is almost everywhere differentiable and $g = f'$ almost everywhere.

Now we show that the function K_1 defined by

$$K_1(x, y) := 1 + \min(1 - x, 1 - y) \tag{3.5}$$

is the reproducing kernel of $\mathcal{H}_1$. For fixed $y \in [0, 1]$ we have

$$K_1(x, y) = \begin{cases} 2 - x & \text{if } x > y, \\ 2 - y & \text{if } x \le y, \end{cases}$$

and

$$\frac{\partial K_1(x, y)}{\partial x} = \begin{cases} -1 & \text{if } x > y, \\ 0 & \text{if } x \le y. \end{cases}$$

Thus with $g : [0, 1] \to \mathbb{R}$ defined by

$$g(x) := \begin{cases} -1 & \text{if } x > y, \\ 0 & \text{if } x \le y, \end{cases}$$

we obtain

$$K_1(1, y) - \int_x^1 g(t)\,dt = 1 + \int_{\max(x,y)}^1 dt = 1 + \min(1 - x, 1 - y) = K_1(x, y).$$

Therefore the function $K_1(\cdot, y)$ is absolutely continuous. In particular, $K_1(\cdot, y)$ is also integrable. Furthermore,

$$\langle K_1(\cdot, y), K_1(\cdot, y)\rangle_1 = K_1(1, y)^2 + \int_0^1 \left(\frac{\partial K_1(x, y)}{\partial x}\right)^2 dx$$

$$= 1 + \int_y^1 dx = 2 - y < \infty,$$

and hence $\|K_1(\cdot, y)\|_{1,2} < \infty$. This implies that $K_1(\cdot, y) \in \mathcal{H}_1$. It remains to check the reproducing property of K_1. We have

$$\langle f, K_1(\cdot, y)\rangle_1 = f(1)K_1(1, y) + \int_0^1 f'(x)\frac{\partial K_1(x, y)}{\partial x} dx$$

$$= f(1) - \int_y^1 f'(x) dx = f(y).$$

Altogether, we have shown that K_1 is the reproducing kernel of the space $\mathcal{H}_1$.

Let us consider QMC integration for the Hilbert space $\mathcal{H}_1$. Using the reproducing Kernel K_1 we can write the integration functional as

$$\int_0^1 f(y) \, dy = \int_0^1 \langle f, K_1(\cdot, y)\rangle_1 \, dy$$

$$= \int_0^1 \left(f(1)K_1(1, y) + \int_0^1 f'(x)\frac{\partial K_1(x, y)}{\partial x} dx\right) dy$$

$$= f(1) \int_0^1 K_1(1, y) \, dy + \int_0^1 f'(x) \int_0^1 \frac{\partial K_1(x, y)}{\partial x} \, dy \, dx$$

$$= f(1) \int_0^1 K_1(1, y) \, dy + \int_0^1 f'(x)\frac{d}{dx} \left(\int_0^1 K_1(x, y) \, dy\right) dx$$

$$= \left\langle f, \int_0^1 K_1(\cdot, y) \, dy\right\rangle_1,$$

where we used Fubini's theorem for changing the order of integration, and where the interchange of integration and differentiation can be justified by direct calculation of the terms $\int_0^1 \frac{\partial K_1(x,y)}{\partial x} \, dy$ and $\frac{d}{dx}\left(\int_0^1 K_1(x, y) \, dy\right)$. In a similar manner we can write the QMC functional $Q_{N,1}(f)$ as

$$Q_{N,1}(f) = \frac{1}{N}\sum_{n=0}^{N-1} f(x_n) = \frac{1}{N}\sum_{n=0}^{N-1}\langle f, K_1(\cdot, x_n)\rangle_1 = \left\langle f, \frac{1}{N}\sum_{n=0}^{N-1} K_1(\cdot, x_n)\right\rangle_1.$$

Put

$$h(x) := \int_0^1 K_1(x, y) \, dy - \frac{1}{N}\sum_{n=0}^{N-1} K_1(x, x_n).$$

Note that $h \in \mathcal{H}_1$, since $K_1(\cdot, y) \in \mathcal{H}_1$ as well as $\int_0^1 K_1(\cdot, y)\,\mathrm{d}y \in \mathcal{H}_1$. Now we have

$$e(f, \mathcal{P}) = \int_0^1 f(y)\,\mathrm{d}y - Q_{N,1}(f) = \langle f, h \rangle_1$$

and hence, using the Cauchy-Schwarz inequality,

$$|e(f, \mathcal{P})| = \left|\langle f, h \rangle_1\right| \leq \|f\|_{1,2}\,\|h\|_{1,2}\,. \tag{3.6}$$

In (3.6) we have equality if $f(x) = h(x)$.

Due to the linearity of the inner product we obtain for $f \in \mathcal{H}_1$ with $\|f\|_{1,2} \neq 0$ that

$$\frac{|e(f, \mathcal{P})|}{\|f\|_{1,2}} = \left| e\left(\frac{f}{\|f\|_{1,2}}, \mathcal{P}\right)\right| \leq \|h\|_{1,2} = \frac{|\langle h, h \rangle_1|}{\|h\|_{1,2}} = \frac{|e(h, \mathcal{P})|}{\|h\|_{1,2}}$$

with equality if $f = h$. Consequently,

$$e(\mathcal{H}_1, \mathcal{P}) = \frac{|e(h, \mathcal{P})|}{\|h\|_{1,2}} = \frac{|\langle h, h \rangle_1|}{\|h\|_{1,2}} = \|h\|_{1,2}.$$

Observe that for $x \notin \mathcal{P}$

$$\Delta_{\mathcal{P},N}(x) = \frac{\mathrm{d}}{\mathrm{d}x}\left(\int_0^1 K_1(x, y)\,\mathrm{d}y - \frac{1}{N}\sum_{n=0}^{N-1} K_1(x, x_n)\right) = h'(x)$$

(a proof of this fact is left as Exercise 3.5) and hence

$$L_{2,N}(\mathcal{P}) = \|h\|_{1,2}\,.$$

This implies that

$$e(\mathcal{H}_1, \mathcal{P}) = L_{2,N}(\mathcal{P}), \tag{3.7}$$

i.e., the worst-case error is exactly the L_2 discrepancy of the underlying point set. Furthermore, among all functions from the unit ball of $\mathcal{H}_1$ the function $h/\|h\|_{1,2}$ is the hardest to integrate (using the point set $\mathcal{P}$), since for h we have equality in (3.6).

3.3   The Worst-Case Error in Reproducing Kernel Hilbert Spaces

Let $\mathcal{H}$ be a Hilbert space of integrable functions $f : [0, 1]^s \to \mathbb{C}$, equipped with inner product $\langle \cdot, \cdot \rangle$ and corresponding norm $\|\cdot\| = \langle \cdot, \cdot \rangle^{1/2}$.

Recall from the general theory of normed spaces that a linear functional T on $\mathcal{H}$ is called *bounded* if there exists $M < \infty$ such that $|T(f)| \leq M$ for all f satisfying $\|f\| \leq 1$, and that for linear functionals boundedness is equivalent to continuity.

Recall further from Definition 3.2 that the worst-case error of a QMC rule based on a point set $\mathcal{P} = \{x_0, \ldots, x_{N-1}\}$ in $[0, 1)^s$ in $\mathcal{H}$ is defined as

$$e(\mathcal{H}, \mathcal{P}) := \sup_{f \in \mathcal{H}, \|f\| \leq 1} |e(f, \mathcal{P})|,$$

where

$$e(f, \mathcal{P}) := \int_{[0,1]^s} f(x)\, dx - \frac{1}{N} \sum_{n=0}^{N-1} f(x_n)$$

for $f \in \mathcal{H}$. A priori it is not clear that $e(\mathcal{H}, \mathcal{P})$ is finite or, in other words, that the linear functional $e(\cdot, \mathcal{P})$ is continuous. In the following we shall find conditions under which this is the case.

For $y \in [0, 1]^s$ consider the linear functional T_y which evaluates $f \in \mathcal{H}$ at y, i.e.:

$$T_y(f) = f(y) \quad \text{for} \ f \in \mathcal{H},$$

where T_y is called the *evaluation functional* in y. If T_y is continuous for all $y \in [0, 1]^s$, then so is every QMC rule.

It turns out that continuity of the evaluation functionals is equivalent to the existence of a reproducing kernel. For the proof of this remarkable fact we use a well-known result for continuous linear functionals from functional analysis, the Fréchet-Riesz representation theorem.

Theorem 3.4 (Fréchet-Riesz Representation Theorem) *Let $\mathcal{X}$ be a Hilbert space equipped with inner product $\langle \cdot, \cdot \rangle$ and let $T : \mathcal{X} \to \mathbb{C}$ be a continuous linear functional. Then there exists exactly one element $z \in \mathcal{X}$ with the property that*

$$T(x) = \langle x, z \rangle \ \text{for all} \ x \in \mathcal{X}.$$

Theorem 3.5 *Let $\mathcal{H}$ be a Hilbert space of functions on $[0, 1]^s$. Then $\mathcal{H}$ is a reproducing kernel Hilbert space on $[0, 1]^s$ if and only if the evaluation functionals*

$$T_y(f) = f(y) \ \text{for} \ f \in \mathcal{H}, \ y \in [0, 1]^s$$

are continuous.

Proof If the evaluation functionals are continuous, then the Fréchet-Riesz representation theorem guarantees, for every y, the existence of a uniquely determined function $k_y \in \mathcal{H}$ with the property that

$$T_y(f) = \langle f, k_y \rangle \quad \text{for all} \ f \in \mathcal{H}.$$

If we now define $K(x, y) := k_y(x)$ for $x, y \in [0, 1]^s$, then K satisfies the properties **K1** and **K2** and thus $\mathcal{H}$ is a reproducing kernel Hilbert space with reproducing kernel K.

Conversely, assume that K is a reproducing kernel for $\mathcal{H}$ and let $y \in [0, 1]^s$. Using the Cauchy-Schwarz inequality, we get for every $f \in \mathcal{H}$

$$|T_y(f)| = |f(y)| = |\langle f, K(\cdot, y)\rangle| \le \|f\|\,\|K(\cdot, y)\|.$$

From the reproducing property we get $\|K(\cdot, y)\|^2 = \langle K(\cdot, y), K(\cdot, y)\rangle = K(y, y)$, so that $|T_y(f)| \le M$ for every f with $\|f\| \le 1$, where $M = \sqrt{K(y, y)}$. That means that T_y is continuous.
$\qquad\square$

Next consider the integration functional $I(f) := \int_{[0,1]^s} f(x)\,dx$. If $\mathcal{H}$ has a reproducing kernel K, then for any $f \in \mathcal{H}$ with $\|f\| \le 1$

$$\left| \int_{[0,1]^s} f(y)\,dy \right| = \left| \int_{[0,1]^s} T_y(f)\,dy \right| \le \int_{[0,1]^s} |T_y(f)|\,dy \le \int_{[0,1]^s} \sqrt{K(y, y)}\,dy.$$

Hence if the kernel K satisfies the condition

C: $\int_{[0,1]^s} \sqrt{K(y, y)}\,dy < \infty,$

then the integration functional I is continuous.

We can summarize: if $\mathcal{H}$ has a reproducing kernel K which satisfies condition **C**, then function evaluation and integration are continuous linear functionals, and so is $e(\cdot, \mathcal{P})$ for any point set $\mathcal{P}$. Moreover, under these conditions $e(\mathcal{H}, \mathcal{P})$ is a well-defined finite number.

Remark 3.6 After studying a couple of examples of reproducing kernel Hilbert spaces, the property that the evaluation functionals are continuous becomes increasingly natural. So one may justly ask the question whether there exist any examples at all of Hilbert spaces of functions for which function evaluation is not continuous.

Such an example can indeed be constructed, see [11, Chap. 1, Exercise 14], however this example makes use of the fact that any linear subspace of a vector space has an algebraic complement. Thus the example is rather non-constructive and we may state informally that all "natural" examples of Hilbert spaces of functions have continuous evaluation functionals.

The example cited above also serves as an example of a Hilbert space of functions where the integration functional is defined on the whole space but is not continuous.

Remark 3.7 Condition **C** is not strictly necessary for continuity of the integration functional. In fact, under our global assumption that we have a Hilbert space of integrable functions, continuity of the evaluation functionals is sufficient for continuity of the integration functional. The proof for this uses the closed graph theorem and can be found in [120, Sect. 23.4]. In the notes on the same section, one can find an example of a reproducing kernel Hilbert space for which $K(\cdot, y)$ is integrable for every y, but which contains functions that are not integrable.

In the last section we used the important property that for the specific kernel $K_1(x, y) = 1 + \min(1 - x, 1 - y)$ one may interchange the order of integration and the inner product, i.e.:

$$\int_0^1 \langle f, K_1(\cdot, y)\rangle_1 \, \mathrm{d}y = \left\langle f, \int_0^1 K_1(\cdot, y) \, \mathrm{d}y \right\rangle_1. \tag{3.8}$$

There, this interchange reduced to a change of the order of integration and of the order of integration and differentiation. Property (3.8) is essential for our error analysis, and in the following we show that it holds for any reproducing kernel Hilbert space for which the integration functional is continuous.

Lemma 3.8 *Let $\mathcal{H}$ be a reproducing kernel Hilbert space with reproducing kernel K and inner product $\langle \cdot, \cdot \rangle$. Assume that the mapping*

$$I(f) := \int_{[0,1]^s} f(\boldsymbol{y}) \, \mathrm{d}\boldsymbol{y} \quad \text{for } f \in \mathcal{H}$$

is a continuous linear functional on $\mathcal{H}$. Then we have

$$\int_{[0,1]^s} \langle f, K(\cdot, \boldsymbol{y})\rangle \, \mathrm{d}\boldsymbol{y} = \left\langle f, \int_{[0,1]^s} K(\cdot, \boldsymbol{y}) \, \mathrm{d}\boldsymbol{y} \right\rangle.$$

Proof Since I is continuous, the Fréchet-Riesz representation theorem guarantees the existence of a unique function $R \in \mathcal{H}$, such that

$$\int_{[0,1]^s} f(\boldsymbol{y}) \, \mathrm{d}\boldsymbol{y} = I(f) = \langle f, R \rangle \quad \text{for all } f \in \mathcal{H}.$$

Since $R \in \mathcal{H}$, the evaluation of R at some $\boldsymbol{x}$ can be expressed in terms of the inner product with the kernel function $K(\cdot, \boldsymbol{x})$ and hence

$$R(\boldsymbol{x}) = \langle R, K(\cdot, \boldsymbol{x})\rangle = \overline{\langle K(\cdot, \boldsymbol{x}), R\rangle} = \overline{\int_{[0,1]^s} K(\boldsymbol{y}, \boldsymbol{x}) \, \mathrm{d}\boldsymbol{y}}.$$

Thus we have

$$\int_{[0,1]^s} \langle f, K(\cdot, \boldsymbol{y})\rangle \, \mathrm{d}\boldsymbol{y} = \int_{[0,1]^s} f(\boldsymbol{y}) \, \mathrm{d}\boldsymbol{y} = \langle f, R\rangle$$

$$= \left\langle f, \overline{\int_{[0,1]^s} K(\boldsymbol{y}, \cdot) \, \mathrm{d}\boldsymbol{y}} \right\rangle = \left\langle f, \int_{[0,1]^s} K(\cdot, \boldsymbol{y}) \, \mathrm{d}\boldsymbol{y} \right\rangle.$$

$\square$

From now on we will always tacitly assume that K satisfies condition **C**.

With this assumption we can proceed in our worst-case error analysis as in the special case from the previous section. We have

$$I(f) = \int_{[0,1]^s} \langle f, K(\cdot, \boldsymbol{y}) \rangle \, \mathrm{d}\boldsymbol{y} = \left\langle f, \int_{[0,1]^s} K(\cdot, \boldsymbol{y}) \, \mathrm{d}\boldsymbol{y} \right\rangle$$

and

$$Q_{N,s}(f) = \frac{1}{N} \sum_{n=0}^{N-1} \langle f, K(\cdot, \boldsymbol{x}_n) \rangle = \left\langle f, \frac{1}{N} \sum_{n=0}^{N-1} K(\cdot, \boldsymbol{x}_n) \right\rangle.$$

Therefore, the integration error of the QMC rule $Q_{N,s}$ in $\mathcal{H}$ can again be expressed as an inner product

$$e(f, \mathcal{P}) = \langle f, h \rangle, \tag{3.9}$$

where

$$h(\boldsymbol{x}) := \int_{[0,1]^s} K(\boldsymbol{x}, \boldsymbol{y}) \, \mathrm{d}\boldsymbol{y} - \frac{1}{N} \sum_{n=0}^{N-1} K(\boldsymbol{x}, \boldsymbol{x}_n).$$

This function is often called the *representer* of the integration error. Taking the absolute value and applying the Cauchy-Schwarz inequality leads to

$$|e(f, \mathcal{P})| \le \|f\| \, \|h\|.$$

Moreover it follows from (3.9) that among all functions in the unit ball of $\mathcal{H}$ the normalized representer $h/\|h\|$ is the hardest to integrate. Therefore, the worst-case error can be written as

$$e(\mathcal{H}, \mathcal{P}) = \|h\|.$$

From this result we obtain the following formula for the squared worst-case error $e^2(\mathcal{H}, \mathcal{P}) = \langle h, h \rangle$, which is very convenient for the calculation of the worst-case error of QMC rules in specific reproducing kernel Hilbert spaces.

Theorem 3.9 *Let $\mathcal{H}$ be a reproducing kernel Hilbert space with reproducing kernel K that satisfies condition* **C** *and let $\mathcal{P} = \{\boldsymbol{x}_0, \dots, \boldsymbol{x}_{N-1}\}$ be an N-element point set in $[0, 1)^s$. Then*

$$e^2(\mathcal{H}, \mathcal{P}) = \int_{[0,1]^s} \int_{[0,1]^s} K(\boldsymbol{x}, \boldsymbol{y}) \, \mathrm{d}\boldsymbol{x} \, \mathrm{d}\boldsymbol{y} - \frac{2}{N} \sum_{n=0}^{N-1} \int_{[0,1]^s} K(\boldsymbol{x}_n, \boldsymbol{y}) \, \mathrm{d}\boldsymbol{y}$$

$$+ \frac{1}{N^2} \sum_{n,m=0}^{N-1} K(\boldsymbol{x}_n, \boldsymbol{x}_m).$$

In the next section, we apply the theory developed here to show some classical results for the QMC integration error.

3.4 Koksma-Hlawka-Type Inequalities

The Koksma-Hlawka inequality is the fundamental error estimate for QMC rules which separates the influence of the integrand f and of the underlying integration nodes on the integration error. It was proved by J.F. Koksma in 1942 for dimension $s = 1$ and later, in 1961, generalized by E. Hlawka to arbitrary dimensions $s \in \mathbb{N}$. The Koksma-Hlawka inequality is a very general error estimate valid for all functions of finite variation in the sense of Hardy and Krause, which reduces to the total variation in dimension $s = 1$.

We consider the reproducing kernel $K_s : [0, 1]^s \times [0, 1]^s \to \mathbb{R}$ given by

$$K_s(\boldsymbol{x}, \boldsymbol{y}) := \prod_{j=1}^{s} K_1(x_j, y_j) = \prod_{j=1}^{s} (1 + \min(1 - x_j, 1 - y_j)), \qquad (3.10)$$

where $\boldsymbol{x} = (x_1, \ldots, x_s) \in [0, 1]^s$ and $\boldsymbol{y} = (y_1, \ldots, y_s) \in [0, 1]^s$, and where K_1 is defined as in (3.5). The corresponding inner product is given by

$$\langle f, g \rangle_s = \sum_{\mathfrak{u} \subseteq [s]} \int_{[0,1]^{|\mathfrak{u}|}} \frac{\partial^{|\mathfrak{u}|} f}{\partial \boldsymbol{x}_{\mathfrak{u}}} (\boldsymbol{x}_{\mathfrak{u}}, \boldsymbol{1}) \frac{\partial^{|\mathfrak{u}|} g}{\boldsymbol{x}_{\mathfrak{u}}} (\boldsymbol{x}_{\mathfrak{u}}, \boldsymbol{1}) \, \mathrm{d}\boldsymbol{x}_{\mathfrak{u}}, \qquad (3.11)$$

where $[s] := \{1, \ldots, s\}$. For $\mathfrak{u} \subseteq [s]$ and $\boldsymbol{x} = (x_1, \ldots, x_s) \in [0, 1]^s$ we put $(\boldsymbol{x}_{\mathfrak{u}}, \boldsymbol{1}) := (z_1, \ldots, z_s)$, where

$$z_j = \begin{cases} x_j & \text{if } j \in \mathfrak{u}, \\ 1 & \text{if } j \notin \mathfrak{u}. \end{cases}$$

Furthermore $\partial^{|\mathfrak{u}|} / \partial \boldsymbol{x}_{\mathfrak{u}}$ denotes the mixed first partial derivative with respect to the components of $\boldsymbol{x}$ whose index belongs to $\mathfrak{u}$.

Now let $\mathcal{H}_s$ be the reproducing kernel Hilbert space with reproducing kernel K_s and norm $\| \cdot \|_{s,2} = \langle \cdot, \cdot \rangle_s^{1/2}$. We call $\mathcal{H}_s$ an *anchored Sobolev space* with anchor in 1. This function space contains all functions f on $[0, 1]^s$ whose mixed partial derivatives $\partial^{|\mathfrak{u}|} f / \partial \boldsymbol{x}_{\mathfrak{u}}$ up to order one in each variable belong to $L_2([0, 1]^s)$ and that are expressible in the form:

$$f(\boldsymbol{x}) = \langle f, K_s(\cdot, \boldsymbol{x}) \rangle_s$$

$$= \sum_{\mathfrak{u} \subseteq [s]} (-1)^{|\mathfrak{u}|} \int_{[0,1]^{|\mathfrak{u}|}} \frac{\partial^{|\mathfrak{u}|} f}{\partial \boldsymbol{y}_{\mathfrak{u}}} (\boldsymbol{y}_{\mathfrak{u}}, \boldsymbol{1}) \chi_{[0,(\boldsymbol{y}_{\mathfrak{u}},\boldsymbol{1}))}(\boldsymbol{x}) \, \mathrm{d}\boldsymbol{y}_{\mathfrak{u}} \quad \text{for } \boldsymbol{x} \in [0, 1]^s.$$

In Sect. 3.2 we already considered the one-dimensional case $\mathcal{H}_1$ and learned that it consists of all absolutely continuous functions f with square integrable first derivative. For $s > 1$ one can show that $\mathcal{H}_s$ is the s-fold tensor product of the spaces $\mathcal{H}_1$, i.e.:

$$\mathcal{H}_s = \underbrace{\mathcal{H}_1 \otimes \ldots \otimes \mathcal{H}_1}_{s\text{-fold}} = \text{clos span}\left\{ \boldsymbol{x} \mapsto \prod_{j=1}^{s} f_j(x_j) \ : \ f_j \in \mathcal{H}_1 \right\},$$

where $\boldsymbol{x} = (x_1, \ldots, x_s)$ and where the closure clos is taken with respect to the norm $\|\cdot\|_{s,2}$.

We have

$$\int_{[0,1]^s} \sqrt{K_s(\boldsymbol{y}, \boldsymbol{y})}\, d\boldsymbol{y} = \left(\int_0^1 \sqrt{2-y}\, dy \right)^s = \left(\frac{2(\sqrt{8}-1)}{3} \right)^s < \infty$$

and hence K_s satisfies condition **C**. Thus it follows from (3.9) that

$$e(f, \mathcal{P}) = \langle f, h \rangle_s$$

for all $f \in \mathcal{H}_s$ and all N-element point sets $\mathcal{P}$ in $[0,1)^s$, where h is the representer of the integration error in $\mathcal{H}_s$. We have

$$h(\boldsymbol{x}) = \int_{[0,1]^s} K_s(\boldsymbol{x}, \boldsymbol{y})\, d\boldsymbol{y} - \frac{1}{N} \sum_{n=0}^{N-1} K_s(\boldsymbol{x}, \boldsymbol{x}_n)$$

$$= \prod_{j=1}^{s} \left[\int_0^1 (1 + \min(1-x_j, 1-y_j))\, dy_j \right]$$

$$- \frac{1}{N} \sum_{n=0}^{N-1} \prod_{j=1}^{s} (1 + \min(1-x_j, 1-x_{n,j}))$$

$$= \prod_{j=1}^{s} \frac{3-x_j^2}{2} - \frac{1}{N} \sum_{n=0}^{N-1} \prod_{j=1}^{s} (1 + \min(1-x_j, 1-x_{n,j})).$$

For $\mathfrak{u} \subseteq [s]$ we have

$$\frac{\partial^{|\mathfrak{u}|}}{\partial \boldsymbol{x}_{\mathfrak{u}}} h(\boldsymbol{x}_{\mathfrak{u}}, \mathbf{1}) = (-1)^{|\mathfrak{u}|} \left(\prod_{j \in \mathfrak{u}} x_j - \frac{1}{N} \sum_{n=0}^{N-1} \prod_{j \in \mathfrak{u}} \chi_{[0,x_j)}(x_{n,j}) \right)$$

$$= (-1)^{|\mathfrak{u}|} \left(\prod_{j \in \mathfrak{u}} x_j - \frac{1}{N} \sum_{n=0}^{N-1} \chi_{[\mathbf{0},(\boldsymbol{x}_{\mathfrak{u}},\mathbf{1}))}(\boldsymbol{x}_n) \right)$$

$$= (-1)^{|\mathfrak{u}|+1} \Delta_{\mathcal{P},N}(\boldsymbol{x}_{\mathfrak{u}}, \mathbf{1}),$$

where $\boldsymbol{x}_n = (x_{n,1}, \ldots, x_{n,s})$ and where $\Delta_{\mathcal{P},N}$ denotes the local discrepancy of the N-element point set $\mathcal{P}$. Note that $\Delta_{\mathcal{P},N}(\boldsymbol{x}_\emptyset, \mathbf{1}) = \Delta_{\mathcal{P},N}(\mathbf{1}) = 0$.

From these considerations we obtain a formula for the integration error in $\mathcal{H}_s$ which is known as *Hlawka's identity* or as *Zaremba's identity*.

Theorem 3.10 (Hlawka's identity, Zaremba's identity) *For $f \in \mathcal{H}_s$ and $\mathcal{P}$ in $[0, 1)^s$ we have*

$$e(f, \mathcal{P}) = \sum_{\emptyset \neq u \subseteq [s]} (-1)^{|u|+1} \int_{[0,1]^{|u|}} \frac{\partial^{|u|} f}{\partial x_u}(x_u, 1) \Delta_{\mathcal{P},N}(x_u, 1) \, dx_u.$$

We use Theorem 3.10 as basis for developing the fundamental error estimates for the QMC method according to Koksma and Hlawka. These error estimates hold for subclasses of $\mathcal{H}_s$ consisting of functions with a certain finite norm, which depends on a parameter $q \in [1, \infty]$.

At first we introduce the semi-norm

$$\|f\|_{s,q}^* := \left(\int_{[0,1]^s} \left| \frac{\partial^s f}{\partial x}(x) \right|^q dx \right)^{1/q}$$

if $q \in [1, \infty)$ and

$$\|f\|_{s,\infty}^* := \sup_{x \in [0,1]^s} \left| \frac{\partial^s f}{\partial x}(x) \right|$$

if $q = \infty$, where $\partial^s f / \partial x := \partial^s f / (\partial x_1 \partial x_2 \ldots \partial x_s)$, and consider the space

$$\mathcal{F}_{s,q}^* := \{f \in \mathcal{H}_s \ : \ f(x) = 0 \text{ if } x_j = 1 \text{ for some } j \in [s], \text{ and } \|f\|_{s,q}^* < \infty\}.$$

Since we have boundary conditions for functions in $\mathcal{F}_{s,q}^*$, Hlawka's and Zaremba's identity boils down to

$$e(f, \mathcal{P}) = (-1)^{s+1} \int_{[0,1]^s} \frac{\partial^s f}{\partial x}(x) \, \Delta_{\mathcal{P},N}(x) \, dx \quad \text{for } f \in \mathcal{F}_{s,q}^*.$$

Applying the absolute value and Hölder's inequality, we obtain a first variant of a Koksma-Hlawka-type inequality.

Theorem 3.11 *Let $\mathcal{P}$ be an N-element point set in $[0, 1)^s$. Let $p, q \in [1, \infty]$ with $1/p + 1/q = 1$. Then for all $f \in \mathcal{F}_{s,q}^*$ we have*

$$|e(f, \mathcal{P})| \leq \|f\|_{s,q}^* L_{p,N}(\mathcal{P}).$$

Next, we aim at losing the boundary conditions in the definition of $\mathcal{F}_{s,q}^*$. To this end we put

$$\|f\|_{s,q} := \left(\sum_{u \subseteq [s]} \int_{[0,1]^{|u|}} \left| \frac{\partial^{|u|} f}{\partial x_u}(x_u, 1) \right|^q dx_u \right)^{1/q}$$

for $q \in [1, \infty)$ and

$$\|f\|_{s,\infty} := \max_{u \subseteq [s]} \sup_{x \in [0,1]^s} \left| \frac{\partial^{|u|} f}{\partial x_u}(x_u, 1) \right|$$

for $q = \infty$, where in both cases the term for $\mathfrak{u} = \emptyset$ corresponds to $|f(\mathbf{1})|$, and consider the corresponding function class

$$\mathcal{F}_{s,q} := \{ f \in \mathcal{H}_s \ : \ \|f\|_{s,q} < \infty \}.$$

Now we need to introduce a more general concept of discrepancy, which takes also projections of point sets onto faces of the unit cube into account.

Definition 3.12 For an N-element point set $\mathcal{P}$ in $[0, 1)^s$ and for $p \in [1, \infty]$ the *combined L_p discrepancy* is given by

$$L_{p,N,1}(\mathcal{P}) := \left(\sum_{\emptyset \neq u \subseteq [s]} \int_{[0,1]^{|u|}} |\Delta_{\mathcal{P},N}(\boldsymbol{t}_{\mathfrak{u}}, \mathbf{1})|^p \, \mathrm{d}\boldsymbol{t}_{\mathfrak{u}} \right)^{1/p},$$

if $p \in [1, \infty)$ and

$$L_{\infty,N,1}(\mathcal{P}) := \max_{\emptyset \neq u \subseteq [s]} \sup_{\boldsymbol{t}_{\mathfrak{u}} \in [0,1]^{|u|}} |\Delta_{\mathcal{P},N}(\boldsymbol{t}_{\mathfrak{u}}, \mathbf{1})| = D_N^*(\mathcal{P}),$$

if $p = \infty$.

We continue with the error analysis of QMC rules in $\mathcal{F}_{s,q}$. Applying Hölder's inequality to Hlawka's and Zaremba's identity from Theorem 3.10 again, now both for integrals and for sums, we conclude the following.

Theorem 3.13 *Let $\mathcal{P}$ be an N-element point set in $[0, 1)^s$, and let $p, q \in [1, \infty]$ be such that $1/p + 1/q = 1$. Then for all $f \in \mathcal{F}_{s,q}$ we have*

$$|e(f, \mathcal{P})| \leq \|f\|_{s,q} L_{p,N,1}(\mathcal{P}).$$

By choosing $q = 1$ and $p = \infty$ we get a "weak" form of the classical Koksma-Hlawka inequality as an immediate corollary.

Corollary 3.14 (Koksma-Hlawka inequality) *Let $\mathcal{P}$ be an N-element point set in $[0, 1)^s$. Then, for all functions f defined on $[0, 1]^s$ with $\|f\|_{s,1} < \infty$, we have*

$$|e(f, \mathcal{P})| \leq \|f\|_{s,1} D_N^*(\mathcal{P}).$$

All presented error estimates separate the influence of the function f and of the integration notes $\mathcal{P}$ on the integration error. Such results are usually called Koksma-Hlawka-type inequalities. The Koksma-Hlawka inequality in its original, more general form states that $|e(f, \mathcal{P})| \leq V(f) D_N^*(\mathcal{P})$, where $V(f)$ denotes the variation in the sense of Hardy and Krause. We do not give the definition of the latter notion here, but we mention that if all partial mixed derivatives of f up to order one in each variable are continuous on $[0, 1]^s$, then we have

$$V(f) = \sum_{\emptyset \neq u \subseteq [s]} \int_{[0,1]^{|u|}} \left| \frac{\partial^{|u|} f}{\partial \boldsymbol{x}_u}(\boldsymbol{x}_u, \boldsymbol{1}) \right| d\boldsymbol{x}_u = \|f\|_{s,1} - f(\boldsymbol{1}).$$

3.5 The Weighted Koksma-Hlawka Inequality

In the Koksma-Hlawka-type inequalites from the last section, it is obvious that all variables and groups of variables of the integrands are considered equally important. However, this may not adequately reflect the situation in many practical applications, where it may occur that not all variables or groups of variables have the same influence on the integral of a function. An extreme example is a function $f : [0, 1]^s \to \mathbb{R}$ of the form $f(x_1, \ldots, x_s) = g(x_1)$ with a univariate function $g : [0, 1] \to \mathbb{R}$. So f is nominally an s-variate function, but in fact it only depends on x_1 with no dependence on $x_2, \ldots, x_s$ at all. Or, less extreme, $f(x_1, \ldots, x_s) = \sum_{1 \leq i < j \leq s} g_{i,j}(x_i, x_j)$ with functions $g_{i,j} : [0, 1]^2 \to \mathbb{R}$. Again, f is nominally an s-variate function, but in fact it is a sum of just two-variate functions and the whole integration problem boils down to a sum of two-dimensional integration problems. Similar phenomena can occur in practical problems, though usually to a lesser extent than in our extreme examples. Such observations motivated I.H. Sloan and H. Woźniakowski in 1998 to introduce additional parameters—so-called *coordinate weights*—in the definition of the norms of function spaces studied in integration problems. Behind that idea is the intuition that if a problem depends on many variables, of which only few have significant contribution to the integral, it is natural to expect that the problem will be easier to solve than a problem where all variables have the same influence. It should behave more like a lower dimensional problem.

Now let $\gamma = \{\gamma_u : u \subseteq [s]\}$ be a set of positive reals which are called *coordinate weights* or simply *weights*. Usually, one uses the convention $\gamma_\emptyset := 1$ and we will do so as well. A prominent example of weights are so-called *product weights* which are of the form $\gamma_u = \prod_{j \in u} \gamma_j$ for a given sequence $(\gamma_j)_{j \geq 1}$ of positive reals. With a slight abuse of nomenclature, we also use the name *coordinate weights* for the elements γ_j of such a sequence.

Now the starting point for developing a weighted Koksma-Hlawka inequality is Theorem 3.10. Let $f \in \mathcal{H}_s$ and $\mathcal{P}$ in $[0, 1)^s$. Rewriting the identity of Hlawka and Zaremba by multiplying and dividing by γ_u we obtain

$$e(f, \mathcal{P}) = \sum_{\emptyset \neq u \subseteq [s]} (-1)^{|u|+1} \gamma_u \frac{1}{\gamma_u} \int_{[0,1]^{|u|}} \frac{\partial^{|u|} f}{\partial \boldsymbol{x}_u}(\boldsymbol{x}_u, \boldsymbol{1}) \Delta_{\mathcal{P},N}(\boldsymbol{x}_u, \boldsymbol{1}) d\boldsymbol{x}_u.$$

Then, applying Hölder's inequality for integrals and sums with $p, q \in [1, \infty]$ satisfying $1/p + 1/q = 1$ we obtain

$$
|e(f, \mathcal{P})| \leq \left(\sum_{\emptyset \neq \mathfrak{u} \subseteq [s]} \gamma_{\mathfrak{u}}^p \int_{[0,1]^{|\mathfrak{u}|}} |\Delta_{\mathcal{P},N}(\boldsymbol{x}_{\mathfrak{u}}, \mathbf{1})|^p \, d\boldsymbol{x}_{\mathfrak{u}} \right)^{1/p}
$$

$$
\times \left(\sum_{\emptyset \neq \mathfrak{u} \subseteq [s]} \frac{1}{\gamma_{\mathfrak{u}}^q} \int_{[0,1]^{|\mathfrak{u}|}} \left| \frac{\partial^{|\mathfrak{u}|} f}{\partial \boldsymbol{x}_{\mathfrak{u}}}(\boldsymbol{x}_{\mathfrak{u}}, \mathbf{1}) \right|^q \, d\boldsymbol{x}_{\mathfrak{u}} \right)^{1/q},
$$

with the usual adaptions if $p = \infty$ or $q = \infty$.

Similar to the above, we put

$$
\|f\|_{s,\gamma,q} := \left(\sum_{\mathfrak{u} \subseteq [s]} \frac{1}{\gamma_{\mathfrak{u}}^q} \int_{[0,1]^{|\mathfrak{u}|}} \left| \frac{\partial^{|\mathfrak{u}|} f}{\partial \boldsymbol{x}_{\mathfrak{u}}}(\boldsymbol{x}_{\mathfrak{u}}, \mathbf{1}) \right|^q \, d\boldsymbol{x}_{\mathfrak{u}} \right)^{1/q}
$$

if $q \in [1, \infty)$ and

$$
\|f\|_{s,\gamma,\infty} := \max_{\mathfrak{u} \subseteq [s]} \frac{1}{\gamma_{\mathfrak{u}}} \sup_{\boldsymbol{x} \in [0,1]^s} \left| \frac{\partial^{|\mathfrak{u}|} f}{\partial \boldsymbol{x}_{\mathfrak{u}}}(\boldsymbol{x}_{\mathfrak{u}}, \mathbf{1}) \right|
$$

if $q = \infty$, where in both cases the term for $\mathfrak{u} = \emptyset$ corresponds to $|f(\mathbf{1})|$, and consider the corresponding (weighted) function class

$$
\mathcal{F}_{s,\gamma,q} := \{ f \in \mathcal{H}_s \ : \ \|f\|_{s,\gamma,q} < \infty \}.
$$

Furthermore, we also introduce weighted versions of L_p discrepancy.

Definition 3.15 For an N-element point set $\mathcal{P}$ in $[0, 1)^s$ and for $p \in [1, \infty]$ the *weighted L_p discrepancy* is given by

$$
L_{p,N,\gamma}(\mathcal{P}) := \left(\sum_{\emptyset \neq \mathfrak{u} \subseteq [s]} \gamma_{\mathfrak{u}}^p \int_{[0,1]^{|\mathfrak{u}|}} |\Delta_{\mathcal{P},N}(\boldsymbol{t}_{\mathfrak{u}}, \mathbf{1})|^p \, d\boldsymbol{t}_{\mathfrak{u}} \right)^{1/p},
$$

if $p \in [1, \infty)$, and

$$
L_{\infty,N,\gamma}(\mathcal{P}) := \max_{\emptyset \neq \mathfrak{u} \subseteq [s]} \gamma_{\mathfrak{u}} \sup_{\boldsymbol{t}_{\mathfrak{u}} \in [0,1]^{|\mathfrak{u}|}} |\Delta_{\mathcal{P},N}(\boldsymbol{t}_{\mathfrak{u}}, \mathbf{1})|, \tag{3.12}
$$

if $p = \infty$. Traditionally, the weighted L_∞-discrepancy is called *weighted star discrepancy* and is denoted by $D_{N,\gamma}^*$.

Now we can state a weighted version of the Koksma-Hlawka inequality.

Theorem 3.16 *Let $\gamma = \{\gamma_{\mathfrak{u}} : \mathfrak{u} \subseteq [s]\}$ be a set of positive coordinate weights. Let $\mathcal{P}$ be an N-element point set in $[0, 1)^s$, and let $p, q \in [1, \infty]$ be such that $1/p + 1/q = 1$. Then for all $f \in \mathcal{F}_{s,\gamma,q}$ we have*

$$|e(f, \mathcal{P})| \leq \|f\|_{s,\gamma,q} L_{p,N,\gamma}(\mathcal{P}).$$

What is now the meaning of the weights in this context? To explain this let us switch to the worst-case error

$$e(\mathcal{F}_{s,\gamma,q}, \mathcal{P}) = \sup_{f \in \mathcal{F}_{s,\gamma,q}, \|f\|_{s,\gamma,q} \leq 1} |e(f, \mathcal{P})|.$$

Then the weighted Koksma-Hlawka inequality shows that this worst-case error is dominated by the weighted L_p discrepancy and in fact it can be shown (since the Hölder inequality is sharp) that we even have equality

$$e(\mathcal{F}_{s,\gamma,q}, \mathcal{P}) = L_{p,N,\gamma}(\mathcal{P}). \tag{3.13}$$

Now the role of the coordinate weights in the space $\mathcal{F}_{s,\gamma,q}$ is the following. Since the error of QMC algorithms is defined over the unit ball of $\mathcal{F}_{s,\gamma,q}$, a small weight $\gamma_{\mathfrak{u}}$ means that the L_q-norm of the partial derivative $\frac{\partial^{|\mathfrak{u}|} f}{\partial x_{\mathfrak{u}}}(x_{\mathfrak{u}}, 1)$ must also be small. In fact, we can even permit that the $\gamma_{\mathfrak{u}}$ are zero. In that case we adopt the convention that $0/0 = 0$ in the definition of the norm $\| \cdot \|_{s,\gamma,q}$; hence, $\gamma_{\mathfrak{u}} = 0$ implies that the functions must be constant with respect to $(x_{\mathfrak{u}}, 1)$.

It is clear that the unit ball of $\mathcal{F}_{s,\gamma,q}$ shrinks for small $\gamma_{\mathfrak{u}}$, and so it makes multivariate integration easier. This is reflected in the weighted discrepancy, which becomes also smaller for small weights compared to the combined discrepancy.

The weighted setting plays an important role in the study of the dependence of the integration error on the dimension s. Many problems are known to suffer from the curse of dimensionality in the unweighted setting, but often the curse can be broken by suitably chosen small weights. See also Sect. 4.6 and Chap. 6 for a discussion of this issue.

Further Reading

This chapter is mainly based on [33, Chap. 2] to which we refer for more detailed information. A good overview can be also found in the survey by Dick, Kuo, and Sloan [29]. For an introduction into the theory of reproducing kernel Hilbert spaces, we refer to the paper [5] of Aronszajn. The foundations of the theory of Hilbert spaces and of absolutely continuous functions can be found, for example, in [128]. Hickernell [57] was the first who introduced reproducing kernel Hilbert spaces in the context of QMC. A proof of the Koksma-Hlawka inequality in its general form can be found in the original paper by Hlawka [67] and in the book by Kuipers and Niederreiter [85]. The idea of introducing coordinate weights and to study a weighted Koksma-

Hlawka inequality stems from Sloan and Woźniakowski [137]. Nowadays, QMC rules are commonly studied in the weighted setting and it is natural to ask what the minimal conditions on the weights γ are in order to break the curse of dimensionality by means of QMC rules. For general information in this direction, we refer to the trilogy [118–120] by Novak and Woźniakowski. Concrete results for the (weighted) L_p discrepancy can be found in [100, 116, 117, 137] and the references therein.

Exercises

3.1 Let $x_0, x_1, \ldots, x_{N-1} \in [0, 1)$ and let τ be an arbitrary real number. Construct two continuous functions $f, g : [0, 1] \to \mathbb{R}$ for which we have $f(x_n) = g(x_n)$ for all $n = 0, 1, \ldots, N - 1$, but $\int_0^1 f(x)\,\mathrm{d}x = \tau + \int_0^1 g(x)\,\mathrm{d}x$. Can you find f, g that are infinitely smooth but still have that property?

3.2 Let

$$\mathcal{B}_r := \{p(x) = a_0 + a_1 x + \cdots + a_r x^r \ : \ a_0, \ldots, a_r \in \mathbb{R}, \ x \in [0, 1]\}$$

be the space of all polynomials on $[0, 1]$ of degree at most $r \in \mathbb{N}_0$. Define the inner product of $p(x) = a_0 + a_1 x + \cdots + a_r x^r$ and $q(x) = b_0 + b_1 x + \cdots + b_r x^r$ in $\mathcal{B}_r$ by

$$\langle p, q \rangle := a_0 b_0 + \cdots + a_r b_r.$$

Determine the reproducing kernel $K(x, y)$ for this space and show directly that K is symmetric, uniquely determined, and positive semi-definite.

3.3 Let

$$\mathcal{C}_r := \{a_0 + a_1 \exp(2\pi \mathrm{i}x) + \cdots + a_r \exp(2\pi \mathrm{i}rx) \ : \ a_0, \ldots, a_r \in \mathbb{C}, \ x \in [0, 1]\}$$

be the space of all trigonometric polynomials on $[0, 1]$ of degree at most $r \in \mathbb{N}_0$. Define the inner product of $f(x) = a_0 + a_1 \exp(2\pi \mathrm{i}x) + \cdots + a_r \exp(2\pi \mathrm{i}rx)$ and $g(x) = b_0 + b_1 \exp(2\pi \mathrm{i}x) + \cdots + b_r \exp(2\pi \mathrm{i}rx)$ in $\mathcal{C}_r$ by

$$\langle f, g \rangle := a_0 \overline{b_0} + \cdots + a_r \overline{b_r}.$$

Determine the reproducing kernel $K(x, y)$ for this space and show directly that K is symmetric, uniquely determined, and positive semi-definite.

3.4 Show that the function $K(x, y) = \min(1 - x, 1 - y)$ is the reproducing kernel for the space

$$\mathcal{H} = \{f : [0, 1] \to \mathbb{R} \ : \ f \text{ absolutely continuous, } \|f\| < \infty \text{ and } f(1) = 0\},$$

where $\|f\| = \langle f, f \rangle^{1/2}$ and $\langle f, g \rangle = \int_0^1 f'(x) g'(x)\,\mathrm{d}x$.

3.5 For an integer $b \geq 2$ let $\omega_b = \exp(2\pi \mathrm{i}/b)$. Let $k \in \mathbb{N}_0$ with b-adic expansion $k = \kappa_0 + \kappa_1 b + \cdots + \kappa_{a-1} b^{a-1}$. The kth b-adic Walsh function $_b\mathrm{wal}_k : \mathbb{R} \to \mathbb{C}$, periodic with period one, is, for $x \in [0, 1)$, defined as

$$_b\mathrm{wal}_k(x) := \omega_b^{\kappa_0\xi_1+\kappa_1\xi_2+\cdots+\kappa_{a-1}\xi_a},$$

where $x = \xi_1 b^{-1} + \xi_2 b^{-2} + \xi_3 b^{-3} + \cdots$ is the b-adic expansion of x (unique in the sense that infinitely many of the digits ξ_i must be different from $b - 1$).

The system $\{\,_b\mathrm{wal}_k : k \in \mathbb{N}_0\}$ is a complete orthonormal basis of $L_2([0, 1])$ and it is called the *b-adic Walsh function system*.

For $\alpha > 1$ we define the Hilbert space $\mathcal{H}_{\mathrm{wal},b,\alpha} \subseteq L_1([0, 1])$ consisting of all functions f with absolutely convergent Walsh series

$$f(x) = \sum_{k=0}^{\infty} \widehat{f}_{\mathrm{wal}}(h)\,_b\mathrm{wal}_k(x) \quad \text{where} \quad \widehat{f}_{\mathrm{wal}}(h) = \int_0^1 f(x)\overline{_b\mathrm{wal}(x)}\,\mathrm{d}x$$

and with finite norm $\|f\|_{\mathrm{wal},b,\alpha} = \langle f, f\rangle_{\mathrm{wal},b,\alpha}^{1/2}$, where the inner product is

$$\langle f, g\rangle_{\mathrm{wal},b,\alpha} = \sum_{k=0}^{\infty} r_{\mathrm{wal},b,\alpha}(h)\widehat{f}(h)\overline{\widehat{g}(h)}$$

and $r_{b,\alpha}(0) = 1$ and, for $k \in \mathbb{N}$, $r_{b,\alpha}(k) = b^{-\alpha(1+\lfloor \log_b k\rfloor)}$.

Show that $\mathcal{H}_{\mathrm{wal},b,\alpha}$ is a reproducing kernel Hilbert space with kernel

$$K_{\mathrm{wal}}(x, y) = \sum_{k=0}^{\infty} r_{b,\alpha}(k)\,_b\mathrm{wal}_k(x)\overline{_b\mathrm{wal}_k(y)} \quad \text{for} \quad x, y \in [0, 1).$$

3.6 Show that the worst-case error of QMC integration in $\mathcal{B}_r$ from Exercise 3.5 based on a point set $\mathcal{P} = \{x_0, \ldots, x_{N-1}\}$ in $[0, 1)$ is given by

$$e^2(\mathcal{B}_r, \mathcal{P}) = \sum_{l=0}^{r} \left(\frac{1}{l+1} - \frac{1}{N} \sum_{n=0}^{N-1} x_n^l \right)^2.$$

Assume that $\mathcal{P}$ is given by $x_n = n/N$ for $n = 0, \ldots, N - 1$. Show that then $e(\mathcal{B}_r, \mathcal{P}) \asymp_r 1/N$.

3.7 Show that the worst-case error of QMC integration in $\mathcal{C}_r$ from Exercise 3.5 based on a point set $\mathcal{P} = \{x_0, \ldots, x_{N-1}\}$ in $[0, 1)$ is given by

$$e^2(\mathcal{C}_r, \mathcal{P}) = \sum_{l=1}^{r} \left| \frac{1}{N} \sum_{n=0}^{N-1} \exp(2\pi \mathrm{i} l x_n) \right|^2.$$

Let $N > r$ be a prime number and assume that $\mathcal{P}$ is given by $x_n = n/N$ for $n = 0, \ldots, N - 1$. Show that then $e(\mathcal{C}_r, \mathcal{P}) = 0$.

3.8 Let $K_1(x, y) = 1 + \min(1 - x, 1 - y)$. Show that the local discrepancy of a point set $\mathcal{P} = \{x_0, \ldots, x_{N-1}\}$ satisfies

$$\Delta_{\mathcal{P},N}(y) = \frac{\mathrm{d}}{\mathrm{d}y}\left(\int_0^1 K_1(x, y)\,\mathrm{d}x - \frac{1}{N}\sum_{n=0}^{N-1} K_1(x_n, y) \right) \tag{3.14}$$

for all $y \notin \mathcal{P}$, such that in particular (3.14) holds for almost all $y \in [0, 1]$.

3.9 For $\boldsymbol{x} = (x_1, \ldots, x_s)$ and $\boldsymbol{y} = (y_1, \ldots, y_s)$ in $[0, 1]^s$ define

$$K(\boldsymbol{x}, \boldsymbol{y}) = \prod_{j=1}^{s} \min(1 - x_j, 1 - y_j)$$

and

$$\langle f, g \rangle = \int_{[0,1]^s} \frac{\partial^s f}{\partial \boldsymbol{x}}(\boldsymbol{x}) \frac{\partial^s g}{\partial \boldsymbol{x}}(\boldsymbol{x})\,\mathrm{d}\boldsymbol{x}.$$

Let $\mathcal{H}$ be the reproducing kernel Hilbert space with kernel K. Show that for every N-element point set $\mathcal{P}$ in $[0, 1)^s$ we have $e(\mathcal{H}, \mathcal{P}) = L_{2,N}(\mathcal{P})$.

3.10 Use Theorem 3.9 to compute $e(\mathcal{H}, \mathcal{P})$ from Exercise 3.5 and show in this way the formula from Exercise 2.4.

3.11 Let $\mathcal{H}_s$ be the reproducing kernel Hilbert space with reproducing kernel (3.10) and inner product (3.11). Show that

$$e^2(\mathcal{H}_s, \mathcal{P}) = \sum_{\emptyset \neq \mathfrak{u} \subseteq [s]} (L_{2,N}(\mathcal{P}_\mathfrak{u}))^2$$

$$= \frac{4^s}{3^s} - \frac{2}{N}\sum_{n=1}^{N}\prod_{j=1}^{s}\frac{3 - x_{n,j}^2}{2} + \frac{1}{N^2}\sum_{n,m=1}^{N}\prod_{j=1}^{s}[1 + \min(1 - x_{n,j}, 1 - x_{m,j})],$$

where $\mathcal{P}_\mathfrak{u}$ stands for the projection of the points in $\mathcal{P}$ onto the coordinates in $\mathfrak{u}$ and $L_{2,N}(\mathcal{P}_\mathfrak{u})$ stands for the L_2 discrepancy of $\mathcal{P}_\mathfrak{u}$. *Remark*: Note that this result is the multi-dimensional version of (3.7).

3.12 Let $\mathcal{H}$ be a Hilbert space of functions $f : [0, 1]^s \to \mathbb{C}$ with inner product $\langle \cdot, \cdot \rangle$ and norm $\|\cdot\| = \langle \cdot, \cdot \rangle^{1/2}$, and with a reproducing kernel K.

For every $i = 1, \ldots, k$ let $\mathcal{P}_i$ be an N_i-element point set in $[0, 1)^s$. Let $\mathcal{P}$ be the superposition of the $\mathcal{P}_i$, where repeated elements are allowed. Let $N = N_1 + \cdots + N_k$. Show the so-called *triangle inequality for the worst-case error*:

$$e(\mathcal{H}, \mathcal{P}) \le \sum_{i=1}^{k} \frac{N_i}{N} e(\mathcal{H}, \mathcal{P}_i).$$

3.13 Let K be a reproducing kernel. Show that condition $\mathbf{C}$ implies that

$$\int_{[0,1]^{2s}} K(\boldsymbol{x}, \boldsymbol{y})\,\mathrm{d}\boldsymbol{x}\,\mathrm{d}\boldsymbol{y} < \infty.$$

3.14 Let $\mathcal{H}$ be a Hilbert space of functions $f : [0, 1] \to \mathbb{C}$ with inner product $\langle \cdot, \cdot \rangle$ and norm $\| \cdot \| = \langle \cdot, \cdot \rangle^{1/2}$, and with a reproducing kernel K that satisfies condition **C**. The so-called *initial error* is defined as

$$e(\mathcal{H}, 0) := \sup_{f \in \mathcal{H}, \|f\| \leq 1} \left| \int_{[0,1]^s} f(x) \, dx \right| .$$

Show that

$$e(\mathcal{H}, 0) = \int_{[0,1]^{2s}} K(x, y) \, dx \, dy .$$

Remark: The initial error is often used as a reference value for the worst-case error.

3.15 Let $\mathcal{H}$ be a Hilbert space of functions $f : [0, 1] \to \mathbb{C}$ with inner product $\langle \cdot, \cdot \rangle$ and norm $\| \cdot \| = \langle \cdot, \cdot \rangle^{1/2}$, and with a reproducing kernel K satisfying condition **C**.

Rather than QMC rules we now consider more general so-called *linear integration rules* of the form

$$Q_{\mathcal{P}, w}(f) := \sum_{n=0}^{N-1} w_n f(x_n),$$

where $w = (w_0, \ldots, w_{N-1}) \in \mathbb{R}^N$ and $\mathcal{P} = \{x_0, \ldots, x_{N-1}\} \subseteq [0, 1)^s$ are given. Note that a QMC rule is obtained by choosing $w = (N^{-1}, \ldots, N^{-1})$. The worst-case integration error for a linear rule is defined as

$$e(\mathcal{H}, \mathcal{P}, w) := \sup_{f \in \mathcal{H}, \|f\| \leq 1} \left| \int_{[0,1]^s} f(x) \, dx - Q_{\mathcal{P}, w}(f) \right| .$$

Show that

$$e^2(\mathcal{H}, \mathcal{P}, w) = \int_{[0,1]^2} K(x, y) \, dx \, dy - 2 \sum_{n=0}^{N-1} w_n \int_{[0,1]^s} K(x_n, y) \, dy$$

$$+ \sum_{n,m=0}^{N-1} w_n w_m K(x_n, x_m) .$$

3.16 Let $\mathcal{H}$ be a reproducing kernel Hilbert space with kernel K satisfying

$$\int_{[0,1]^s} K(x, x) \, dx < \infty .$$

As integration nodes $\mathcal{P}$ choose realizations of N independent and uniformly distributed random variables $X_0, \ldots, X_{N-1}$ in $[0, 1]^s$. Show that the QMC mean square worst-case error $\mathbb{E}[e^2(\mathcal{H}, \mathcal{P})]$ is given by

$$\mathbb{E}[e^2(\mathcal{H}, \mathcal{P})] = \frac{1}{N}\left[\int_{[0,1]^s} K(\boldsymbol{x},\boldsymbol{x})\,\mathrm{d}\boldsymbol{x} - \int_{[0,1]^{2s}} K(\boldsymbol{x},\boldsymbol{y})\,\mathrm{d}\boldsymbol{x}\,\mathrm{d}\boldsymbol{y}\right].$$

3.17 We give an elementary proof of the identity of Hlawka and Zaremba: Let $f : [0,1]^s \to \mathbb{R}$ be such that for every $\mathfrak{u} \subseteq [s]$ the partial derivatives $\frac{\partial^{|\mathfrak{u}|}}{\partial \boldsymbol{x}_{\mathfrak{u}}} f$ exist and are continuous. Let further $\mathcal{P}$ be an N-point set in $[0,1)^s$. Then

$$e(f,\mathcal{P}) = \sum_{\emptyset \neq \mathfrak{u} \subseteq [s]} (-1)^{|\mathfrak{u}|+1} \int_{[0,1]^{|\mathfrak{u}|}} \frac{\partial^{|\mathfrak{u}|} f}{\partial \boldsymbol{x}_{\mathfrak{u}}}(\boldsymbol{x}_{\mathfrak{u}}, \mathbf{1})\Delta_{\mathcal{P},N}(\boldsymbol{x}_{\mathfrak{u}}, \mathbf{1})\,\mathrm{d}\boldsymbol{x}_{\mathfrak{u}}. \quad (3.15)$$

(a) Show that it suffices to prove (3.15) for one-element point sets.

(b) Let $\boldsymbol{t} = (t_1, \ldots, t_s) \in [0,1)^s$. Prove by induction on s that

$$\int_{[0,1]^s} \frac{\partial^s}{\partial \boldsymbol{x}} f(\boldsymbol{x}) \prod_{j=1}^{s} \chi_{[0,x_j)}(t_j)\,\mathrm{d}\boldsymbol{x} = \sum_{\mathfrak{v} \subseteq [s]} (-1)^{|\mathfrak{v}|} f(\boldsymbol{t}_{\mathfrak{v}}, \mathbf{1}).$$

(c) Prove by induction on s that

$$\int_{[0,1]^s} \frac{\partial^s}{\partial \boldsymbol{x}} f(\boldsymbol{x}) \prod_{j=1}^{s} x_j\,\mathrm{d}\boldsymbol{x} = \int_{[0,1]^s} \sum_{\mathfrak{v} \subseteq [s]} (-1)^{|\mathfrak{v}|} f(\boldsymbol{x}_{\mathfrak{v}}, \mathbf{1})\,\mathrm{d}\boldsymbol{x}.$$

(d) Prove by induction on s that

$$\sum_{\mathfrak{u} \subseteq [s]} (-1)^{|\mathfrak{u}|+1} \sum_{\mathfrak{v} \subseteq \mathfrak{u}} (-1)^{|\mathfrak{v}|}(f(\boldsymbol{t}_{\mathfrak{v}}, \mathbf{1}) - f(\boldsymbol{x}_{\mathfrak{v}}, \mathbf{1})) = f(\boldsymbol{x}) - f(\boldsymbol{t}).$$

(e) Deduce (3.15).

3.18 In literature, the weighted star discrepancy is sometimes defined as

$$D^*_{N,\boldsymbol{\gamma}}(\mathcal{P}) = \sup_{\boldsymbol{t} \in [0,1]^s} \max_{\emptyset \neq \mathfrak{u} \subseteq [s]} \gamma_{\mathfrak{u}}|\Delta_{\mathcal{P},N}(\boldsymbol{t}_{\mathfrak{u}}, \mathbf{1})|.$$

Show that this definition is in accordance with the definition given in Definition 3.15, where the order of sup and max is interchanged.

3.19 Prove that for the weighted star discrepancy of an N-element point set $\mathcal{P}$ in $[0,1)^s$ we have

$$D^*_{N,\boldsymbol{\gamma}}(\mathcal{P}) \leq \sum_{\emptyset \neq \mathfrak{u} \subseteq [s]} \gamma_{\mathfrak{u}} D^*_N(\mathcal{P}_{\mathfrak{u}}),$$

where $\mathcal{P}_{\mathfrak{u}} \subseteq [0,1)^{|\mathfrak{u}|}$ consists of the points from $\mathcal{P}$, projected onto the coordinates with index $j \in \mathfrak{u}$, and D^*_N is the usual (unweighted) star discrepancy.

Lattice Point Sets

4.1 Definition and Discrepancy Estimates

We have shown in Proposition 2.6 that the infinite sequence $(\{n\boldsymbol{\alpha}\})_{n\in\mathbb{N}_0}$ is uniformly distributed modulo one under a certain condition on the vector $\boldsymbol{\alpha} \in \mathbb{R}^s$. In this chapter we consider "finite" versions of such sequences which are referred to as lattice point sets.

Definition 4.1 Let $s, N \in \mathbb{N}$, $N \geq 2$, and let $\boldsymbol{g} \in \mathbb{Z}^s$. The point set $\mathcal{P}(\boldsymbol{g}, N) = \{\boldsymbol{x}_0, \ldots, \boldsymbol{x}_{N-1}\}$, with

$$\boldsymbol{x}_n := \left\{ \frac{n}{N} \boldsymbol{g} \right\} \quad \text{for} \quad n = 0, 1, \ldots, N - 1,$$

where the fractional part $\{\cdot\}$ is applied component-wise, is called a *lattice point set*. The vector $\boldsymbol{g}$ is called the *generating vector* of the lattice point set.

A lattice point set is obtained by successive addition modulo one of the vector $(1/N)\boldsymbol{g}$. This is illustrated in Fig. 4.1. Another example, consisting of $N = 987$ elements, is illustrated in Fig. 4.2.

Every element of a lattice point set $\mathcal{P}(\boldsymbol{g}, N) = \{\boldsymbol{x}_0, \ldots, \boldsymbol{x}_{N-1}\}$ with generating vector $\boldsymbol{g} \in \mathbb{Z}^s$ is of the form $\boldsymbol{x}_n = \{\boldsymbol{y}_n/N\}$ with $\boldsymbol{y}_n = n\boldsymbol{g} \in \mathbb{Z}^s$. Hence, for a discrepancy estimate we can apply Theorem 2.30. In this way we obtain

$$D_N(\mathcal{P}(\boldsymbol{g}, N)) \leq 1 - \left(1 - \frac{1}{N}\right)^s + \sum_{\boldsymbol{h} \in C_s^*(N)} \frac{1}{r(\boldsymbol{h}, N)} \left| \frac{1}{N} \sum_{n=0}^{N-1} \exp(2\pi \mathrm{i} n \boldsymbol{h} \cdot \boldsymbol{g}/N) \right|.$$

© The Author(s), under exclusive license to Springer Nature Switzerland AG 2026
G. Leobacher and F. Pillichshammer, *Introduction to Quasi-Monte Carlo Integration and Applications*, Compact Textbooks in Mathematics,
https://doi.org/10.1007/978-3-032-05446-3_4

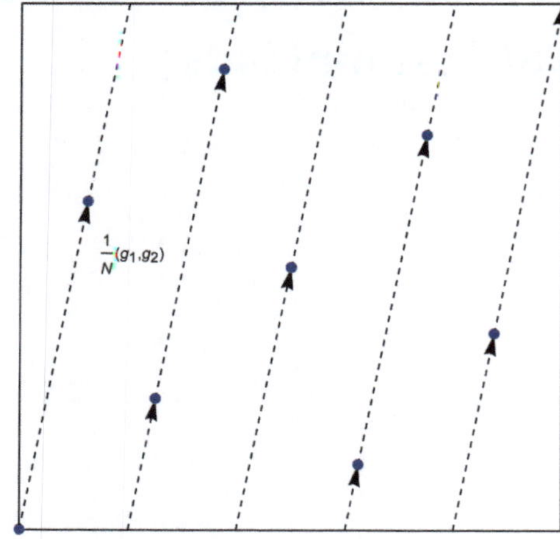

Fig. 4.1 A two-dimensional lattice point set with $N = 8$ elements and generating vector $\boldsymbol{g} = (1, 5)$. The point set is obtained by successive addition modulo one of the vector $(1/8, 5/8)$

Fig. 4.2 A two-dimensional lattice point set with $N = 987$ elements and generating vector $\boldsymbol{g} = (1, 610)$

Using the formula for geometric sums we obtain the following important property of lattice point sets:

$$\sum_{n=0}^{N-1} \exp(2\pi\,\mathrm{i}n\boldsymbol{h} \cdot \boldsymbol{g}/N) = \begin{cases} N & \text{if } \boldsymbol{h} \cdot \boldsymbol{g} \equiv 0 \pmod{N}, \\ 0 & \text{if } \boldsymbol{h} \cdot \boldsymbol{g} \not\equiv 0 \pmod{N}. \end{cases} \tag{4.1}$$

Definition 4.2 The set

$$\mathcal{L}(\boldsymbol{g}, N) := \{\boldsymbol{h} \in \mathbb{Z}^s \,:\, \boldsymbol{h} \cdot \boldsymbol{g} \equiv 0 \pmod{N}\}$$

is called the *dual lattice* of the lattice point set $\mathcal{P}(\boldsymbol{g}, N)$.

Hence the above discrepancy estimate for lattice point sets reduces to

$$D_N(\mathcal{P}(\boldsymbol{g}, N)) \leq 1 - \left(1 - \frac{1}{N}\right)^s + \sum_{\boldsymbol{h} \in C_s^*(N) \cap \mathcal{L}(\boldsymbol{g}, N)} \frac{1}{r(\boldsymbol{h}, N)}. \qquad (4.2)$$

For $h \in \mathbb{Z}$ we define $r_1(h) = \max(1, |h|)$ and for $\boldsymbol{h} = (h_1, \ldots, h_s) \in \mathbb{Z}^s$ we define $r_1(\boldsymbol{h}) = \prod_{j=1}^{s} r_1(h_j)$. Then for any $\boldsymbol{h} \in C_s^*(N)$ we have $r(\boldsymbol{h}, N) \geq 2r_1(\boldsymbol{h})$. This inequality follows from the fact that $\sin(\pi t) \geq 2t$ for $0 \leq t \leq 1/2$. Now we define

$$R_N(\boldsymbol{g}) := \sum_{\boldsymbol{h} \in C_s^*(N) \cap \mathcal{L}(\boldsymbol{g}, N)} \frac{1}{r_1(\boldsymbol{h})}. \qquad (4.3)$$

Summing up we obtain the following estimate for the discrepancy of a lattice point set.

Proposition 4.3 *For the discrepancy of a lattice point set* $\mathcal{P}(\boldsymbol{g}, N)$ *in* $[0, 1)^s$ *with* $\boldsymbol{g} \in \mathbb{Z}^s$ *and* $N \geq 2$ *we have*

$$D_N(\mathcal{P}(\boldsymbol{g}, N)) \leq \frac{s}{N} + \frac{1}{2} R_N(\boldsymbol{g}).$$

Now, for given N, we aim at finding a generating vector $\boldsymbol{g}$ with low value for the quantity R_N. Such generating vectors yield a low discrepancy for the corresponding lattice point set.

One way of finding generating vectors with small value of R_N is a computer search. Note that it follows from the definition of R_N that it is enough to search for $\boldsymbol{g}$ within the finite set $\{0, \ldots, N-1\}^s$. Nevertheless, apart from small values of N and s, an exhaustive search for $\boldsymbol{g}$ is practically not feasible, since one still has to check N^s integer vectors. At this point we use an idea that was first used by N.M. Korobov in the 1960s and which was later re-invented by I.H. Sloan and A.V. Reztsov in 2002. This basic idea is to construct a generating vector component-by-component (CBC). That is, for given N, one starts with a sufficiently good one-dimensional generator g_1. To this generator one appends a second component g_2 which is chosen as to minimize a desired figure of merit, in our case, the quantity R_N. In a next step, one appends to the now two-dimensional generator (g_1, g_2) a third component g_3 which is again chosen as to minimize the desired figure of merit. This procedure is repeated until one obtains an s-dimensional generating vector. In each of the s steps the search space $\{0, \ldots, N-1\}$ has cardinality N and hence the overall search space for the component-by-component method reduces to a size of sN. Therefore, this provides a feasible way of finding a generating vector which is (hopefully) "good" with respect to the involved figure of merit.

It is obvious that we may exclude the element 0 from the search space. Let in the following $G_N := \{1, \ldots, N-1\}$.

Algorithm 4.4 (*CBC algorithm*) Let $s, N \in \mathbb{N}$.

1. Choose $g_1 = 1$.
2. For $d = 1, 2, \ldots, s - 1$, choose $g_{d+1} \in G_N$ to minimize $R_N((g_1, \ldots, g_d, z))$ as a function of $z \in G_N$.

Having formulated a construction algorithm, there is still the question whether the resulting generating vector is good in the sense that it yields a sufficiently small value for the figure of merit R_N, and hence also for the discrepancy of the corresponding lattice point set. This question is answered in the affirmative in Theorem 4.5. For technical reasons we restrict ourselves to integers N which are prime numbers.

Theorem 4.5 *Let N be a prime number. If the lattice point $\boldsymbol{g} = (g_1, \ldots, g_s)$ is constructed using Algorithm 4.4, then we have*

$$R_N(\boldsymbol{g}) \leq \frac{1}{N-1}(1 + S_N)^s,$$

where $S_N := \sum_{h \in C_1^*(N)} |h|^{-1}$.

Proof We prove

$$R_N(\boldsymbol{g}_d) \leq \frac{1}{N-1}(1 + S_N)^d, \tag{4.4}$$

where $\boldsymbol{g}_d = (g_1, \ldots, g_d)$ for $d \in \{1, \ldots, s\}$ by induction on d.

Since N is a prime number it follows that $R_N(z) = 0$ for all $z \in G_N$, which, in particular, applies for $g_1 = 1$.

Let $d \in \{1, \ldots, s - 1\}$ and assume that (4.4) holds. Now, with some abuse of notation, we consider the $(d + 1)$-dimensional lattice point $(\boldsymbol{g}_d, g_{d+1}) := (g_1, \ldots, g_d, g_{d+1})$. According to Algorithm 4.4, g_{d+1} is a minimizer of the function $R_N((\boldsymbol{g}_d, \cdot))$ over G_N. Therefore,

$$R_N((\boldsymbol{g}_d, g_{d+1})) \leq \frac{1}{N-1} \sum_{g_{d+1}=1}^{N-1} \sum_{\substack{(\boldsymbol{h},h_{d+1}) \in C_{d+1}^*(N) \\ \boldsymbol{h} \cdot \boldsymbol{g}_d + h_{d+1}g_{d+1} \equiv 0 \pmod{N}}} \frac{1}{r_1(\boldsymbol{h})} \frac{1}{r_1(h_{d+1})}$$

$$= \sum_{(\boldsymbol{h},h_{d+1}) \in C_{d+1}^*(N)} \frac{1}{r_1(\boldsymbol{h})} \frac{1}{r_1(h_{d+1})} \frac{1}{N-1} \sum_{\substack{g_{d+1} \in G_N \\ \boldsymbol{h} \cdot \boldsymbol{g}_d + h_{d+1}g_{d+1} \equiv 0 \pmod{N}}} 1,$$

where we just interchanged the order of summation. Decomposing the range of summation according to $C_{d+1}^*(N) = (C_d^*(N) \times \{\boldsymbol{0}\}) \cup (C_d(N) \times C_1^*(N))$ we obtain

$$R_N((\boldsymbol{g}_d, g_{d+1}))$$

$$\leq R_N(\boldsymbol{g}_d) + \sum_{\boldsymbol{h} \in C_d(N)} \frac{1}{r_1(\boldsymbol{h})} \sum_{h_{d+1} \in C_1^*(N)} \frac{1}{r_1(h_{d+1})} \frac{1}{N-1} \sum_{\substack{g_{d+1} \in G_N \\ h_{d+1}g_{d+1} \equiv -\boldsymbol{h} \cdot \boldsymbol{g}_d \pmod{N}}} 1.$$

Now we use that N is a prime number. In this case we have that the congruence $h_{d+1}g_{d+1} \equiv -\boldsymbol{h} \cdot \boldsymbol{g}_d \pmod{N}$ has exactly one solution $g_{d+1} \in G_N$ whenever $\boldsymbol{h} \cdot \boldsymbol{g}_d \not\equiv 0 \pmod{N}$ and no solution in G_N whenever $\boldsymbol{h} \cdot \boldsymbol{g}_d \equiv 0 \pmod{N}$. This leads to

$$
\begin{aligned}
R_N((\boldsymbol{g}_d, g_{d+1})) &\leq R_N(\boldsymbol{g}_d) + \frac{1}{N-1} \sum_{\boldsymbol{h} \in C_d(N)} \frac{1}{r_1(\boldsymbol{h})} \sum_{h_{d+1} \in C_1^*(N)} \frac{1}{r_1(h_{d+1})} \\
&= R_N(\boldsymbol{g}_d) + \frac{S_N}{N-1} \sum_{\boldsymbol{h} \in C_d(N)} \frac{1}{r_1(\boldsymbol{h})} \\
&= R_N(\boldsymbol{g}_d) + \frac{S_N}{N-1}(1 + S_N)^d \\
&\leq \frac{1}{N-1}(1 + S_N)^d + \frac{S_N}{N-1}(1 + S_N)^d \\
&= \frac{1}{N-1}(1 + S_N)^{d+1},
\end{aligned}
$$

where from line three to four of this displayed formula we applied the induction hypothesis for $R_N(\boldsymbol{g}_d)$. Hence the desired result follows by induction on d. $\qquad\square$

Using a straightforward estimate we have

$$
S_N \leq 2 \sum_{h=1}^{\lfloor N/2 \rfloor} \frac{1}{h} \leq 2 \left(1 + \int_1^{N/2} \frac{\mathrm{d}t}{t}\right) = 2(\log N + 1 - \log 2). \tag{4.5}
$$

Hence Algorithm 4.4 yields lattice points $\boldsymbol{g} \in G_N^s$ with

$$
\begin{aligned}
R_N(\boldsymbol{g}) &\leq \frac{1}{N-1}(1 + 2(\log N + 1 - \log 2))^s \\
&\leq \frac{2^{s+1}}{N}(\log N + 1)^s = O_s\left(\frac{(\log N)^s}{N}\right).
\end{aligned}
$$

This estimate is best possible in the order of magnitude in N as shown by G. Larcher in 1987, and stated in the following theorem.

Theorem 4.6 (Larcher) *For every dimension $s \geq 2$ there exists a $c_s > 0$, depending only on s, such that for all $\boldsymbol{g} \in \mathbb{Z}^s$ and every $N \in \mathbb{N}$ we have*

$$
R_N(\boldsymbol{g}) > c_s \frac{\ell(\log(N/\ell))^s}{N}
$$

with $\ell = \gcd(g_1, g_2, \ldots, g_s, N)$.

The sophisticated proof of this result is based on the theory of continued fractions, which is beyond the scope of this introductory book.

Combining Proposition 4.3 and Theorem 4.5 we obtain the following estimate for the discrepancy of lattice point sets whose generating vector is constructed by Algorithm 4.4.

Corollary 4.7 *Let N be a prime number. If the lattice point $g = (g_1, \ldots, g_s)$ is constructed with Algorithm 4.4, then we have*

$$D_N(\mathcal{P}(g, N)) \leq \frac{s}{N} + \frac{2^s}{N} \, (\log N + 1)^s \, .$$

Hence the CBC algorithm yields lattice points $g \in G_N^s$ for which the corresponding lattice point set $\mathcal{P}(g, N)$ satisfies

$$D_N(\mathcal{P}(g, N)) = O_s \left(\frac{(\log N)^s}{N} \right) .$$

Using a different approach, it was shown by V.A. Bykovskii in 2012, that, for every $s, N \in \mathbb{N}$, $s \geq 2$ and $N \geq 3$, there exists a generating vector $g \in \mathbb{Z}^s$ for which

$$D_N(\mathcal{P}(g, N)) = O_s \left(\frac{(\log N)^{s-1} \log \log N}{N} \right) . \tag{4.6}$$

For dimension $s = 2$ this result was already obtained by G. Larcher in 1986. It is currently the best result for the discrepancy of lattice point sets. In contrast to Corollary 4.7, it is a pure existence result.

4.2   The Fast CBC Construction

In the last section we learned that, for given N and s, we can construct a generating vector g with low discrepancy component-by-component. So far we did not think about the construction cost of the CBC algorithm. Is the CBC algorithm reasonably fast to construct generating vectors also for large N and large dimensions s? We will answer this question in the affirmative within this section.

First, let us think about the computation of the figure of merit R_N which is required in each step of the algorithm. Using (4.1) we have

$$R_N(\boldsymbol{g}) = \sum_{\boldsymbol{h} \in C_s^*(N) \cap \mathcal{L}(\boldsymbol{g},N)} \frac{1}{r_1(\boldsymbol{h})}$$

$$= \sum_{\boldsymbol{h} \in C_s^*(N)} \frac{1}{r_1(\boldsymbol{h})} \frac{1}{N} \sum_{n=0}^{N-1} \exp(2\pi \mathrm{i} n \boldsymbol{h} \cdot \boldsymbol{g}/N)$$

$$= -1 + \frac{1}{N} \sum_{n=0}^{N-1} \prod_{j=1}^{s} \left[\sum_{h \in C(N)} \frac{1}{r_1(h)} \exp(2\pi \mathrm{i} n h g_j/N) \right] \qquad (4.7)$$

$$= -1 + \frac{1}{N} \sum_{n=0}^{N-1} \prod_{j=1}^{s} \phi\left(\frac{ng_j}{N}\right),$$

where

$$\phi(x) := \sum_{h \in C(N)} \frac{1}{r_1(h)} \exp(2\pi \mathrm{i} h x).$$

Observe that $\phi(ng/N)$ for $n = 0, 1, \ldots, N-1$ and $g \in G_N$ takes on at most N different values, $\phi(ng/N) = \phi(k/N)$ for some $k \in \{0, 1, \ldots, N-1\}$. The possible values for ϕ can be precomputed and stored, and the computational cost for this is at most $O(N^2)$.

Once we have available the values $\phi(k/N)$ for $k \in \{0, 1, \ldots, N-1\}$, then we can compute the quantity $R_N(\boldsymbol{g})$ for a vector $\boldsymbol{g} \in G_N^s$ at a cost of at most $O(sN)$ elementary operations. Algorithm 4.4 needs at most sN evaluations of the quantity R_N in order to find the minimizer in G_N for all dimensions $d = 1, 2, \ldots, s$. Therefore we find that the construction cost of Algorithm 4.4 is at most of order $O(s^2 N^2)$. The order in which s appears can in fact easily be reduced to 1: After each step of the CBC construction we may store the values $\prod_{j=1}^{d} \phi(\frac{ng_j}{N})$, for all $n \in \{0, 1, \ldots, N-1\}$, which are needed for the computation of $R_N((g_1, \ldots, g_d, z))$, $z \in \{0, 1, \ldots, N-1\}$ in the next step. Thus the computation of the values $\prod_{j=1}^{d+1} \phi(\frac{ng_j}{N})$ for all $n \in \{0, 1, \ldots, N-1\}$ needs only N multiplications for every value of g_{d+1}, so that the total cost for computing the next component is $O(N^2)$. Note that updating the values $\prod_{j=1}^{d} \phi(\frac{ng_j}{N})$ to $\prod_{j=1}^{d+1} \phi(\frac{ng_j}{N})$, for all $n \in \{0, 1, \ldots, N-1\}$ only needs another N operations. Therefore the cost of computing a generating vector using Algorithm 4.4 is $O((s+1)N^2 + sN) = O(sN^2)$.

The quadratic occurrence of N in the cost means that the CBC algorithm in the form of Algorithm 4.4 can only be used for moderate values of N. However, we are interested in lattice point sets with a large number of points. Thus we need to reduce the factor N^2 in the construction cost to get an applicable construction method also for large values of N.

A breakthrough regarding this problem was obtained by D. Nuyens and R. Cools in 2006 using fast Fourier transform (FFT) methods for the construction of lattice point sets. This way it is possible to construct, for a given prime number N, an s-

dimensional generating vector $\boldsymbol{g}$ in $O(sN \log N)$ operations, compared to $O(s^2 N^2)$ operations for the usual CBC algorithm. In the following we explain this so-called *fast CBC construction*. Throughout we assume that N is an odd prime number.

First, the values $\phi(k/N)$ for $k \in \{0, 1, \dots, N-1\}$ can be computed using only $O(N \log N)$ operations. Consider the vector

$$\boldsymbol{\phi}_N := \left(\phi\left(\frac{0}{N}\right), \phi\left(\frac{1}{N}\right), \dots, \phi\left(\frac{N-1}{N}\right) \right)^\top,$$

where

$$\phi\left(\frac{k}{N}\right) = \sum_{h \in C(N)} \frac{1}{r_1(h)} \exp\left(2\pi \mathrm{i} \frac{hk}{N}\right)$$

$$= \sum_{h=0}^{\frac{N-1}{2}} \frac{1}{r_1(h)} \exp\left(2\pi \mathrm{i} \frac{hk}{N}\right) + \sum_{h=\frac{N+1}{2}}^{N-1} \frac{1}{r_1(h-N)} \exp\left(2\pi \mathrm{i} \frac{hk}{N}\right)$$

for $k = 0, 1, \dots, N-1$ (recall that N is an odd number).

For $m \in \mathbb{N}$, let $\omega_m := \exp(2\pi \mathrm{i}/m)$ and let $F_m := \frac{1}{\sqrt{m}} \left(\omega_m^{kl}\right)_{k,l=0}^{m-1}$ be the Fourier matrix of order m. Note that F_m is symmetric and that $F_m \overline{F}_m = I_m$, where I_m denotes the $m \times m$ identity matrix and where $\overline{F}_m$ denotes the complex conjugate of the matrix F_m.

With this notation we can write

$$\boldsymbol{\phi}_N = \sqrt{N} F_N \boldsymbol{x},$$

where

$$\boldsymbol{x} = \left(r_1(0)^{-1}, \dots, r_1\left(\frac{N-1}{2}\right)^{-1}, r_1\left(\frac{N-1}{2}\right)^{-1}, \dots, r_1(1)^{-1} \right)^\top.$$

Using FFT the matrix-vector product $F_N \boldsymbol{x}$ for the computation of $\boldsymbol{\phi}_N$ requires only $O(N \log N)$ operations. The procedure of the FFT will be explained later on in this section (see Theorem 4.10). Note that for the storage of the vector $\boldsymbol{\phi}_N$ we require a memory space of size $O(N)$.

Now we turn to the actual CBC algorithm. Using Algorithm 4.4 we construct, component-by-component, a generating vector $\boldsymbol{g} = (g_1, \dots, g_s) \in G_N^s$ such that $g_1 = 1$ and for all $d = 1, 2, \dots, s-1$ the quantity $R_N((g_1, \dots, g_d, z))$ is minimized with respect to $z \in G_N$ for fixed $g_1, \dots, g_d$.

Thus, assume that the components $g_1, \dots, g_d$ have already been constructed, and that the values $\eta_d(n) := \prod_{j=1}^{d} \phi(\frac{ng_j}{N})$ have been computed for all $n \in \{0, 1, \dots, N-1\}$.

We have to find $z \in G_N$ which minimizes

$$R_N(z) := R_N((g_1, \ldots, g_d, z)) = -1 + \frac{1}{N} \sum_{n=0}^{N-1} \phi\left(\frac{nz}{N}\right) \eta_d(n).$$

For $n = 0$ we have $\phi\left(\frac{ng}{N}\right) = \phi(0)$, which does not depend on z. Hence we can write

$$R_N(z) = -1 + \frac{\phi(0)^d}{N} + \frac{1}{N} \sum_{n=1}^{N-1} \phi\left(\frac{nz}{N}\right) \eta_d(n). \tag{4.8}$$

In formula (4.8) only the terms under the sum depend on z, so minimizing $R_N(z)$ is equivalent to minimizing

$$T_{N,d+1}(z) := \sum_{n=1}^{N-1} \phi\left(\frac{nz}{N}\right) \eta_d(n). \tag{4.9}$$

Now a key observation is that formula (4.9) does express the column vector $T_{N,d+1} = (T_{N,d+1}(1), \ldots, T_{N,d+1}(N-1))^\top$ as the product of the $(N-1) \times (N-1)$ matrix

$$\Omega_N := \left(\phi\left(\frac{nz}{N}\right)\right)_{\substack{z=1,\ldots,N-1 \\ n=1,\ldots,N-1}}$$

with the vector $\boldsymbol{\eta}_d := (\eta_d(1), \ldots, \eta_d(N-1))^\top$:

$$\boldsymbol{T}_{N,d+1} = \Omega_N \boldsymbol{\eta}_d.$$

Now, while the cost of general matrix-vector multiplication is quadratic in the dimension, it can be much lower for special matrices. Fortunately, it turns out that Ω_N is such a special matrix, and we will show in the remainder of this section that the cost for computation of $\boldsymbol{T}_{N,d+1} = \Omega_N \boldsymbol{\eta}_d$ is $O(N \log N)$.

Algorithm 4.8 (*Fast CBC algorithm*) Let $s \in \mathbb{N}$ and $N > 2$ be a prime number.

1. Compute the vector $\boldsymbol{\phi}_N$.
2. Set $g_1 = 1$ and $\boldsymbol{\eta}_1 = \boldsymbol{\phi}_N$.
3. For $d = 1, \ldots, s - 1$ do:

    (i) compute $\Omega_N \boldsymbol{\eta}_d =: \boldsymbol{T}_{N,d+1} = (T_{N,d+1}(1), \ldots, T_{N,d+1}(N-1))^\top$;
    (ii) set $g_{d+1} := \text{argmin}_{z \in G_N} T_{N,d+1}(z)$;
    (iii) compute $\eta_{d+1}(n) := \eta_d(n)\phi\left(\frac{ng_{d+1}}{N}\right)$ for all $n \in \{1, \ldots, N-1\}$.
    (iv) If $d < s - 1$ increase d by 1, otherwise exit.

It is now evident that, provided $T_{N,d+1} = \Omega_N \eta_d$ can be computed in $O(N \log N)$ operations, the total cost of computing $g = (g_1, \ldots, g_s)$ using Algorithm 4.8 is $O(sN \log N)$.

Now let us investigate the matrix-vector product $\Omega_N \eta$. The entries of this matrix are of the form $\phi\left(\frac{nz}{N}\right)$, where the product of the nonzero integers z and n has to be evaluated modulo N. Since N is a prime number, there exists a primitive root q modulo N, that is, there is a $q \in \{1, \ldots, N-1\}$ such that $\{q^k \pmod{N} : k = 0, \ldots, N-2\} = \{1, \ldots, N-1\}$. Thus, any product of nonzero integers z and n can be written—modulo N—as a power of q. Now we aim at permuting the rows of Ω_N by the positive powers of the primitive root q and the columns by the negative powers of q.

We describe this procedure, which is often called *Rader transform*, in detail. Let q be a primitive root modulo N. Define an $(N-1) \times (N-1)$ matrix $\Pi(q) = (\pi_{k,l}(q))_{k,l=1,\ldots,N-1}$ by

$$\pi_{k,l}(q) := \begin{cases} 1 \text{ if } k \equiv q^l \pmod{N}, \\ 0 \text{ otherwise.} \end{cases}$$

Since q is a primitive root modulo N, it follows that each row and each column of $\Pi(q)$ has exactly one entry which is 1 and the remaining entries are 0. Further, $\Pi(q)\Pi(q)^{\top} = I_{N-1}$, the $(N-1) \times (N-1)$ identity matrix. In fact, the matrix $\Pi(q)$ is a permutation matrix. That is, for any $(N-1) \times (N-1)$ matrix C, $\Pi(q)C$ just changes the order of the rows of C and $C\Pi(q)$ only changes the order of the columns of C.

Let $C = (c_{k,l})_{k,l=1,\ldots,N-1}$, where

$$C = \Pi(q)\Omega_N\Pi(q^{-1}).$$

Then

$$c_{k,l} = \sum_{u,v=1}^{N-1} \pi_{u,k}(q)\phi\left(\frac{uv}{N}\right)\pi_{v,l}(q^{-1}) = \phi\left(\frac{q^k q^{-l}}{N}\right).$$

Let $c_r = \phi(q^r/N)$. Note that $c_r = c_{r'}$ for all $r, r' \in \mathbb{Z}$ with $r \equiv r' \pmod{N-1}$, since $q^{N-1} \equiv 1 \pmod{N}$. Then $c_{k,l} = c_{k-l}$ and therefore we obtain

$$C = \begin{pmatrix} c_0 & c_{-1} & \cdots & \cdots & c_2 & c_1 \\ c_1 & c_0 & c_{-1} & \cdots & \cdots & c_2 \\ \vdots & \ddots & \ddots & \ddots & & \vdots \\ \vdots & & \ddots & \ddots & \ddots & \vdots \\ c_{-2} & \cdots & \cdots & c_1 & c_0 & c_{-1} \\ c_{-1} & c_{-2} & \cdots & \cdots & c_1 & c_0 \end{pmatrix}. \tag{4.10}$$

Matrices of the form C are called *circulant*. In general, a circulant matrix $C_m = \text{circ}(c)$ of order m is an $m \times m$ matrix defined by the m elements of a vector $c =$

$(c_0, c_1, \ldots, c_{m-1})^\top$ as

$$
C_m = \begin{pmatrix}
c_0 & c_{m-1} & \cdots & \cdots & c_2 & c_1 \\
c_1 & c_0 & c_{m-1} & \cdots & \cdots & c_2 \\
\vdots & \ddots & \ddots & \ddots & & \vdots \\
\vdots & & & \ddots & \ddots & \vdots \\
c_{m-2} & \cdots & & \cdots & c_1 & c_0 & c_{m-1} \\
c_{m-1} & c_{m-2} & \cdots & \cdots & c_1 & c_0
\end{pmatrix}.
$$

For such a matrix we set $c_{k'} = c_k$ for all $k, k' \in \mathbb{Z}$ such that $k \equiv k'$ (mod m).

Note that the circulant matrix C_m is fully determined by its first column c. Such matrices have a similarity transform which has the Fourier matrix as its eigenvectors. We shall prove this in the next lemma. Before we do so, recall that the Fourier matrix of order m is defined as $F_m = \frac{1}{\sqrt{m}} \left(\omega_m^{kl}\right)_{k,l=0}^{m-1}$ with $\omega_m = \exp(2\pi \mathrm{i}/m)$. Let $\mathrm{diag}(a_1, \ldots, a_m)$ be the $m \times m$ diagonal matrix $(A_{i,j})_{i,j=1}^{m}$ with $A_{i,i} = a_i$ for $i = 1, \ldots, m$ and $A_{i,j} = 0$ for $i \neq j$.

Lemma 4.9 *A circulant matrix* $C_m = \mathrm{circ}(c)$ *of order* m, *with first column* $c = (c_0, c_{m-1}, c_{m-2}, \ldots, c_1)^\top$, *has a similarity transform* $C_m = \overline{F}_m D F_m$, *with a diagonal matrix* $D = \mathrm{diag}(p_c(1), p_c(\omega_m), \ldots, p_c(\omega_m^{m-1}))$, *where* $p_c(z) := c_0 + c_1 z + \cdots + c_{m-1} z^{m-1}$.

Proof Let $D = (d_{k,l})_{k,l=0,\ldots,m-1}$ be given by $D = F_m C_m \overline{F}_m$. Then

$$
d_{k,l} = \frac{1}{m} \sum_{u,v=0}^{m-1} \omega_m^{ku} c_{u-v} \omega_m^{-lv} = \frac{1}{m} \sum_{u=0}^{m-1} \omega_m^{u(k-l)} \sum_{v=0}^{m-1} c_{u-v} \omega_m^{l(u-v)}.
$$

We have $\sum_{v=0}^{m-1} c_{u-v} \omega_m^{l(u-v)} = p_c(\omega_m^l)$, and therefore

$$
d_{k,l} = p_c(\omega_m^l) \frac{1}{m} \sum_{u=0}^{m-1} \omega_m^{u(l-k)}.
$$

The result now follows since $\frac{1}{m} \sum_{u=0}^{m-1} \omega_m^{u(l-k)} = 1$ if $l = k$ and 0 otherwise. $\square$

Thus, we have shown that

$$
\Omega_N = \Pi(q)^\top \overline{F}_{N-1} D F_{N-1} \Pi(q^{-1})^\top,
$$

where $\Pi(q)^\top$, $\Pi(q^{-1})^\top$ are permutation matrices, F_{N-1} is a Fourier matrix, $\overline{F}_{N-1}$ its complex conjugate, and D is a diagonal matrix.

For any vector $x = (x_1, \ldots, x_{N-1})^\top$ with $x_j \in \mathbb{C}$ for $j = 1, \ldots, N-1$ the matrix-vector multiplications $\Pi(g)^\top x$, Dx, and $\Pi(g^{-1})^\top x$ can be done in at most $O(N)$ operations. Hence it only remains to show that $F_{N-1} x$ and $\overline{F}_{N-1} x$ can be computed in $O(N \log N)$ operations. Since $\overline{F}_{N-1} x = \overline{F_{N-1} \overline{x}}$, it is enough to show that $F_{N-1} x$ can be computed in $O(N \log N)$ operations. Again this can be done using the FFT.

We illustrate this in the following. The Fourier matrix considered here is of size $N - 1$ (for the computation of the vector ϕ_N it is of size N—but this case can be handled analogously). The FFT is most efficient for Fourier matrices whose size is a power of 2. Hence we first aim at transforming the matrix-vector product of size $N - 1$ into a related one of size 2^{h+1}.

Note that the matrix C given in (4.10) is a $(N - 1) \times (N - 1)$ matrix. Let $h \in \mathbb{N}$ such that $2^{h-1} < N - 1 \leq 2^h$ and let $k \in \{0, \dots, 2^{h-1} - 1\}$ with $N - 1 + k = 2^h$. We extend C by k rows and k columns to obtain the $2^h \times 2^h$ matrix

$$
T = \begin{pmatrix}
c_0 & c_{-1} & \cdots & \cdots & c_2 & c_1 & 0 & 0 & \cdots & 0 \\
c_1 & c_0 & c_{-1} & \cdots & \cdots & c_2 & c_1 & 0 & \cdots & \ddots \\
\vdots & \ddots & \ddots & \ddots & & \vdots & c_2 & c_1 & \ddots & \ddots \\
\vdots & & & \ddots & \ddots & \ddots & \vdots & \vdots & \vdots & \ddots \\
c_{-2} & \cdots & \cdots & c_1 & c_0 & c_{-1} & \vdots & \vdots & \ddots & \\
c_{-1} & c_{-2} & \cdots & \cdots & c_1 & c_0 & c_{-1} & c_{-2} & \ddots & \ddots \\
0 & c_{-1} & c_{-2} & \cdots & \cdots & c_1 & c_0 & c_{-1} & \ddots & \ddots \\
0 & 0 & c_{-1} & c_{-2} & \cdots & \cdots & c_1 & c_0 & \ddots & \ddots \\
\vdots & \ddots & \ddots & & & & & \ddots & \ddots & \ddots \\
0 & \ddots & \ddots & \ddots & & & & \ddots & c_1 & c_0
\end{pmatrix}.
$$

T is no longer a circulant matrix, but it is still a Toeplitz matrix. To obtain a circulant matrix again, let

$$
R = \begin{pmatrix}
0 & \cdots & 0 & c_{-1} & \cdots & \cdots & c_2 & c_1 \\
\vdots & \ddots & \ddots & \ddots & \ddots & \ddots & \ddots & \ddots \\
0 & \ddots & 0 & \ddots & 0 & c_{-1} & \ddots & \ddots \\
c_1 & \ddots & \ddots & \ddots & \ddots & \ddots & \ddots & \ddots \\
\vdots & \ddots & \ddots & \ddots & \ddots & \ddots & \ddots & c_{-1} \\
\vdots & \ddots & c_1 & \ddots & \ddots & \ddots & \ddots & 0 \\
c_{-2} & \ddots & \ddots & \ddots & \ddots & \ddots & \ddots & \ddots \\
c_{-1} & c_{-2} & \ddots & \ddots & c_1 & 0 & \ddots & 0
\end{pmatrix}
$$

of size $2^h \times 2^h$. Then the matrix

$$
C' = \begin{pmatrix} T & R \\ R & T \end{pmatrix}
$$

is a circulant matrix of size $2^{h+1} \times 2^{h+1}$.

Let $x = (x_1, \ldots, x_{N-1})^\top$ be a complex column vector of length $N-1$ and assume we want to multiply the $(N-1) \times (N-1)$ matrix C given by (4.10) by x. Then we can do so by multiplying the matrix C' by the vector

$$x' = (x_1, \ldots, x_{N-1}, \underbrace{0, \ldots, 0}_{k+2^h \text{ zeros}})^\top$$

of length 2^{h+1}. Let $y' = (y_1, \ldots, y_{2^{h+1}})^\top = C'x'$. Then $y = (y_1, \ldots, y_{N-1})^\top = Cx$. Hence we can use Lemma 4.9 with $n = 2^{h+1}$ (rather than $n = N - 1$). This simplifies the FFT algorithm. The following result was shown by J.W. Cooley and J.W. Tukey in 1965.

Theorem 4.10 (Cooley and Tukey) *Let* $F_{2^{h+1}} = 2^{-(h+1)/2}(\omega_{2^{h+1}}^{kl})_{k,l=0,\ldots,2^{h+1}-1}$ *be a Fourier matrix. Let* $u = (u_0, \ldots, u_{2^{m+1}-1})^\top$ *be a given complex vector of length* 2^{h+1}. *Then the matrix-vector product* $F_{2^{h+1}}u$ *can be computed in* $O((h+1)2^{h+1})$ *operations.*

Proof Let $z = (z_0, \ldots, z_{2^{h+1}-1})^\top = F_{2^{h+1}}u$, where $u = (u_0, \ldots, u_{2^{h+1}-1})^\top$. Then

$$z_k = 2^{-\frac{h+1}{2}} \sum_{l=0}^{2^{h+1}-1} \omega_{2^{h+1}}^{kl} u_l \quad \text{for } k = 0, \ldots, 2^{h+1} - 1.$$

We compute the above sum recursively. Let $k = \kappa_0 + \kappa_1 2 + \cdots + \kappa_h 2^h$ and $l = \lambda_0 + \lambda_1 2 + \cdots + \lambda_h 2^h$, with dyadic digits $\kappa_0, \ldots, \kappa_h, \lambda_0, \ldots, \lambda_h \in \{0, 1\}$. Put

$$G_0(\lambda_0, \ldots, \lambda_{h-1}, \kappa_0) := \sum_{\lambda_h=0}^{1} \omega_2^{\kappa_0 \lambda_h} u_{\lambda_0 + \lambda_1 2 + \cdots + \lambda_h 2^h}, \tag{4.11}$$

and, for $r \in \{1, \ldots, h\}$, put

$$G_r(\lambda_0, \ldots, \lambda_{h-r-1}, \kappa_0, \ldots, \kappa_r) \tag{4.12}$$

$$:= \sum_{\lambda_{h-r}=0}^{1} \omega_{2^{r+1}}^{(\kappa_0 + \cdots + \kappa_r 2^r)\lambda_{h-r}} G_{r-1}(\lambda_0, \ldots, \lambda_{h-r}, \kappa_0, \ldots, \kappa_{r-1}).$$

Now we show by induction on r that for all $r = 1, \ldots, h$ we have

$$G_r(\lambda_0, \ldots, \lambda_{h-r-1}, \kappa_0, \ldots, \kappa_r)$$

$$= \sum_{\lambda_{h-r}=0}^{1} \cdots \sum_{\lambda_h=0}^{1} \omega_{2^{r+1}}^{(\kappa_0 + \cdots + \kappa_r 2^r)(\lambda_{h-r} + \cdots + \lambda_h 2^r)} u_{\lambda_0 + \cdots + \lambda_h 2^h}. \tag{4.13}$$

For $r = 1$ we have

$$G_1(\lambda_0, \ldots, \lambda_{h-2}, \kappa_0, \kappa_1) = \sum_{\lambda_{h-1}=0}^{1} \omega_4^{(\kappa_0+\kappa_1 2)\lambda_{h-1}} \sum_{\lambda_h=0}^{1} \omega_2^{\kappa_0\lambda_h} u_{\lambda_0+\cdots+\lambda_h 2^h}.$$

Now note that $\omega_2^{\kappa_0\lambda_h} = \omega_4^{2\kappa_0\lambda_h} = \omega_4^{2\kappa_0\lambda_h+4\kappa_1\lambda_h} = \omega_4^{(\kappa_0+2\kappa_1)2\lambda_h}$, and hence

$$G_1(\lambda_0, \ldots, \lambda_{h-2}, \kappa_0, \kappa_1) = \sum_{\lambda_{h-1}=0}^{1} \sum_{\lambda_h=0}^{1} \omega_4^{(\kappa_0+\kappa_1 2)(\lambda_{h-1}+\lambda_h 2)} u_{\lambda_0+\cdots+\lambda_h 2^h}.$$

Thus (4.13) holds for $r = 1$. Assume that (4.13) holds for the index r. Then

$$G_{r+1}(\lambda_0, \ldots, \lambda_{h-r-2}, \kappa_0, \ldots, \kappa_{r+1})$$

$$= \sum_{\lambda_{h-r-1}=0}^{1} \omega_{2^{r+2}}^{(\kappa_0+\cdots+\kappa_{r+1}2^{r+1})\lambda_{h-r-1}} G_r(\lambda_0, \ldots, \lambda_{h-r-1}, \kappa_0, \ldots, \kappa_r)$$

$$= \sum_{\lambda_{h-r-1}=0}^{1} \omega_{2^{r+2}}^{(\kappa_0+\cdots+\kappa_{r+1}2^{r+1})\lambda_{h-r-1}}$$

$$\times \sum_{\lambda_{h-r}=0}^{1} \cdots \sum_{\lambda_h=0}^{1} \omega_{2^{r+1}}^{(\kappa_0+\cdots+\kappa_r 2^r)(\lambda_{h-r}+\cdots+\lambda_h 2^r)} u_{\lambda_0+\cdots+\lambda_h 2^h}.$$

Since

$$\omega_{2^{r+1}}^{(\kappa_0+\cdots+\kappa_r 2^r)(\lambda_{h-r}+\cdots+\lambda_h 2^r)} = \omega_{2^{r+1}}^{(\kappa_0+\cdots+\kappa_r 2^r+\kappa_{r+1}2^{r+1})(\lambda_{h-r}+\cdots+\lambda_h 2^r)}$$

$$= \omega_{2^{r+2}}^{(\kappa_0+\cdots+\kappa_{r+1}2^{r+1})(\lambda_{h-r}2+\cdots+\lambda_h 2^{r+1})},$$

we obtain

$$G_{r+1}(\lambda_0, \ldots, \lambda_{h-r-2}, \kappa_0, \ldots, \kappa_{r+1})$$

$$= \sum_{\lambda_{h-r-1}=0}^{1} \cdots \sum_{\lambda_h=0}^{1} \omega_{2^{r+2}}^{(\kappa_0+\cdots+\kappa_{r+1}2^{r+1})(\lambda_{h-r-1}+\cdots+\lambda_h 2^{r+1})} u_{\lambda_0+\cdots+\lambda_h 2^h}$$

and hence (4.13) is shown.

In particular, (4.13) leads with the choice $r = h$ to the desired formula

$$z_k = 2^{-\frac{h+1}{2}} G_h(\kappa_0, \ldots, \kappa_h),$$

for all $k = \kappa_0 + \cdots + \kappa_h 2^h \in \{0, \ldots, 2^{h+1} - 1\}$.

For the computation of the z_k we compute G_r recursively. For $r = 0, \ldots, h$ compute

$$G_r(\lambda_0, \ldots, \lambda_{m-r-1}, \kappa_0, \ldots, \kappa_r)$$

for all $\lambda_0, \ldots, \lambda_{h-r-1}, \kappa_0, \ldots, \kappa_r \in \{0, 1\}$ using (4.11) and the recursion formula (4.12). For each r this requires $O(2^{h+1})$ operations.

Overall we require $O((h + 1)2^{h+1})$ operations to compute G_h. Hence, we can compute z in $O((h + 1)2^{h+1})$ operations. $\square$

Recall that we have chosen h such that $2^{h-1} < N - 1 \leq 2^h$ and hence $h < 1 + \log_2 N$. Thus we obtain the following result.

Corollary 4.11 *Let Ω_N be defined as above and let $\boldsymbol{x}$ be a complex column vector of length $N - 1$. Then the matrix-vector product $\Omega_N \boldsymbol{x}$ can be computed in $O(N \log N)$ operations.*

Therefore Algorithm 4.8 requires only $O(sN \log N)$ operations by using $O(N)$ memory space (compared to $O(sN^2)$ operations). This is a significant speed-up compared to a straightforward implementation of Algorithm 4.8. Only through this reduction of the construction cost does the CBC algorithm become applicable for the generation of lattice point sets with reasonably large cardinality.

4.3   The Weighted Star Discrepancy of Lattice Point Sets

Next, we consider the weighted star discrepancy of lattice point sets $\mathcal{P}(\boldsymbol{g}, N)$. Recall from Sect. 3.5, Eq. (3.13), that the weighted star discrepancy equals the worst-case error in the weighted space $\mathcal{F}_{s,\boldsymbol{\gamma},1}$.

Obviously, the projection of the elements of $\mathcal{P}(\boldsymbol{g}, N)$ onto the components with indices in $\mathfrak{u} \subseteq [s]$, $\mathfrak{u} \neq \emptyset$, yields the rank-1 lattice point set $\mathcal{P}(\boldsymbol{g}_\mathfrak{u}, N)$ with the $|\mathfrak{u}|$-dimensional generating vector $\boldsymbol{g}_\mathfrak{u}$ consisting of the components of $\boldsymbol{g}$ whose indices belong to $\mathfrak{u}$. With this observation we obtain directly from the definition of the weighted star discrepancy and Proposition 4.3 that

$$D_{N,\boldsymbol{\gamma}}^*(\mathcal{P}(\boldsymbol{g}, N)) \leq \max_{\emptyset \neq \mathfrak{u} \subseteq [s]} \gamma_\mathfrak{u} \left(\frac{|\mathfrak{u}|}{N} + \frac{1}{2} R_N(\boldsymbol{g}_\mathfrak{u}) \right).$$

From this estimate we can deduce the following result, which can be considered as the weighted version of Proposition 4.3.

Proposition 4.12 *The weighted star discrepancy for product weights $\boldsymbol{\gamma} = \{\gamma_\mathfrak{u} = \prod_{j \in \mathfrak{u}} \gamma_j : \mathfrak{u} \subseteq [s]\}$ with coordinate weights $(\gamma_j)_{j \geq 1}$ of a lattice point set $\mathcal{P}(\boldsymbol{g}, N)$ in $[0, 1)^s$ with $\boldsymbol{g} \in \mathbb{Z}^s$ and $N \geq 2$ satisfies*

$$D_{N,\boldsymbol{\gamma}}^*(\mathcal{P}(\boldsymbol{g}, N)) \leq \frac{1}{N} \prod_{j=1}^{s} (1 + 2\gamma_j) + \frac{1}{2} \widetilde{R}_{N,\boldsymbol{\gamma}}(\boldsymbol{g}), \tag{4.14}$$

where

$$\widetilde{R}_{N,\boldsymbol{\gamma}}(\boldsymbol{g}) := \sum_{\boldsymbol{h} \in C_s^*(N) \cap \mathcal{L}(\boldsymbol{g},N)} \frac{1}{\widetilde{r}_{1,\boldsymbol{\gamma}}(\boldsymbol{h})},$$

where, for a generic $\gamma > 0$ and $h \in \mathbb{Z}$,

$$\widetilde{r}_{1,\gamma}(h) := \begin{cases} (1+\gamma)^{-1} & \text{if } h = 0, \\ r_1(h)\gamma^{-1} & \text{if } h \neq 0, \end{cases}$$

and $\widetilde{r}_{1,\boldsymbol{\gamma}}(\boldsymbol{h}) := \prod_{j=1}^{s} \widetilde{r}_{1,\gamma_j}(h_j)$ for $\boldsymbol{h} = (h_1,\ldots,h_s) \in \mathbb{Z}^s$.

Proof First of all, for $\mathfrak{u} \neq \emptyset$ the trivial estimate $|\mathfrak{u}| \leq 2^{|\mathfrak{u}|}$ implies

$$\max_{\emptyset \neq \mathfrak{u} \subseteq [s]} \gamma_{\mathfrak{u}} \frac{|\mathfrak{u}|}{N} \leq \frac{1}{N} \max_{\emptyset \neq \mathfrak{u} \subseteq [s]} \prod_{j \in \mathfrak{u}} (2\gamma_j) \leq \frac{1}{N} \sum_{\mathfrak{u} \subseteq [s]} \prod_{j \in \mathfrak{u}} (2\gamma_j) = \frac{1}{N} \prod_{j=1}^{s} (1 + 2\gamma_j).$$

Next, we estimate

$$\max_{\emptyset \neq \mathfrak{u} \subseteq [s]} \gamma_{\mathfrak{u}} R_N(\boldsymbol{g}_{\mathfrak{u}}) \leq \sum_{\emptyset \neq \mathfrak{u} \subseteq [s]} \gamma_{\mathfrak{u}} R_N(\boldsymbol{g}_{\mathfrak{u}}).$$

From the form of R_N in (4.7) we obtain

$$R_N(\boldsymbol{g}_{\mathfrak{u}}) = -1 + \frac{1}{N} \sum_{n=0}^{N-1} \prod_{j \in \mathfrak{u}} \left[\sum_{h \in C(N)} \frac{1}{r_1(h)} \exp(2\pi \mathrm{i} nhg_j/N) \right].$$

Hence

$$\sum_{\emptyset \neq \mathfrak{u} \subseteq [s]} \gamma_{\mathfrak{u}} R_N(\boldsymbol{g}_{\mathfrak{u}})$$

$$= - \sum_{\emptyset \neq \mathfrak{u} \subseteq [s]} \gamma_{\mathfrak{u}} + \sum_{\emptyset \neq \mathfrak{u} \subseteq [s]} \frac{1}{N} \sum_{n=0}^{N-1} \prod_{j \in \mathfrak{u}} \gamma_j \left[\sum_{h \in C(N)} \frac{1}{r_1(h)} \exp(2\pi \mathrm{i} nhg_j/N) \right]$$

$$= 1 - \prod_{j=1}^{s} (1 + \gamma_j)$$

$$+ \frac{1}{N} \sum_{n=0}^{N-1} \left(-1 + \prod_{j=1}^{s} \left(1 + \gamma_j \sum_{h \in C(N)} \frac{1}{r_1(h)} \exp(2\pi \mathrm{i} nhg_j/N) \right) \right)$$

$$= - \prod_{j=1}^{s} (1 + \gamma_j) + \frac{1}{N} \sum_{n=0}^{N-1} \prod_{j=1}^{s} \left(\sum_{h \in C(N)} \frac{1}{\widetilde{r}_{1,\gamma_j}(h)} \exp(2\pi \mathrm{i} nhg_j/N) \right),$$

with $\widetilde{r}_{1,\gamma}(h)$ as defined in the statement of the proposition. Then we obtain

$$
\sum_{\emptyset \neq \mathfrak{u} \subseteq [s]} \gamma_{\mathfrak{u}} R_N(g_{\mathfrak{u}}) = \sum_{h \in C_s^*(N)} \frac{1}{\widetilde{r}_{1,\gamma}(h)} \frac{1}{N} \sum_{n=0}^{N-1} \exp(2\pi \mathrm{i} n h \cdot g / N)
$$

$$
= \sum_{h \in C_s^*(N) \cap \mathcal{L}(g,N)} \frac{1}{\widetilde{r}_{1,\gamma}(h)}
$$

$$
= \widetilde{R}_{N,\gamma}(g),
$$

with $\widetilde{R}_{N,\gamma}(g)$ as defined in the statement of the proposition. Altogether, we obtain
(4.14). $\qquad\square$

The quantity $\widetilde{R}_{N,\gamma}(g)$ in Theorem 4.14 is the weighted version of the quantity
$R_N(g)$ from (4.3). Both are of a very similar form, the only difference is in the
coefficients, which are $1/\widetilde{r}_{1,\gamma}(h)$ in the weighted case and $1/r_1(h)$ in the unweighted
case. One can now proceed in the same way as in the previous sections in order to
construct a good generating vector with respect to the weighted star discrepancy.

Algorithm 4.13 (*CBC algorithm*) Let $s, N \in \mathbb{N}$.

1. Choose $g_1 = 1$.
2. For $d = 1, 2, \ldots, s - 1$, choose $g_{d+1} \in G_N$ to minimize $\widetilde{R}_{N,\gamma}((g_1, \ldots, g_d, z))$
   as a function of $z \in G_N$.

With the same method as in the proof of Theorem 4.5 we obtain the following
result. The proof is left as an exercise.

Theorem 4.14 *Let N be a prime number. If the lattice point $g = (g_1, \ldots, g_s)$ is
constructed using Algorithm 4.13, then we have*

$$
\widetilde{R}_{N,\gamma}(g) \leq \frac{1}{N-1} \prod_{j=1}^{s} (1 + \gamma_j S_N),
$$

where $S_N := \sum_{h \in C_1^*(N)} |h|^{-1}$.

We can now apply Theorem 4.14 to Proposition 4.12 to get a result for the weighted
star discrepancy.

Corollary 4.15 *Let N be a prime number. If the lattice point $g = (g_1, \ldots, g_s)$ is
constructed using Algorithm 4.13, then we have*

$$
D_{N,\gamma}^*(\mathcal{P}(g, N)) \leq \frac{2}{N} \prod_{j=1}^{s} (1 + 3\gamma_j \log N).
$$

Proof The estimates in (4.14) and Theorem 4.14 yield

$$D^*_{N,\boldsymbol{\gamma}}(\mathcal{P}(\boldsymbol{g},N)) \le \frac{1}{N}\prod_{j=1}^{s}(1+2\gamma_j) + \frac{1}{2(N-1)}\prod_{j=1}^{s}(1+\gamma_j S_N).$$

From (4.5) we know that $S_N \le 2(\log N + 1 - \log 2) \le 3\log N$, for $N \ge 2$. Hence

$$D^*_{N,\boldsymbol{\gamma}}(\mathcal{P}(\boldsymbol{g},N)) \le \frac{1}{N}\prod_{j=1}^{s}(1+2\gamma_j) + \frac{1}{2(N-1)}\prod_{j=1}^{s}(1+3\gamma_j \log N)$$

$$\le \frac{2}{N}\prod_{j=1}^{s}(1+3\gamma_j \log N),$$

since N is a prime number and therefore $N \ge 2$. $\qquad\square$

If the coordinate weights $(\gamma_j)_{j\ge 1}$ are sufficiently small one can prove a bound on the weighted star discrepancy whose convergence rate in N is essentially best possible and which holds uniformly for any dimension s.

Corollary 4.16 *Let N be a prime number, let $s \in \mathbb{N}$. Assume that the coordinate weights $(\gamma_j)_{j\ge 1}$ are summable, i.e., $\sum_{j=1}^{\infty}\gamma_j < \infty$. Then for every $\delta > 0$ there exists a quantity $C_{\delta,\boldsymbol{\gamma}} > 0$, independent of s and N, such that the weighted star discrepancy of the lattice point set $\mathcal{P}(\boldsymbol{g},N)$ whose generating vector $\boldsymbol{g}$ is constructed using Algorithm 4.13 satisfies*

$$D^*_{N,\boldsymbol{\gamma}}(\mathcal{P}(\boldsymbol{g},N)) \le \frac{C_{\delta,\boldsymbol{\gamma}}}{N^{1-\delta}}.$$

Proof Let $U(\boldsymbol{\gamma},N) := \prod_{j=1}^{\infty}(1+3\gamma_j \log N)$ and $\sigma_k := 3\sum_{j=k+1}^{\infty}\gamma_j$ for $k \in \mathbb{N}_0$. Since $\sum_{j=1}^{\infty}\gamma_j < \infty$, it is clear that $\lim_{k\to\infty}\sigma_k = 0$. We have

$$\log U(\boldsymbol{\gamma},N) = \sum_{j=1}^{\infty}\log(1+3\gamma_j \log N)$$

$$\le \sum_{j=1}^{k}\log(1+\sigma_d^{-1}+3\gamma_j \log N) + \sum_{j=k+1}^{\infty}\log(1+3\gamma_j \log N)$$

$$\le k\log(1+\sigma_k^{-1}) + \sum_{j=1}^{k}\log(1+3\sigma_k\gamma_j \log N) + \sum_{j=k+1}^{\infty}\log(1+3\gamma_j \log N)$$

$$\le \sum_{j=1}^{k}\log(1+\sigma_k^{-1}+3\gamma_j \log N) + \sum_{j=k+1}^{\infty}\log(1+3\gamma_j \log N)$$

$$\le k\log(1+\sigma_k^{-1}) + 3\sigma_k(\log N)\sum_{j=1}^{k}\gamma_j + \sigma_k \log N$$

$$\le k\log(1+\sigma_k^{-1}) + \sigma_k(\sigma_0+1)\log N.$$

Thus,

$$U(\boldsymbol{\gamma}, N) \le (1 + \sigma_k^{-1})^k N^{(\sigma_0 + 1)\sigma_k}.$$

For $\delta > 0$ choose k large enough to make $\sigma_k \le \delta/(\sigma_0 + 1)$. Hence

$$U(\boldsymbol{\gamma}, N) \le C'_{\delta, \boldsymbol{\gamma}} N^{\delta},$$

with a suitable quantity $C'_{\delta, \boldsymbol{\gamma}} > 0$. Now the result follows by applying this estimate to the discrepancy bound in Corollary 4.15. $\qquad\square$

4.4 Numerical Integration in Korobov Spaces

If we use a lattice point set as constructed by means of Algorithm 4.4 as underlying sample points of a QMC rule for functions $f : [0, 1]^s \to \mathbb{R}$ with bounded variation in the sense of Hardy and Krause, then Corollary 4.7 together with the inequality of Koksma and Hlawka guarantees a convergence rate of order $O((\log N)^s / N)$. For the Koksma-Hlawka inequality no additional information on the smoothness of the integrand was required. On the other hand, for functions of higher smoothness we would certainly expect a better convergence rate of the integration error. This, however, is not reflected in the error estimate via the Koksma-Hlawka inequality.

Originally, lattice point sets were introduced for the integration of smooth and one-periodic functions. Such functions will be considered in the following.

Definition 4.17 A QMC rule that uses a lattice point set as underlying sample nodes is called a *lattice rule*.

We will see that lattice rules are particularly suitable for the integration of periodic functions of sufficient smoothness. We define reproducing kernel Hilbert spaces of smooth and one-periodic functions and analyze the worst-case error of lattice rules in these function spaces.

We begin with the one-dimensional case. The smoothness in this function space will be described by the so-called smoothness parameter $\alpha > 1$, which controls the decay of the Fourier coefficients of the functions. Let $\gamma > 0$ be a real number. The value of γ does not have any influence on the one-dimensional analysis and may be chosen equal to one in that case. However, it will play a crucial role in the multi-dimensional case, where every dimension will be assigned its own γ. We define the Hilbert space $\mathcal{H}_{\alpha,\gamma} \subseteq L_1([0, 1])$ consisting of all one-periodic functions f with absolutely convergent Fourier series

$$f(x) = \sum_{h \in \mathbb{Z}} \widehat{f}(h) \exp(2\pi \mathrm{i} h x)$$

with Fourier coefficients

$$\widehat{f}(h) = \int_0^1 f(x)\exp(-2\pi\,\mathrm{i}hx)\,\mathrm{d}x$$

and with finite norm $\|f\|_{\alpha,\gamma} = \langle f, f\rangle_{\alpha,\gamma}^{1/2}$, where the inner product is defined by

$$\langle f, g\rangle_{\alpha,\gamma} := \sum_{h\in\mathbb{Z}} r_{\alpha,\gamma}(h)\widehat{f}(h)\overline{\widehat{g}(h)}$$

and

$$r_{\alpha,\gamma}(h) := \begin{cases} 1 & \text{if } h = 0, \\ \gamma^{-1}|h|^\alpha & \text{if } h \neq 0. \end{cases}$$

Definition 4.18 The space $\mathcal{H}_{\alpha,\gamma}$ is called the *Korobov space of smoothness* α (with weight γ).

We now show that the reproducing kernel of $\mathcal{H}_{\alpha,\gamma}$ is given by

$$K_{\alpha,\gamma}(x, y) = \sum_{h\in\mathbb{Z}} \frac{1}{r_{\alpha,\gamma}(h)}\exp(2\pi\,\mathrm{i}h(x - y)). \tag{4.15}$$

First, note that due to $\alpha > 1$ the series in (4.15) is uniformly convergent, so $K_{\alpha,\gamma}(\cdot, y)$ is integrable and

$$\widehat{K}_{\alpha,\gamma}(\cdot, y)(h) = \int_0^1 K_{\alpha,\gamma}(x, y)\exp(-2\pi\,\mathrm{i}hx)\,\mathrm{d}x$$

$$= \sum_{k\in\mathbb{Z}} \frac{1}{r_{\alpha,\gamma}(h)}\exp(-2\pi\,\mathrm{i}ky)\int_0^1 \exp(2\pi\,\mathrm{i}(k - h)x)\,\mathrm{d}x$$

$$= \frac{\exp(-2\pi\,\mathrm{i}hy)}{r_{\alpha,\gamma}(h)}.$$

Thus

$$\|K_{\alpha,\gamma}(\cdot, y)\|_\alpha^2 = \langle K_{\alpha,\gamma}(\cdot, y), K_{\alpha,\gamma}(\cdot, y)\rangle_{\alpha,\gamma} = \sum_{h\in\mathbb{Z}} r_{\alpha,\gamma}(h)\frac{1}{r_{\alpha,\gamma}^2(h)} = 1 + 2\gamma\zeta(\alpha),$$

where $\zeta(\alpha) = \sum_{h=1}^{\infty} h^{-\alpha}$ denotes the Riemann Zeta function. For $\alpha > 1$ we have $\zeta(\alpha) < \infty$ and therefore we obtain $K_{\alpha,\gamma}(\cdot, y) \in \mathcal{H}_{\alpha,\gamma}$. Furthermore, we have

$$\langle f, K_{\alpha,\gamma}(\cdot, y)\rangle_{\alpha,\gamma} = \sum_{h\in\mathbb{Z}} r_{\alpha,\gamma}(h)\widehat{f}(h)\frac{\exp(2\pi\,\mathrm{i}hy)}{r_{\alpha,\gamma}(h)} = f(y)$$

and hence also the reproducing property of $K_{\alpha,\gamma}$ is satisfied. This shows that $K_{\alpha,\gamma}$ is indeed the reproducing kernel of $\mathcal{H}_{\alpha,\gamma}$.

The smoothness parameter α is linked to the actual smoothness of functions in the sense that a periodic function f which is sufficiently often continuously differentiable belongs to the space $\mathcal{H}_{\alpha,\gamma}$.

Proposition 4.19 *Let $\alpha \geq 2$ be an integer and assume that $f : [0, 1] \to \mathbb{R}$ is one periodic and $f \in C^{\alpha}([0, 1])$. Then we have $f \in \mathcal{H}_{\alpha,\gamma}$.*

Proof The function f and its derivatives $f^{(1)}, f^{(2)}, \ldots, f^{(\alpha-1)}$ are continuously differentiable and one periodic, and the derivative $f^{(\alpha)}$ is continuous and one periodic. Let $h \neq 0$ be an integer. Integration by parts then yields

$$
\begin{aligned}
\widehat{f}(h) &= \int_0^1 f(x) \exp(-2\pi \mathrm{i} h x) \, \mathrm{d}x \\
&= f(x) \frac{\exp(-2\pi \mathrm{i} h x)}{-2\pi \mathrm{i} h} \Big|_0^1 + \frac{1}{2\pi \mathrm{i} h} \int_0^1 f^{(1)}(x) \exp(-2\pi \mathrm{i} h x) \, \mathrm{d}x \\
&= \frac{1}{2\pi \mathrm{i} h} \int_0^1 f^{(1)}(x) \exp(-2\pi \mathrm{i} h x) \, \mathrm{d}x \\
&= \frac{1}{2\pi \mathrm{i} h} \widehat{f^{(1)}}(h).
\end{aligned}
$$

Now we can repeat this with $\widehat{f^{(1)}}(h)$ in place of $\widehat{f}(h)$ and obtain

$$
\widehat{f}(h) = \frac{1}{(2\pi \mathrm{i} h)^2} \widehat{f^{(2)}}(h).
$$

After α steps we finally obtain

$$
\widehat{f}(h) = \frac{1}{(2\pi \mathrm{i} h)^{\alpha}} \widehat{f^{(\alpha)}}(h).
$$

Consequently,

$$
|\widehat{f}(h)| \leq \frac{1}{(2\pi |h|)^{\alpha}} \int_0^1 |f^{(\alpha)}(x)| \, \mathrm{d}x = \frac{c}{|h|^{\alpha}},
$$

where $c := \frac{1}{(2\pi)^{\alpha}} \int_0^1 |f^{(\alpha)}(x)| \, \mathrm{d}x < \infty$, since $f^{(\alpha)}$ is continuous. Thus we obtain

$$
\|f\|_{\alpha,\gamma}^2 = \langle f, f \rangle_{\alpha,\gamma} = \sum_{h \in \mathbb{Z}} r_{\alpha,\gamma}(h) |\widehat{f}(h)|^2 \leq |\widehat{f}(0)|^2 + 2\gamma^{-1} c \zeta(\alpha) < \infty
$$

and this implies $f \in \mathcal{H}_{\alpha,\gamma}$. $\qquad\square$

On the other hand, every function in $\mathcal{H}_{\alpha,\gamma}$ has some smoothness which will be shown in Proposition 4.22 for arbitrary dimension $s \in \mathbb{N}$.

Now we turn to the s-dimensional case where $s \in \mathbb{N}$. Let $\boldsymbol{\gamma} = (\gamma_j)_{j \in \mathbb{N}}$ be a sequence of positive weights. We consider the s-fold tensor product $\mathcal{H}_{s,\alpha,\gamma}$ of the spaces $\mathcal{H}_{\alpha,\gamma_1}, \mathcal{H}_{\alpha,\gamma_2}, \ldots, \mathcal{H}_{\alpha,\gamma_s}$, i.e.:

$$\mathcal{H}_{s,\alpha,\boldsymbol{\gamma}} = \mathcal{H}_{\alpha,\gamma_1} \otimes \mathcal{H}_{\alpha,\gamma_2} \otimes \cdots \otimes \mathcal{H}_{\alpha,\gamma_s}$$

$$= \text{clos span} \left\{ \boldsymbol{x} \mapsto \prod_{j=1}^{s} f_j(x_j) \ : \ f_j \in \mathcal{H}_{\alpha,\gamma_j} \right\},$$

where $\boldsymbol{x} = (x_1, \ldots, x_s)$ and where the closure clos is taken with respect to the norm induced by the inner product

$$\langle f, g \rangle_{s,\alpha,\boldsymbol{\gamma}} = \sum_{\boldsymbol{h} \in \mathbb{Z}^s} r_{\alpha,\boldsymbol{\gamma}}(\boldsymbol{h}) \widehat{f}(\boldsymbol{h}) \overline{\widehat{g}(\boldsymbol{h})},$$

and where $r_{\alpha,\boldsymbol{\gamma}}(\boldsymbol{h}) := \prod_{j=1}^{s} r_{\alpha,\gamma_j}(h_j)$. The reproducing kernel $K_{s,\alpha,\boldsymbol{\gamma}}$ of $\mathcal{H}_{s,\alpha,\boldsymbol{\gamma}}$ is the s-fold product of the one-dimensional kernels K_{α,γ_j}, for $j = 1, \ldots, s$, i.e.:

$$K_{s,\alpha,\boldsymbol{\gamma}}(\boldsymbol{x}, \boldsymbol{y}) = \prod_{j=1}^{s} K_{\alpha,\gamma_j}(x_j, y_j),$$

where $\boldsymbol{x} = (x_1, \ldots, x_s)$ and $\boldsymbol{y} = (y_1, \ldots, y_s)$. Then we have

$$K_{s,\alpha,\boldsymbol{\gamma}}(\boldsymbol{x}, \boldsymbol{y}) = \sum_{h_1,\ldots,h_s \in \mathbb{Z}} \frac{\exp(2\pi \mathrm{i} h_1(x_1 - y_1)) \cdots \exp(2\pi \mathrm{i} h_s(x_s - y_s))}{r_{\alpha,\gamma_1}(h_1) \cdots r_{\alpha,\gamma_s}(h_s)}$$

$$= \sum_{\boldsymbol{h} \in \mathbb{Z}^s} \frac{\exp(2\pi \mathrm{i} \boldsymbol{h} \cdot (\boldsymbol{x} - \boldsymbol{y}))}{r_{\alpha,\boldsymbol{\gamma}}(\boldsymbol{h})}.$$

The proof that $K_{s,\alpha,\boldsymbol{\gamma}}$ is the reproducing kernel of $\mathcal{H}_{\alpha,s,\boldsymbol{\gamma}}$ follows along the same lines as in the one-dimensional case.

Definition 4.20 The space $\mathcal{H}_{s,\alpha,\boldsymbol{\gamma}}$ is called a *weighted Korobov space of smoothness* α *with weights* $\boldsymbol{\gamma}$.

We present a multivariate version of Proposition 4.19, but without a proof. For $\boldsymbol{\beta} = (\beta_1, \beta_2, \ldots, \beta_s) \in \mathbb{N}_0^s$ denote by

$$D^{\boldsymbol{\beta}} f := \frac{\partial^{|\boldsymbol{\beta}|}}{\partial x_1^{\beta_1} \partial x_2^{\beta_2} \cdots \partial x_s^{\beta_s}} f$$

the operator of partial differentiation, where $|\boldsymbol{\beta}| = \beta_1 + \beta_2 + \cdots + \beta_s$.

Proposition 4.21 *Let* $\alpha \geq 2$ *be an integer and assume that* $f : [0, 1]^s \to \mathbb{R}$ *is one periodic in each of its s variables and that* $D^{\boldsymbol{\beta}} f$ *exists and is continuous for all* $\boldsymbol{\beta} \in \{0, \ldots, \alpha\}^s$. *Then* $f \in \mathcal{H}_{s,\alpha,\boldsymbol{\gamma}}$.

On the other hand, every function in $\mathcal{H}_{s,\alpha,\gamma}$ has some smoothness.

Proposition 4.22 *If* $f \in \mathcal{H}_{s,\alpha,\gamma}$, *then* $D^{\boldsymbol{\beta}} f$ *exists and is continuous for all* $\boldsymbol{\beta} \in \{0, 1, \ldots, \lceil \frac{\alpha-1}{2} \rceil - 1\}^s$.

Proof We have

$$D^{\boldsymbol{\beta}} f(\boldsymbol{x}) = \sum_{\boldsymbol{h} \in \mathbb{Z}^s} \left[\widehat{f}(\boldsymbol{h}) \, (2\pi \mathrm{i})^{|\boldsymbol{\beta}|} \prod_{j=1}^{s} h_j^{\beta_j} \right] \exp(2\pi \mathrm{i} \boldsymbol{h} \cdot \boldsymbol{x}),$$

where, by convention, we take $0^0 = 1$. The last series is convergent as long as $\max_{1 \le j \le s} \beta_j < \frac{\alpha-1}{2}$. Indeed,

$$|D^{\boldsymbol{\beta}} f(\boldsymbol{x})| = \left| \sum_{\boldsymbol{h} \in \mathbb{Z}^s} \left[\widehat{f}(\boldsymbol{h}) r_{\alpha,\gamma}^{1/2}(\boldsymbol{h}) \right] \left[r_{\alpha,\gamma}^{-1/2}(\boldsymbol{h}) (2\pi \mathrm{i})^{|\boldsymbol{\beta}|} \prod_{j=1}^{s} h_j^{\beta_j} \right] \exp(2\pi \mathrm{i} \boldsymbol{h} \cdot \boldsymbol{x}) \right|$$

$$\le \|f\|_{s,\alpha,\gamma} \left[\sum_{\boldsymbol{h} \in \mathbb{Z}^s} (2\pi)^{2|\boldsymbol{\beta}|} \prod_{j=1}^{s} \frac{|h_j|^{2\beta_j}}{r_{\alpha,\gamma_j}(h_j)} \right]^{1/2}$$

$$= \|f\|_{s,\alpha,\gamma} \left[(2\pi)^{2|\boldsymbol{\beta}|} \prod_{j=1}^{s} \sum_{h_j \in \mathbb{Z}} \frac{|h_j|^{2\beta_j}}{r_{\alpha,\gamma_j}(h_j)} \right]^{1/2}$$

$$= \|f\|_{s,\alpha,\gamma} \left[(2\pi)^{2|\boldsymbol{\beta}|} \prod_{j=1}^{s} 2\gamma_j \sum_{h_j=1}^{\infty} \frac{h_j^{2\beta_j}}{h_j^{\alpha}} \right]^{1/2}$$

$$= (2\pi)^{|\boldsymbol{\beta}|} 2^{s/2} \|f\|_{s,\alpha,\gamma} \left[\prod_{j=1}^{s} \gamma_j \zeta(\alpha - 2\beta_j) \right]^{1/2} < \infty. \qquad \square$$

Now we study QMC integration in $\mathcal{H}_{s,\alpha,\gamma}$ for $\alpha > 1$. We have

$$K_{s,\alpha,\gamma}(\boldsymbol{y}, \boldsymbol{y}) = \prod_{j=1}^{s} (1 + 2\gamma_j \zeta(\alpha)) < \infty.$$

Hence condition **C** is satisfied and we can therefore use the formula for the worst-case error from Theorem 3.9. We have

$$\int_{[0,1]^s} \int_{[0,1]^s} K_{s,\alpha,\gamma}(\boldsymbol{x}, \boldsymbol{y}) \, \mathrm{d}\boldsymbol{x} \, \mathrm{d}\boldsymbol{y} = \prod_{j=1}^{s} \int_0^1 \int_0^1 K_{\alpha,\gamma_j}(x_j, y_j) \, \mathrm{d}x_j \, \mathrm{d}y_j$$

$$= \prod_{j=1}^{s} \left(\sum_{h \in \mathbb{Z}} \frac{1}{r_{\alpha,\gamma_j}(h)} \int_0^1 \int_0^1 \exp(2\pi \mathrm{i} h(x_j - y_j)) \, \mathrm{d}x_j \, \mathrm{d}y_j \right) = 1.$$

In the same way one can show that for any fixed $\boldsymbol{x}_n \in [0, 1)^s$ we have

$$\int_{[0,1]^s} K_{s,\alpha,\gamma}(\boldsymbol{x}_n, \boldsymbol{y}) \, \mathrm{d}\boldsymbol{y} = 1.$$

Let $\mathcal{P} = \{\boldsymbol{x}_0, \ldots, \boldsymbol{x}_{N-1}\}$ be a point set in $[0, 1)^s$. From Theorem 3.9 we now obtain

$$e^2(\mathcal{H}_{s,\alpha,\gamma}, \mathcal{P}) = -1 + \frac{1}{N^2} \sum_{n,m=0}^{N-1} \sum_{\boldsymbol{h} \in \mathbb{Z}^s} \frac{\exp(2\pi \mathrm{i} \boldsymbol{h} \cdot (\boldsymbol{x}_n - \boldsymbol{x}_m))}{r_{\alpha,\gamma}(\boldsymbol{h})}$$

$$= \sum_{\boldsymbol{h} \in \mathbb{Z}^s \setminus \{\boldsymbol{0}\}} \frac{1}{r_{\alpha,\gamma}(\boldsymbol{h})} \left| \frac{1}{N} \sum_{n=0}^{N-1} \exp(2\pi \mathrm{i} \boldsymbol{h} \cdot \boldsymbol{x}_n) \right|^2. \tag{4.16}$$

For lattice point sets $\mathcal{P}(\boldsymbol{g}, N)$, i.e., $\boldsymbol{x}_n = \{n\boldsymbol{g}/N\}$ where $\boldsymbol{g} \in \mathbb{Z}^s$, the above formula yields

$$e^2(\mathcal{H}_{s,\alpha,\gamma}, \mathcal{P}(\boldsymbol{g}, N)) = \sum_{\boldsymbol{h} \in \mathbb{Z}^s \setminus \{\boldsymbol{0}\}} \frac{1}{r_{\alpha,\gamma}(\boldsymbol{h})} \left| \frac{1}{N} \sum_{n=0}^{N-1} \exp(2\pi \mathrm{i} n \boldsymbol{h} \cdot \boldsymbol{g}/N) \right|^2. \tag{4.17}$$

Now, using property (4.1) and the notion $\mathcal{L}(\boldsymbol{g}, N)$ of the dual lattice of $\mathcal{P}(\boldsymbol{g}, N)$ as given in Definition 4.2 we obtain the following result.

Theorem 4.23 *The worst-case error of a lattice rule in the weighted Korobov space* $\mathcal{H}_{s,\alpha,\gamma}$ *is given by*

$$e^2(\mathcal{H}_{s,\alpha,\gamma}, \mathcal{P}(\boldsymbol{g}, N)) = \sum_{\boldsymbol{h} \in \mathcal{L}(\boldsymbol{g},N) \setminus \{\boldsymbol{0}\}} \frac{1}{r_{\alpha,\gamma}(\boldsymbol{h})}.$$

Remark 4.24 If $\alpha \geq 2$ is an even integer, then the Bernoulli polynomial B_α of degree α has the Fourier expansion

$$B_\alpha(x) = \frac{(-1)^{(\alpha+2)/2}\alpha!}{(2\pi)^\alpha} \sum_{h \in \mathbb{Z} \setminus \{0\}} \frac{\exp(2\pi \mathrm{i} h x)}{|h|^\alpha} \qquad \text{for all } x \in [0, 1).$$

Hence in this case we obtain

$$e^2(\mathcal{H}_{s,\alpha,\gamma}, \mathcal{P}(\boldsymbol{g}, N)) = -1 + \frac{1}{N} \sum_{k=0}^{N-1} \prod_{j=1}^{s} \left(1 + \gamma_j \frac{(-1)^{(\alpha+2)/2}(2\pi)^\alpha}{\alpha!} B_\alpha \left(\left\{ \frac{k g_j}{N} \right\} \right) \right),$$

so that $e^2(\mathcal{H}_{s,\alpha,\gamma}, \mathcal{P}(\boldsymbol{g}, N))$ can be calculated in $O(Ns)$ operations.

4.5   Existence Results for the Unweighted Case

In this section we restrict ourselves to the unweighted case where all weights are equal to 1. In this case we omit the weights in our notations from the previous section and just write $\mathcal{H}_{s,\alpha}$, $K_{s,\alpha}$, r_α, etc. Note that in the unweighted case we have

$$r_\alpha(h) := \max(1, |h|^\alpha) = r_1(h)^\alpha.$$

First we show a result which brings the quantity R_N as defined in (4.3) in relation to the worst-case error of a lattice rule in the Korobov space.

Lemma 4.25 *Let $\alpha > 1$ and let N be a prime number. Then for all $\boldsymbol{g} \in G_N^s$ we have*

$$e^2(\mathcal{H}_{s,\alpha}, P(\boldsymbol{g}, N)) \le (1 + 2\zeta(\alpha))^s \left(\frac{1}{N^\alpha} + R_N(\boldsymbol{g})^\alpha \right).$$

Proof By Theorem 4.23,

$$e^2(\mathcal{H}_{s,\alpha}, P(\boldsymbol{g}, N)) = \sum_{\boldsymbol{h} \in \mathcal{L}(\boldsymbol{g},N)\setminus\{\boldsymbol{0}\}} \frac{1}{r_\alpha(\boldsymbol{h})} =: \Sigma_1 + \Sigma_2,$$

where Σ_1 is the sum over all $\boldsymbol{h}$ such that $\boldsymbol{h} = N\boldsymbol{k}$ with $\boldsymbol{k} \in \mathbb{Z}^s \setminus \{\boldsymbol{0}\}$. In this case we obviously have $\boldsymbol{h} \cdot \boldsymbol{g} \equiv 0 \pmod{N}$. In Σ_2 we sum over the remaining $\boldsymbol{h}$ from $\mathcal{L}(\boldsymbol{g}, N) \setminus \{\boldsymbol{0}\}$.

For $\boldsymbol{h} = N\boldsymbol{k}$ with $\boldsymbol{k} \in \mathbb{Z}^s \setminus \{\boldsymbol{0}\}$ we have

$$\frac{1}{r_\alpha(\boldsymbol{h})} = \frac{1}{r_\alpha(N\boldsymbol{k})} \le \frac{1}{N^\alpha} \frac{1}{r_\alpha(\boldsymbol{k})},$$

and so

$$\Sigma_1 \le \frac{1}{N^\alpha} \sum_{\boldsymbol{k} \in \mathbb{Z}^s} \frac{1}{r_\alpha(\boldsymbol{k})} = \frac{1}{N^\alpha} \left(\sum_{k \in \mathbb{Z}} \frac{1}{r_\alpha(k)} \right)^s = \frac{1}{N^\alpha} (1 + 2\zeta(\alpha))^s.$$

The remaining $\boldsymbol{h}$ which appear in Σ_2 can be uniquely represented in the form $\boldsymbol{h} = \boldsymbol{h}^* + N\boldsymbol{k}$, where $\boldsymbol{k} \in \mathbb{Z}^s$ and $\boldsymbol{h}^* = (h_1^*, \ldots, h_s^*) \in C_s^*(N)$ with $\boldsymbol{h}^* \cdot \boldsymbol{g} \equiv 0 \pmod{N}$. Hence

$$\Sigma_2 = \sum_{\boldsymbol{k} \in \mathbb{Z}^s} \sum_{\substack{\boldsymbol{h}^* \in C_s^*(N) \\ \boldsymbol{h}^* \cdot \boldsymbol{g} \equiv 0 \pmod{N}}} \frac{1}{r_\alpha(\boldsymbol{h}^* + N\boldsymbol{k})}.$$

Now we show that

$$r_1(\boldsymbol{h}^* + N\boldsymbol{k}) \ge r_1(\boldsymbol{h}^*)r_1(\boldsymbol{k}). \tag{4.18}$$

To this end it suffices to verify that

$$\max(1, |h_j^* + Nk_j|) \ge \max(1, |h_j^*|) \max(1, |k_j|)$$

for all $j = 1, 2, \ldots, s$. If $h_j^* = 0$ or $k_j = 0$, then this inequality is obviously satisfied. If $h_j^* \neq 0$ and $k_j \neq 0$, then

$$|h_j^* + Nk_j| \geq N|k_j| - |h_j^*| \geq N|k_j| - \frac{N}{2} = \frac{N}{2}(2|k_j| - 1) \geq |h_j^*| \cdot |k_j|,$$

and the inequality follows as well. Hence (4.18) is shown.

From (4.18) we obtain $r_\alpha(\boldsymbol{h}^* + N\boldsymbol{k}) \geq r_\alpha(\boldsymbol{h}^*)r_\alpha(\boldsymbol{k})$. Using this estimate we find that

$$\Sigma_2 \leq \sum_{\boldsymbol{k} \in \mathbb{Z}^s} \frac{1}{r_1(\boldsymbol{k})^\alpha} \sum_{\substack{\boldsymbol{h}^* \in C_s^*(N) \\ \boldsymbol{h}^* \cdot \boldsymbol{g} \equiv 0 \pmod{N}}} \frac{1}{r_1(\boldsymbol{h}^*)^\alpha}$$

$$= (1 + 2\zeta(\alpha))^s \sum_{\substack{\boldsymbol{h}^* \in C_s^*(N) \\ \boldsymbol{h}^* \cdot \boldsymbol{g} \equiv 0 \pmod{N}}} \frac{1}{r_1(\boldsymbol{h}^*)^\alpha}$$

$$\leq (1 + 2\zeta(\alpha))^s R_N(\boldsymbol{g})^\alpha.$$

Now the result follows by adding the estimates for Σ_1 and Σ_2. $\qquad\square$

It should be pointed out that the quantity $R_N(\boldsymbol{g})$ does not depend on the smoothness parameter α. Assume now that $\boldsymbol{g} \in G_N^s$ is constructed with Algorithm 4.4. Then it follows from Theorem 4.5 in combination with (4.5) that

$$R_N(\boldsymbol{g}) \leq 2^{s+1} \frac{(1 + \log N)^s}{N}.$$

Inserting the bound on $R_N(\boldsymbol{g})$ into the bound from Lemma 4.25 yields the following result, which goes back to work by E. Hlawka.

Theorem 4.26 *Let $\alpha > 1$ and let N be a prime number. Assume that $\boldsymbol{g} \in G_N^s$ is constructed by Algorithm 4.4. Then*

$$e^2(\mathcal{H}_{s,\alpha}, \mathcal{P}(\boldsymbol{g}, N)) \leq (1 + 2\zeta(\alpha))^s \frac{1 + 2^{\alpha(s+1)}(1 + \log N)^{\alpha s}}{N^\alpha}.$$

Theorem 4.26 shows the existence of lattice rules for which the convergence rate for the worst-case error in the Korobov space $\mathcal{H}_{s,\alpha}$ is of order

$$e(\mathcal{H}_{s,\alpha}, \mathcal{P}(\boldsymbol{g}, N)) = O_{s,\alpha}\left(\frac{(\log N)^{\alpha s/2}}{N^{\alpha/2}}\right).$$

This bound reflects the smoothness of the problem. Higher smoothness α leads to improved convergence rates for the worst-case integration error. Furthermore, as Algorithm 4.4 is independent of the parameter α, it is clear that the output vector $\boldsymbol{g}$ is independent of α as well. Hence the lattice rule that is constructed by Algorithm 4.4 adjusts itself to the smoothness of a given integrand.

We next present a lower bound on the integration error for numerical integration in the Korobov space which shows that the result from Theorem 4.26 is, up to some $(\log N)$-powers, best possible in the order of magnitude in N. For reasons of simplicity we restrict ourselves to the QMC case and remark that in fact the following result holds even for more general quadrature rules.

Theorem 4.27 *Let $\mathcal{P}$ be an arbitrary N-element point set in $[0, 1)^s$. For any $\alpha > 1$ we have*
$$e^2(\mathcal{H}_{s,\alpha}, \mathcal{P}) \geq C(s, \alpha) \frac{(\log N)^{s-1}}{N^\alpha},$$
where $C(s, \alpha) > 0$ depends only on α and s, but not on N.

The proof of this theorem is technical and can be skipped by beginners.

Proof Let $\mathcal{P} = \{x_0, \ldots, x_{N-1}\}$ be an arbitrary N-element point set in $[0, 1)^s$. Let $f : \mathbb{R} \to [0, 1]$ be an infinitely differentiable function such that $f(x) > 0$ for $x \in (0, 1)$ and $f^{(r)}(x) = 0$ for $x \in \mathbb{R} \setminus (0, 1)$ for all $0 \leq r \leq a := \lceil \alpha/2 \rceil + 1$. For instance choose
$$f(x) = \begin{cases} x^{a+1}(1 - x)^{a+1} & \text{for } x \in (0, 1) \\ 0 & \text{otherwise.} \end{cases} \tag{4.19}$$
For $m \in \mathbb{N}_0$ let $f_m(x) = f(2^{m+2}x)$ and for $\boldsymbol{m} = (m_1, \ldots, m_s) \in \mathbb{N}_0^s$ put
$$f_{\boldsymbol{m}}(\boldsymbol{x}) := \prod_{j=1}^{s} f_{m_j}(x_j),$$
where $\boldsymbol{x} = (x_1, \ldots, x_s)$. Let $\|\boldsymbol{m}\|_1 := m_1 + \cdots + m_s$ and $I(f) := \int_0^1 f(y)\,dy$. Then
$$\widehat{f_{\boldsymbol{m}}}(\boldsymbol{0}) = \prod_{j=1}^{s} \int_0^1 f(2^{m_j+2}x)\,dx = \prod_{j=1}^{s} \left(\frac{1}{2^{m_j+2}} \int_0^1 f(y)\,dy \right) = \frac{1}{2^{\|\boldsymbol{m}\|_1+2s}} I^s(f).$$
$$\tag{4.20}$$
Let t be such that $2N \leq 2^t < 4N$, let $F(y) = \sum_{n=0}^{N-1} f_{\boldsymbol{m}}(\boldsymbol{x}_n - \boldsymbol{y})$, and let
$$B_{\boldsymbol{m}} := \left\{ \boldsymbol{y} \in [0, 1]^s \ : \ F(\boldsymbol{y}) = 0 \right\}.$$

Notice that the support of $f_{\boldsymbol{m}}(\boldsymbol{x}_n - \boldsymbol{y})$ (as a function of $\boldsymbol{y}$) is contained in the interval $\prod_{j=1}^{s}(x_{n,j} - 2^{-m_j-2}, x_{n,j})$ and hence the support of $F(\boldsymbol{y})$ is contained in $\bigcup_{n=0}^{N-1} \prod_{j=1}^{s}(x_{n,j} - 2^{-m_j-2}, x_{n,j})$. Therefore the volume of the support of F is at most $N2^{-\|\boldsymbol{m}\|_1}$. Thus, for all $\boldsymbol{m}$ such that $\|\boldsymbol{m}\|_1 = t$ we have
$$\lambda_s(B_{\boldsymbol{m}}) \geq 1 - \frac{N}{2^{\|\boldsymbol{m}\|_1}} = 1 - \frac{N}{2^t} > \frac{1}{4}.$$

For the QMC rule $Q_{N,s}$ based on $\mathcal{P}$

$$Q_{N,s}(f_m(\cdot - y)) - \widehat{f_m}(\mathbf{0})$$

$$= \sum_{h \in \mathbb{Z}^s \setminus \{0\}} \left(\widehat{f_m}(h) \frac{1}{N} \sum_{n=0}^{N-1} \exp(2\pi i h \cdot x_n) \right) \exp(-2\pi i h \cdot y)$$

and, for $y \in B_m$, $Q_{N,s}(f_m(\cdot - y)) = \frac{1}{N} F(y) = 0$. Therefore

$$\lambda_s(B_m) |\widehat{f_m}(\mathbf{0})|^2 = \int_{B_m} |Q_{N,s}(f_m(\cdot - y)) - \widehat{f_m}(\mathbf{0})|^2 \, dy$$

$$\leq \int_{[0,1]^s} |Q_{N,s}(f_m(\cdot - y)) - \widehat{f_m}(\mathbf{0})|^2 \, dy$$

$$= \sum_{h \in \mathbb{Z}^s \setminus \{0\}} |\widehat{f_m}(h)|^2 \left| \frac{1}{N} \sum_{n=0}^{N-1} \exp(2\pi i h \cdot x_n) \right|^2, \qquad (4.21)$$

where for the last line we applied Parseval's identity.

We have

$$\widehat{f_m}(h) = \int_0^1 f(2^{m+2} x) \exp(-2\pi i h x) \, dx = \frac{1}{2^{m+2}} \widehat{f}(h 2^{-m-2}).$$

Since, by assumption, f is infinitely times differentiable, integration by parts shows that for any $m \in \mathbb{N}_0$ we have

$$|\widehat{f_m}(h)| \leq \frac{1}{2^{m+2}} |\widehat{f}(h 2^{-m-2})| \leq C_a \frac{1}{2^{m+2}} \min \left(1, \frac{2^{a(m+2)}}{h^a} \right),$$

where $C_a > 0$ depends only on a and f. Then for m with $\|m\|_1 = t$ we have

$$|\widehat{f_m}(h)| \leq C(a, s) \prod_{j=1}^{s} \left(\frac{1}{2^{m_j}} \min \left(1, \frac{2^{a m_j}}{r_a(h_j)} \right) \right)$$

$$= C(a, s) 2^{(\alpha/2 - 1)t} \prod_{j=1}^{s} \left(\frac{1}{2^{\alpha m_j/2}} \min \left(1, \frac{2^{a m_j}}{r_a(h_j)} \right) \right).$$

Taking the square and summing over all choices of m with $\|m\|_1 = t$ we obtain

$$\sum_{\substack{m \in \mathbb{N}_0^s \\ \|m\|_1 = t}} |\widehat{f_m}(h)|^2 \leq 2^{(\alpha-2)t} C^2(a, s) \sum_{\substack{m \in \mathbb{N}_0^s \\ \|m\|_1 = t}} \prod_{j=1}^{s} \left(\frac{1}{2^{\alpha m_j}} \min \left(1, \frac{2^{2 a m_j}}{r_a^2(h_j)} \right) \right)$$

$$\leq 2^{(\alpha-2)t} C^2(a, s) \prod_{j=1}^{s} \left(\sum_{m=0}^{\infty} \frac{1}{2^{\alpha m}} \min \left(1, \frac{2^{2 a m}}{r_a^2(h_j)} \right) \right). \qquad (4.22)$$

The last sum can now be estimated by

$$\sum_{m=0}^{\infty} \frac{1}{2^{\alpha m}} \min\left(1, \frac{2^{2am}}{r_a^2(h_j)}\right)$$

$$= \sum_{0 \le m \le (\log_2 r_a(h_j))/a} \frac{2^{(2a-\alpha)m}}{r_a^2(h_j)} + \sum_{m > (\log_2 r_a(h_j))/a} \frac{1}{2^{\alpha m}}$$

$$\le \frac{r_{2a-\alpha}(h_j)2^{2a-\alpha} - 1}{2^{2a-\alpha} - 1} \frac{1}{r_{2a}(h_j)} + \frac{2^{\alpha}}{2^{\alpha} - 1} \frac{1}{r_{\alpha}(h_j)}$$

$$\le \frac{1}{r_{\alpha}(h_j)} \left(1 + \frac{2^{\alpha}}{2^{\alpha} - 1}\right) \le \frac{3}{r_{\alpha}(h_j)}. \tag{4.23}$$

Thus, combining (4.22) and (4.23), we have

$$\frac{C_1(a,s)}{2^{(\alpha-2)t}} \sum_{\substack{\boldsymbol{m} \in \mathbb{N}_0^s \\ \|\boldsymbol{m}\|_1 = t}} |\widehat{f_{\boldsymbol{m}}}(\boldsymbol{h})|^2 \le \frac{1}{r_{\alpha}(\boldsymbol{h})} \tag{4.24}$$

with some suitable $C_1(a,s) > 0$ depending only on a and s. Using the formula for the worst-case error from (4.16), in conjunction with (4.24), (4.21), and (4.20) we finally get

$$e^2(\mathcal{H}_{s,\alpha}, \mathcal{P}) = \sum_{\boldsymbol{h} \in \mathbb{Z}^s \setminus \{\boldsymbol{0}\}} \frac{1}{r_{\alpha}(\boldsymbol{h})} \left| \frac{1}{N} \sum_{n=0}^{N-1} \exp(2\pi i \boldsymbol{h} \cdot \boldsymbol{x}_n) \right|^2$$

$$\ge \frac{C_1(a,s)}{2^{(\alpha-2)t}} \sum_{\substack{\boldsymbol{m} \in \mathbb{N}_0^s \\ \|\boldsymbol{m}\|_1 = t}} \sum_{\boldsymbol{h} \in \mathbb{Z}^s \setminus \{\boldsymbol{0}\}} |\widehat{f_{\boldsymbol{m}}}(\boldsymbol{h})|^2 \left| \frac{1}{N} \sum_{n=0}^{N-1} \exp(2\pi i \boldsymbol{h} \cdot \boldsymbol{x}_n) \right|^2$$

$$\ge \frac{C_1(a,s)}{2^{(\alpha-2)t}} \sum_{\substack{\boldsymbol{m} \in \mathbb{N}_0^s \\ \|\boldsymbol{m}\|_1 = t}} \lambda_s(B_{\boldsymbol{m}}) |\widehat{f_{\boldsymbol{m}}}(\boldsymbol{0})|^2$$

$$\ge C_2(a,s) \frac{2^{2t}}{N^{\alpha}} \sum_{\substack{\boldsymbol{m} \in \mathbb{N}_0^s \\ \|\boldsymbol{m}\|_1 = t}} 2^{-2t-4s} I^{2s}(f)$$

$$\ge C_3(a,s) \frac{1}{N^{\alpha}} \binom{t+s-1}{s-1}$$

$$\ge C_4(a,s) \frac{t^{s-1}}{N^{\alpha}}$$

$$\ge C_5(a,s) \frac{(\log N)^{s-1}}{N^{\alpha}},$$

since $t \ge \log_2 N$, with suitable $C_j(a,s) > 0$ depending only on a and s.

In this section we have shown that the best convergence rate of QMC rules in the Korobov space of smoothness α is $O(N^{-\alpha/2})$, up to some $\log N$ factors, and this rate can be achieved with a lattice rule. However, it is not clear how long we have to wait to see this nice asymptotic behavior especially for large dimension s. In practical applications s can be huge, for example, in the hundreds or even thousands for problems from mathematical finance. The dependence of the worst-case error on the dimension s is the subject of tractability, which is the topic of the next section.

4.6 Tractability

To systematically analyze the dependence of the worst-case integration error on the dimension one considers the following quantities. For an arbitrary reproducing kernel Hilbert space $\mathcal{H}(K)$ we denote by $e(\mathcal{H}(K), \mathcal{P})$ the worst-case error of a QMC rule based on the point set $\mathcal{P}$ in $[0, 1)^s$. The Nth *minimal (worst-case) error* is given by

$$e(N, s) := \inf_{\mathcal{P}} e(\mathcal{H}(K), \mathcal{P}),$$

where the infimum is taken over all N-element point sets $\mathcal{P}$ in $[0, 1)^s$. For $N = 0$ we set the QMC rule to be zero, and define the *initial error* by

$$e(0, s) := \sup_{\substack{f \in \mathcal{H}(K) \\ \|f\|_{\mathcal{H}(K)} \leq 1}} \left| \int_{[0,1]^s} f(x) \, dx \right|.$$

The initial error is used as a reference value.

When studying QMC rules, we do not only want to control how the error depends on N, but also how it depends on the dimension s, which is of particular importance for high-dimensional problems (see also Chap. 6 for a discussion of this topic). To this end, we define, for $\varepsilon \in (0, 1]$ and $s \in \mathbb{N}$, the *information complexity* by

$$N_{\mathcal{H}(K)}(\varepsilon, s) := \min\{N \in \mathbb{N} : e(N, s) \leq \varepsilon \, e(0, s)\},$$

i.e., as the minimal number of information evaluations needed to reduce the initial error by a factor of ε.

The subject of tractability deals with the question in which way the information complexity depends on ε^{-1} and s. Roughly speaking, tractability means that the information complexity lacks a certain disadvantageous dependence on ε^{-1} and s. Usually, one is interested in situations where $N_{\mathcal{H}(K)}(\varepsilon, s)$ depends at most polynomially on s and ε^{-1}, and frequently the following notions of tractability are studied.

Definition 4.28 We say that we have:

- The *curse of dimensionality* if there exist positive numbers C, τ and ε_0 such that

$$N_{\mathcal{H}(K)}(\varepsilon, s) \geq C(1 + \tau)^s \quad \text{for all } \varepsilon \leq \varepsilon_0 \text{ and infinitely many } s.$$

- *Polynomial (QMC) tractability* if there exist non-negative numbers C, τ_1, τ_2 such that

$$N_{\mathcal{H}(K)}(\varepsilon, s) \leq C s^{\tau_1} \varepsilon^{-\tau_2} \quad \text{for all } s \in \mathbb{N}, \ \varepsilon \in (0, 1).$$

- *Strong polynomial (QMC) tractability* if there exist non-negative numbers C and τ such that

$$N_{\mathcal{H}(K)}(\varepsilon, s) \leq C \varepsilon^{-\tau} \quad \text{for all } s \in \mathbb{N}, \ \varepsilon \in (0, 1).$$

The *exponent* τ^* *of strong polynomial tractability* is defined as the infimum over all τ for which strong polynomial tractability holds.

Remark 4.29 Usually in tractability theory one studies the class of arbitrary algorithms using N information evaluations rather than only QMC rules as we do here. For this reason we added in the above notation the term "(QMC)." However, as we only deal with QMC rules, we will omit this term in the following.

It is known that many multivariate problems defined over standard spaces of functions suffer from the curse of dimensionality, as, for example, integration of Lipschitz functions (cf. Example 1.1), of functions from the unweighted Korobov space $\mathcal{H}_{s,\alpha}$, of monotone functions, of convex functions, of smooth functions, etc. The reason for this disadvantageous behavior may be found in the fact that for standard spaces all variables and groups of variables are equally important. As a way out, I.H. Sloan and H. Woźniakowski suggested to consider weighted spaces, in which the importance of successive variables and groups of variables is monitored by corresponding weights, to vanquish the curse of dimensionality and obtain polynomial or even strong polynomial tractability, depending on the decay of the weights.

For this reason now we switch again to the weighted setting and study tractability for the weighted Korobov space $\mathcal{H}_{s,\alpha,\boldsymbol{\gamma}}$.

Theorem 4.30 *For any prime number N, any dimension s, and any $\lambda \in (1/\alpha, 1]$ there exists a generating vector $\boldsymbol{g} \in G_N^s$ such that*

$$e^2(\mathcal{H}_{s,\alpha,\boldsymbol{\gamma}}, \mathcal{P}(\boldsymbol{g}, N)) \leq \frac{1}{(N-1)^{1/\lambda}} \left(-1 + \prod_{j=1}^{s} (1 + 2\gamma_j^\lambda \zeta(\lambda\alpha)) \right)^{1/\lambda}.$$

Proof Let $\mathrm{av}(N, s)$ denote the average squared worst-case error over all $\boldsymbol{g} \in G_N^s$,

$$\mathrm{av}(N, s) := \frac{1}{(N-1)^s} \sum_{\boldsymbol{g} \in G_N^s} e^2(\mathcal{H}_{s,\alpha,\boldsymbol{\gamma}}, \mathcal{P}(\boldsymbol{g}, N)).$$

Then we have

$$\mathrm{av}(N, s) = \frac{1}{(N-1)^s} \sum_{\boldsymbol{g} \in G_N^s} \sum_{\boldsymbol{h} \in \mathcal{L}(g,N) \setminus \{\boldsymbol{0}\}} \frac{1}{r_{\alpha,\boldsymbol{\gamma}}(\boldsymbol{h})}$$

$$= \frac{1}{(N-1)^s} \sum_{\boldsymbol{h} \in \mathbb{Z}^s \setminus \{\boldsymbol{0}\}} \frac{1}{r_{\alpha,\boldsymbol{\gamma}}(\boldsymbol{h})} \# \{ \boldsymbol{g} \in G_N^s \ : \ \boldsymbol{h} \cdot \boldsymbol{g} \equiv 0 \quad (\mathrm{mod}\ N) \}.$$

Let $\boldsymbol{h} = (h_1, \ldots, h_s) \in \mathbb{Z}^s \setminus \{\boldsymbol{0}\}$. Assume first that $h_1 \neq 0$. Then, for arbitrary $g_2, \ldots, g_s \in G_N$, the condition $\boldsymbol{h} \cdot \boldsymbol{g} \equiv 0$ (mod N) is equivalent to

$$h_1 g_1 \equiv -(h_2 g_2 + \cdots + h_s g_s) \quad (\mathrm{mod}\ N).$$

Since N is a prime, there exists at most one $g_1 \in G_N$ which satisfies this congruence. Consequently,

$$\# \{ \boldsymbol{g} \in G_N^s \ : \ \boldsymbol{h} \cdot \boldsymbol{g} \equiv 0 \quad (\mathrm{mod}\ N) \} \leq (N-1)^{s-1}.$$

The same argument applies if $h_j \neq 0$ for any $j \in \{1, 2, \ldots, s\}$. Now it follows that

$$\mathrm{av}(N, s) \leq \frac{1}{N-1} \sum_{\boldsymbol{h} \in \mathbb{Z}^s \setminus \{\boldsymbol{0}\}} \frac{1}{r_{\alpha,\boldsymbol{\gamma}}(\boldsymbol{h})} = \frac{1}{N-1} \left(-1 + \prod_{j=1}^{s} (1 + 2\gamma_j \zeta(\alpha)) \right).$$

However, since the average square worst-case error over G_N^s satisfies this bound, there must exist at least one particular $\boldsymbol{g} \in G_N^s$ for which the worst-case error satisfies this bound as well. Hence we have shown the existence of $\boldsymbol{g} \in G_N^s$ for which we have

$$e^2(\mathcal{H}_{s,\alpha,\boldsymbol{\gamma}}, \mathcal{P}(\boldsymbol{g}, N)) \leq \frac{1}{N-1} \left(-1 + \prod_{j=1}^{s} (1 + 2\gamma_j \zeta(\alpha)) \right). \tag{4.25}$$

Now we use a popular trick to improve the convergence rate of this existence result. To this end we require, for any $\lambda \in (0, 1]$ and non-negative reals a_k, the inequality

$$\left(\sum_k a_k \right)^\lambda \leq \sum_k a_k^\lambda \tag{4.26}$$

which can be proved as follows: We have $0 \le a_j / \left(\sum_k a_k \right) \le 1$ and hence, since $\lambda \in (0, 1]$,

$$\frac{a_j}{\sum_k a_k} \le \left(\frac{a_j}{\sum_k a_k} \right)^{\lambda}.$$

Summation over all j implies

$$1 = \frac{\sum_j a_j}{\sum_k a_k} \le \frac{\sum_j a_j^{\lambda}}{\left(\sum_k a_k \right)^{\lambda}}$$

which finally yields (4.26).

Inequality (4.26) applied to the formula for the square worst-case error yields

$$e^{2\lambda}(\mathcal{H}_{s,\alpha,\boldsymbol{\gamma}}, \mathcal{P}(\boldsymbol{g}, N)) = \left(\sum_{\boldsymbol{h} \in \mathcal{L}(\boldsymbol{g},N) \setminus \{0\}} \frac{1}{r_{\alpha,\boldsymbol{\gamma}}(\boldsymbol{h})} \right)^{\lambda} \le \sum_{\boldsymbol{h} \in \mathcal{L}(\boldsymbol{g},N) \setminus \{0\}} \frac{1}{r_{\alpha,\boldsymbol{\gamma}}(\boldsymbol{h})^{\lambda}},$$

where we have to restrict λ to $\lambda \in (1/\alpha, 1]$ in order to guarantee the convergence of the series $\sum_{h \in \mathbb{Z}} r_{\alpha,\gamma}(h)^{-\lambda}$. For $h \in \mathbb{Z} \setminus \{0\}$ we have $r_{\alpha,\gamma}(h)^{\lambda} = \gamma^{-\lambda} |h|^{\lambda\alpha} = r_{\lambda\alpha,\gamma^{\lambda}}(h)$. Therefore,

$$r_{\alpha,\boldsymbol{\gamma}}(\boldsymbol{h})^{\lambda} = r_{\lambda\alpha,\boldsymbol{\gamma}^{\lambda}}(\boldsymbol{h}),$$

where $\boldsymbol{\gamma}^{\lambda} = (\gamma_j^{\lambda})_{j \in \mathbb{N}}$, and we obtain

$$e^{2\lambda}(\mathcal{H}_{s,\alpha,\boldsymbol{\gamma}}, \mathcal{P}(\boldsymbol{g}, N)) \le \sum_{\boldsymbol{h} \in \mathcal{L}(\boldsymbol{g},N) \setminus \{0\}} \frac{1}{r_{\lambda\alpha,\boldsymbol{\gamma}^{\lambda}}(\boldsymbol{h})} = e^{2}(\mathcal{H}_{s,\lambda\alpha,\boldsymbol{\gamma}^{\lambda}}, \mathcal{P}(\boldsymbol{g}, N)).$$

Applying the existence result (4.25) to $e^2(\mathcal{H}_{s,\lambda\alpha,\boldsymbol{\gamma}^{\lambda}}, \mathcal{P}(\boldsymbol{g}, N))$ we finally obtain the existence of $\boldsymbol{g} \in G_N^s$ such that

$$e^{2\lambda}(\mathcal{H}_{s,\alpha,\boldsymbol{\gamma}}, \mathcal{P}(\boldsymbol{g}, N)) \le \frac{1}{N-1} \left(-1 + \prod_{j=1}^{s} (1 + 2\gamma_j^{\lambda} \zeta(\lambda\alpha)) \right).$$

Note that $\zeta(\lambda\alpha) < \infty$ since $\lambda \in (1/\alpha, 1]$. Hence the desired result follows. $\qquad \square$

Since λ can be chosen arbitrarily close to $1/\alpha$ we obtain a convergence rate of $O(N^{-\frac{\alpha}{2}+\delta})$ for $\delta > 0$. This is in accordance with Theorem 4.26. The result of Theorem 4.30 can even be made explicit with a CBC algorithm; see Sect. 4.7.

If we estimate the bound from Theorem 4.30 further we obtain the following result for the Nth minimal error in the weighted Korobov space $\mathcal{H}_{s,\alpha,\boldsymbol{\gamma}}$:

Corollary 4.31 *Let N be a prime number, let $s \in \mathbb{N}$, $\alpha > 1$, and let $e(N, s)$ be the Nth minimal error in the weighted Korobov space $\mathcal{H}_{s,\alpha,\boldsymbol{\gamma}}$.*

1. *For some $\lambda \in (1/\alpha, 1]$ assume*

$$\Gamma_\lambda := \sum_{j=1}^{\infty} \gamma_j^\lambda < \infty. \tag{4.27}$$

Then we have

$$e(N, s) \le \frac{2^{1/(2\lambda)}}{N^{1/(2\lambda)}} \exp\left(\frac{\zeta(\lambda\alpha)}{\lambda} \Gamma_\lambda\right) \quad \text{for all } s \in \mathbb{N}.$$

This bound is independent of the dimension s and can be achieved by a lattice rule.

2. *Assume that*

$$A := \limsup_{s \to \infty} \frac{\sum_{j=1}^{s} \gamma_j}{\log s} < \infty. \tag{4.28}$$

Then for any $\delta > 0$ there exists a $c_\delta > 0$ such that

$$e(N, s) \le \frac{c_\delta s^{\zeta(\alpha)(A+\delta)}}{\sqrt{N}} \quad \text{for all } s \in \mathbb{N}.$$

This bound depends only polynomially on the dimension s and can be achieved by a lattice rule.

3. *We have $e(0, s) = 1$.*

Proof To begin with, for the bound from Theorem 4.30 we have

$$
\begin{aligned}
e^2(\mathcal{H}_{s,\alpha,\boldsymbol{\gamma}}, \mathcal{P}(\boldsymbol{g}, N)) &\le \frac{2^{1/\lambda}}{N^{1/\lambda}} \exp\left(\frac{1}{\lambda} \log\left(\prod_{j=1}^{s}(1 + 2\gamma_j^\lambda \zeta(\lambda\alpha))\right)\right) \\
&= \frac{2^{1/\lambda}}{N^{1/\lambda}} \exp\left(\frac{1}{\lambda} \sum_{j=1}^{s} \log(1 + 2\gamma_j^\lambda \zeta(\lambda\alpha))\right) \\
&\le \frac{2^{1/\lambda}}{N^{1/\lambda}} \exp\left(\frac{2\zeta(\lambda\alpha)}{\lambda} \sum_{j=1}^{s} \gamma_j^\lambda\right),
\end{aligned}
\tag{4.29}
$$

where we used the estimate $\log(1 + x) \le x$. Now Assertion 1 follows immediately from (4.29) and Condition (4.27) on the weights.

We turn to Assertion 2: Since $A < \infty$, there exists, for any $\delta > 0$, a positive s_δ such that

$$\sum_{j=1}^{s} \gamma_j \le (A + \delta) \log s \quad \text{for all } s \ge s_\delta.$$

From (4.29) with $\lambda = 1$ we obtain for $s \ge s_\delta$ that

$$e(N, s) \le \sqrt{\frac{2}{N}}\, s^{\zeta(\alpha)\frac{\sum_{j=1}^{s}\gamma_j}{\log s}} \le \sqrt{\frac{2}{N}}\, s^{\zeta(\alpha)(A+\delta)}.$$

Hence there exists a $c_\delta > 0$ such that

$$e(N, s) \le \frac{c_\delta s^{\zeta(\alpha)(A+\delta)}}{\sqrt{N}} \quad \text{for all } s \in \mathbb{N}.$$

Finally, to show that $e(0, s) = 1$ is left as an exercise (cf. Exercise 4.8). $\square$

If $\sum_{j=1}^{\infty} \gamma_j < \infty$, then there exists a $\lambda \in (1/\alpha, 1]$ which satisfies (4.27). Put $C_{\alpha,\lambda,\boldsymbol{\gamma}} := 2 \exp\left(2\zeta(\lambda\alpha)\Gamma_\lambda\right)$. For $\varepsilon > 0$, let N be the smallest prime number that is larger or equal to $\lceil C_{\alpha,\lambda,\boldsymbol{\gamma}}\varepsilon^{-2\lambda}\rceil =: M$. Then we have $e(N, s) \le \varepsilon$ and hence

$$N_{\mathcal{H}_{s,\alpha,\boldsymbol{\gamma}}}(\varepsilon, s) \le N < 2M = 2\lceil C_{\alpha,\lambda,\boldsymbol{\gamma}}\varepsilon^{-2\lambda}\rceil,$$

where we used Bertrand's postulate which tells us that $M \le N < 2M$. Hence multivariate integration in $\mathcal{H}_{s,\alpha,\boldsymbol{\gamma}}$ is strongly polynomially tractable with ε-exponent at most $2\lambda_0$ where λ_0 is the infimum over all $\lambda \in (1/\alpha, 1]$ such that (4.27) holds.

In the same way one can show that polynomial tractability holds under Condition (4.28). The proof for this result is left as an exercise (cf. Exercise 4.9).

Now we turn to necessary conditions for (strong) polynomial tractability. Let $\gamma'_j := \min(\gamma_j, 1/(2\zeta(\alpha)))$ and $\boldsymbol{\gamma}' = (\gamma'_j)_{j\in\mathbb{N}}$. From $\gamma'_j \le \gamma_j$ we obtain $\|f\|_{s,\alpha,\boldsymbol{\gamma}'} \ge \|f\|_{s,\alpha,\boldsymbol{\gamma}}$ and hence

$$\{f \in \mathcal{H}_{s,\alpha,\boldsymbol{\gamma}'} : \|f\|_{s,\alpha,\boldsymbol{\gamma}'} \le 1\} \subseteq \{f \in \mathcal{H}_{s,\alpha,\boldsymbol{\gamma}} : \|f\|_{s,\alpha,\boldsymbol{\gamma}} \le 1\}.$$

This implies that integration in $\mathcal{H}_{s,\alpha,\boldsymbol{\gamma}'}$ is no harder than integration in $\mathcal{H}_{s,\alpha,\boldsymbol{\gamma}}$, that is, $e(\mathcal{H}_{s,\alpha,\boldsymbol{\gamma}'}, \mathcal{P}) \le e(\mathcal{H}_{s,\alpha,\boldsymbol{\gamma}}, \mathcal{P})$ for any N-element point set in $[0, 1)^s$. Furthermore, according to the definition of $\boldsymbol{\gamma}'$, we have that $K_{s,\alpha,\boldsymbol{\gamma}'}$ is non-negative.

Now we estimate $e(\mathcal{H}_{s,\alpha,\boldsymbol{\gamma}}, \mathcal{P})$ from below. We have

$$
\begin{aligned}
e^2(\mathcal{H}_{s,\alpha,\boldsymbol{\gamma}}, \mathcal{P}) &\geq e^2(\mathcal{H}_{s,\alpha,\boldsymbol{\gamma}'}, \mathcal{P}) \\
&= -1 + \frac{1}{N^2} \sum_{h,i=0}^{N-1} K_{s,\alpha,\boldsymbol{\gamma}'}(\boldsymbol{x}_h, \boldsymbol{x}_i) \\
&\geq -1 + \frac{1}{N^2} \sum_{h=0}^{N-1} K_{s,\alpha,\boldsymbol{\gamma}'}(\boldsymbol{x}_h, \boldsymbol{x}_h) \\
&= -1 + \frac{1}{N^2} \sum_{h=0}^{N-1} \prod_{j=1}^{s} \left(1 + 2\gamma_j' \zeta(\alpha)\right) \\
&= -1 + \frac{1}{N} \prod_{j=1}^{s} \left(1 + 2\gamma_j' \zeta(\alpha)\right).
\end{aligned}
\tag{4.30}
$$

We summarize the result in the following lemma.

Lemma 4.32 *For any N-element point set $\mathcal{P}$ in $[0,1)^s$ we have*

$$
e^2(\mathcal{H}_{s,\alpha,\boldsymbol{\gamma}}, \mathcal{P}) \geq -1 + \frac{1}{N} \prod_{j=1}^{s} \left(1 + \min(2\zeta(\alpha)\gamma_j, 1)\right)
$$

and hence

$$
N_{\mathcal{H}_{s,\alpha,\boldsymbol{\gamma}}}(s, \varepsilon) \geq \frac{\prod_{j=1}^{s} \left(1 + \min(2\zeta(\alpha)\gamma_j, 1)\right)}{1 + \varepsilon^2}.
$$

Suppose there exists a $\gamma_* > 0$ such that $\gamma_j \geq \gamma_*$ for all $j \in \mathbb{N}$. Then we have

$$
N_{\mathcal{H}_{s,\alpha,\boldsymbol{\gamma}}}(s, \varepsilon) \geq \frac{(1 + \min(2\zeta(\alpha)\gamma_*, 1))^s}{1 + \varepsilon^2}
$$

which implies the curse of dimensionality. This shows that $\lim_{j\to\infty} \gamma_j = 0$ is a necessary condition for tractability.

Now suppose that $\lim_{j\to\infty} \gamma_j = 0$ but $\sum_{j=1}^{\infty} \gamma_j = \infty$. Then for growing s we have

$$
\prod_{j=1}^{s} \left(1 + \min(2\zeta(\alpha)\gamma_j, 1)\right) \geq 1 + \sum_{j=1}^{s} \min(2\zeta(\alpha)\gamma_j, 1) \to \infty.
$$

Hence $\lim_{s\to\infty} N_{\mathcal{H}_{s,\alpha,\boldsymbol{\gamma}}}(s, \varepsilon) = \infty$, which means that we cannot have strong polynomial tractability. Thus $\sum_{j=1}^{\infty} \gamma_j < \infty$ is a necessary condition for strong polynomial tractability.

Finally, suppose that $\lim_{j \to \infty} \gamma_j = 0$ and $\limsup_{s \to \infty} \sum_{j=1}^{s} \gamma_j / \log s = \infty$. Note that for $x \in [0, 2\zeta(\alpha) \sup_j \gamma_j]$ we have $\log(1 + x) \geq cx$ for some $c > 0$. Hence we have

$$\log \prod_{j=1}^{s} \left(1 + \min(2\zeta(\alpha)\gamma_j, 1)\right) = \sum_{j=1}^{s} \log \left(1 + \min(2\zeta(\alpha)\gamma_j, 1)\right) \geq c \sum_{j=1}^{s} \min(2\zeta(\alpha)\gamma_j, 1)$$

and therefore

$$\prod_{j=1}^{s} \left(1 + \min(2\zeta(\alpha)\gamma_j, 1)\right) \geq \exp\left(c \sum_{j=1}^{s} \min(2\zeta(\alpha)\gamma_j, 1)\right) = s^{\frac{c}{\log s} \sum_{j=1}^{s} \min(2\zeta(\alpha)\gamma_j, 1)}.$$

This implies that $N_{\mathcal{H}_{s,\alpha,\gamma}}(s, \varepsilon)$ goes to infinity faster than any power of s and we cannot have polynomial tractability. Hence we have shown that (4.28) is a necessary condition for polynomial tractability.

Summing up we have shown the following result.

Corollary 4.33 *Numerical integration in $\mathcal{H}_{s,\alpha,\gamma}$ is:*

1. *Strongly polynomially tractable if and only if $\sum_{j=1}^{\infty} \gamma_j < \infty$. If λ_0 is the infimum over all $\lambda \in (1/\alpha, 1]$ such that (4.27) holds, then the ε-exponent τ^* of strong polynomial tractability is at most $2\lambda_0$.*
2. *Polynomially tractable if and only if (4.28) holds.*

Remark 4.34 The above if and only if conditions for (strong) polynomial tractability also hold for more general algorithms rather than for QMC rules as considered here. This is clear for the sufficient conditions. For the necessary conditions this was proved by F.J. Hickernell and H. Woźniakowski in 2001.

4.7   The CBC Construction for the Weighted Korobov Space

With Theorem 4.30 we have an important existence result for lattice point sets of excellent quality, achieving essentially the optimal convergence rate of the worst-case error and which even guarantees tractability for sufficiently fast decaying weights, where the sufficient conditions exactly match the necessary conditions for (strong) polynomial tractability.

In this section we discuss how such generating vectors, whose existence is guaranteed by Theorem 4.30, can be constructed in an efficient manner. The suitable approach will again be a CBC construction, which now directly relies on the worst-case error as quality measure, rather than the quantities R_N or $\widetilde{R}_{N,\gamma}$, like in the previous sections. Starting point is the representation of the worst-case error in (4.17) of the form:

$$e^2(\mathcal{H}_{s,\alpha,\boldsymbol{\gamma}}, \mathcal{P}(\boldsymbol{g}, N)) = \sum_{\boldsymbol{h}\in\mathbb{Z}^s\setminus\{0\}} \frac{1}{r_{\alpha,\boldsymbol{\gamma}}(\boldsymbol{h})} \left| \frac{1}{N} \sum_{n=0}^{N-1} \exp(2\pi\,\mathrm{i}n\boldsymbol{h}\cdot\boldsymbol{g}/N) \right|^2,$$

where $\alpha > 1$ and for product weights $\boldsymbol{\gamma} = \{\gamma_{\mathfrak{u}} = \prod_{j\in\mathfrak{u}}\gamma_j \; : \; \mathfrak{u} \subseteq [s]\}$ with a sequence of coordinate weights $(\gamma_j)_{j\geq 1}$. According to (4.1) the term between the absolute value signs is either 0 or 1, which means that we may omit the square. Adding and subtracting the term for $\boldsymbol{h} = \boldsymbol{0}$, which equals 1, and interchanging the order of summation yields

$$\begin{aligned}
e^2(\mathcal{H}_{s,\alpha,\boldsymbol{\gamma}}, \mathcal{P}(\boldsymbol{g}, N)) &= -1 + \frac{1}{N}\sum_{n=0}^{N-1}\prod_{j=1}^{s}\left(\sum_{h\in\mathbb{Z}}\frac{1}{r_{\alpha,\gamma_j}(h)}\exp(2\pi\,\mathrm{i}nhg_j/N)\right) \\
&= -1 + \frac{1}{N}\sum_{n=0}^{N-1}\prod_{j=1}^{s}\left(1 + \gamma_j\phi_\alpha\left(\frac{ng_j}{N}\right)\right),
\end{aligned} \qquad (4.31)$$

where

$$\phi_\alpha(x) := \sum_{h\in\mathbb{Z}\setminus\{0\}}\frac{\exp(2\pi\,\mathrm{i}hx)}{|h|^\alpha}.$$

Thus, for evaluating the worst-case error, it suffices to know the values of ϕ_α for all arguments of the form k/N, where $k \in \{0, 1, \ldots, N-1\}$. These values of ϕ_α can be precomputed and stored in a look-up table of size N. If $\alpha \geq 2$ is an even integer, then, according to Remark 4.24:

$$\phi_\alpha(x) = \phi_\alpha(\{x\}) = \frac{(2\pi)^\alpha}{(-1)^{(\alpha+2)/2}\alpha!}B_\alpha(\{x\}),$$

where B_α is the Bernoulli polynomial of degree α.

Now we formulate the CBC algorithm which uses as its criterion the squared worst-case error $E_s(\boldsymbol{g}) := e^2(\mathcal{H}_{s,\alpha,\boldsymbol{\gamma}}, \mathcal{P}(\boldsymbol{g}, N))$ in the form of (4.31). Let again $G_N := \{1, \ldots, N-1\}$.

Algorithm 4.35 (*CBC algorithm*) Let $s, N \in \mathbb{N}$, $\alpha > 1$ and $\boldsymbol{\gamma}$ be product weights.

1. Choose $g_1 = 1$.
2. For $d = 1, 2, \ldots, s-1$, choose $g_{d+1} \in G_N$ to minimize $E_d((g_1, \ldots, g_d, z))$ as a function of $z \in G_N$.

The error bound for a generating vector obtained from Algorithm (4.35) satisfies (essentially) the error bound from Theorem 4.30, as will be shown in the next theorem. The added benefit is that now we know how to construct the generating vector.

Theorem 4.36 *Let N be a prime number, let $s \in \mathbb{N}$, $\alpha > 1$ and $\boldsymbol{\gamma}$ be product weights. Assume that $\boldsymbol{g} = (g_1, \ldots, g_s)$ has been found by Algorithm 4.35. Then for arbitrary $\lambda \in (1/\alpha, 1]$ we have*

$$e^2(\mathcal{H}_{s,\alpha,\boldsymbol{\gamma}}, \mathcal{P}(\boldsymbol{g}, N)) \leq \frac{2^{1/\lambda}}{N^{1/\lambda}} \prod_{j=1}^{s} \left(1 + 2\gamma_j^\lambda \zeta(\alpha\lambda)\right)^{1/\lambda}.$$

From Theorem 4.36 one can again deduce Corollary 4.31, thus providing an alternative proof to establish the statements about tractability there.

For the proof of Theorem 4.36 we need the following lemma.

Lemma 4.37 *Let N be a prime number and suppose that $\alpha > 1$. Then for $k \in \{0, 1, \ldots, N-1\}$ we have*

$$T_{N,\alpha}(k) := \sum_{g=1}^{N-1} \phi_\alpha\left(\frac{kg}{N}\right) = \begin{cases} 2\zeta(\alpha)(N-1) & \text{if } k = 0, \\ 2\zeta(\alpha)(N^{1-\alpha} - 1) & \text{otherwise.} \end{cases}$$

Furthermore,

$$\frac{1}{N-1} \sum_{k=0}^{N-1} |T_{N,\alpha}(k)| \leq 4\zeta(\alpha).$$

Proof If $k = 0$, then

$$T_{N,\alpha}(0) = \sum_{g=1}^{N-1} \sum_{h \in \mathbb{Z}\setminus\{0\}} \frac{1}{|h|^\alpha} = 2\zeta(\alpha)(N-1).$$

On the other hand, if $k \in \{1, 2, \ldots, N-1\}$, then

$$T_{N,\alpha}(k) = \sum_{g=1}^{N-1} \sum_{h \in \mathbb{Z}\setminus\{0\}} \frac{\exp(2\pi\,\mathrm{i}hkg/N)}{|h|^\alpha}$$

$$= \sum_{h \in \mathbb{Z}\setminus\{0\}} \frac{1}{|h|^\alpha} \sum_{g=1}^{N-1} \exp(2\pi\,\mathrm{i}hkg/N)$$

$$= \sum_{\substack{h \in \mathbb{Z}\setminus\{0\} \\ h \equiv 0 \pmod{N}}} \frac{1}{|h|^\alpha} \sum_{g=1}^{N-1} \exp(2\pi\,\mathrm{i}hkg/N)$$

$$+ \sum_{\substack{h \in \mathbb{Z}\setminus\{0\} \\ h \not\equiv 0 \pmod{N}}} \frac{1}{|h|^\alpha} \sum_{g=1}^{N-1} \exp(2\pi\,\mathrm{i}hkg/N).$$

We treat the two sums in the latter expression separately. First,

$$
T_{N,\alpha}^{(1)}(k) := \sum_{\substack{h \in \mathbb{Z}\setminus\{0\} \\ h \equiv 0 \pmod{N}}} \frac{1}{|h|^\alpha} \sum_{g=1}^{N-1} \exp(2\pi \, \mathrm{i} hkg/N)
$$

$$
= \sum_{h \in \mathbb{Z}\setminus\{0\}} \frac{1}{|Nh|^\alpha} \sum_{g=1}^{N-1} 1
$$

$$
= \frac{N-1}{N^\alpha} 2\zeta(\alpha).
$$

Next, using the fact that

$$
\sum_{g=1}^{N-1} \exp(2\pi \, \mathrm{i} hkg/N) = -1 + \sum_{g=0}^{N-1} \exp(2\pi \, \mathrm{i} hkg/N) = -1
$$

for $h \not\equiv 0 \pmod{N}$, where in the last step we used the geometric sum formula, we obtain

$$
T_{N,\alpha}^{(2)}(k) := \sum_{\substack{h \in \mathbb{Z}\setminus\{0\} \\ h \not\equiv 0 \pmod{N}}} \frac{1}{|h|^\alpha} \sum_{g=1}^{N-1} \exp(2\pi \, \mathrm{i} hkg/N)
$$

$$
= - \sum_{\substack{h \in \mathbb{Z}\setminus\{0\} \\ h \not\equiv 0 \pmod{N}}} \frac{1}{|h|^\alpha}
$$

$$
= \sum_{\substack{h \in \mathbb{Z}\setminus\{0\} \\ h \equiv 0 \pmod{N}}} \frac{1}{|h|^\alpha} - \sum_{h \in \mathbb{Z}\setminus\{0\}} \frac{1}{|h|^\alpha}
$$

$$
= \frac{2\zeta(\alpha)}{N^\alpha} - 2\zeta(\alpha).
$$

Adding $T_{N,\alpha}^{(1)}(k)$ and $T_{N,\alpha}^{(2)}(k)$ thus yields

$$
T_{N,\alpha}(k) = 2\zeta(\alpha)(N^{1-\alpha} - 1).
$$

Note that $T_{N,\alpha}(k)$ is the same for all $k \in \{1, \ldots, N-1\}$. Hence

$$
\frac{1}{N-1} \sum_{k=0}^{N-1} |T_{N,\alpha}(k)| = 2\zeta(\alpha) + 2\zeta(\alpha)\left(1 - \frac{1}{N^{\alpha-1}}\right) \leq 4\zeta(\alpha),
$$

as claimed. $\qquad\square$

We are now in a position to prove Theorem 4.36.

Proof of Theorem   4.36. We prove

$$E_d(\boldsymbol{g}_d) := e^2(\mathcal{H}_{d,\alpha,\gamma}, \mathcal{P}(\boldsymbol{g}_d, N)) \leq \frac{2^{1/\lambda}}{N^{1/\lambda}} \prod_{j=1}^{d} \left(1 + 2\gamma_j^{\lambda}\zeta(\alpha\lambda)\right)^{1/\lambda} \qquad (4.32)$$

for all $\lambda \in (1/\alpha, 1]$, where $\boldsymbol{g}_d = (g_1, \ldots, g_d)$ for $d \in \{1, \ldots, s\}$ by induction on d.
For $d = 1$, we have $g_1 = 1$ according to Algorithm 4.35. Using (4.31) we have

$$E_1(1) = -1 + \frac{1}{N}\sum_{k=0}^{N-1}\left(1 + \gamma_1\phi_\alpha\left(\frac{k}{N}\right)\right)$$

$$= \frac{\gamma_1}{N}\sum_{k=0}^{N-1}\phi_\alpha\left(\frac{k}{N}\right)$$

$$= \frac{\gamma_1}{N}\sum_{h\in\mathbb{Z}\setminus\{0\}}\frac{1}{|h|^\alpha}\sum_{k=0}^{N-1}\exp(2\pi\,\mathrm{i}hk/N).$$

Now we employ (4.1) to obtain

$$E_1(1) = \gamma_1\sum_{\substack{h\in\mathbb{Z}\setminus\{0\}\\ h\equiv 0 \pmod{N}}}\frac{1}{|h|^\alpha} = \frac{\gamma_1}{N^\alpha}\sum_{h\in\mathbb{Z}\setminus\{0\}}\frac{1}{|h|^\alpha} = \frac{2\gamma_1\zeta(\alpha)}{N^\alpha}. \qquad (4.33)$$

Note that $\alpha > 1$ and hence $\zeta(\alpha) < \infty$. Using (4.26) we obtain $\zeta(\alpha)^\lambda \leq \zeta(\alpha\lambda)$ for $\lambda \in (1/\alpha, 1]$ and hence

$$[E_1(1)]^\lambda \leq \frac{2\gamma_1^\lambda\zeta(\alpha\lambda)}{N^{\alpha\lambda}} \leq \frac{1}{N^{\alpha\lambda}}\left(1 + 2\gamma_1^\lambda\zeta(\alpha\lambda)\right).$$

Therefore,

$$E_1(1) \leq \frac{1}{N^\alpha}\left(1 + 2\gamma_1^\lambda\zeta(\alpha\lambda)\right)^{1/\lambda}$$

and this yields the result (4.32) for $d = 1$ and all $\lambda \in (1/\alpha, 1]$.

For the induction step assume that (4.32) holds for all $\lambda \in (1/\alpha, 1]$. Again, with some abuse of notation, we consider the $(d + 1)$-dimensional lattice point $(\boldsymbol{g}_d, g) := (g_1, \ldots, g_d, g)$ for $g \in G_N$. We will now show that the error bound holds if the $(d + 1)$-st component is chosen according to Algorithm 4.35.

First of all, for arbitrary $g \in G_N$ we have

$$
\begin{aligned}
E_{d+1}((\boldsymbol{g}_d, g)) &= -1 + \frac{1}{N} \sum_{k=0}^{N-1} \prod_{j=1}^{d} \left(1 + \gamma_j \sum_{h \in \mathbb{Z} \backslash \{0\}} \frac{\exp(2\pi \mathrm{i} hk g_j/N)}{|h|^\alpha} \right) \\
&\qquad \times \left(1 + \gamma_{d+1} \sum_{h \in \mathbb{Z} \backslash \{0\}} \frac{\exp(2\pi \mathrm{i} hk g/N)}{|h|^\alpha} \right) \\
&= -1 + \frac{1}{N} \sum_{k=0}^{N-1} \prod_{j=1}^{d} \left(1 + \gamma_j \sum_{h \in \mathbb{Z} \backslash \{0\}} \frac{\exp(2\pi \mathrm{i} hk g_j/N)}{|h|^\alpha} \right) \\
&\qquad + \frac{\gamma_{d+1}}{N} \sum_{k=0}^{N-1} \prod_{j=1}^{d} \left(1 + \gamma_j \sum_{h \in \mathbb{Z} \backslash \{0\}} \frac{\exp(2\pi \mathrm{i} hk g_j/N)}{|h|^\alpha} \right) \\
&\qquad \times \left(\sum_{h \in \mathbb{Z} \backslash \{0\}} \frac{\exp(2\pi \mathrm{i} hk g/N)}{|h|^\alpha} \right) \\
&= E_d(\boldsymbol{g}_d) + \theta_{d+1,\alpha,\boldsymbol{\gamma}}(\boldsymbol{g}_d, g),
\end{aligned}
\tag{4.34}
$$

where we write

$$
\begin{aligned}
&\theta_{d+1,\alpha,\boldsymbol{\gamma}}(\boldsymbol{g}_d, g) \\
&:= \frac{\gamma_{d+1}}{N} \sum_{k=0}^{N-1} \prod_{j=1}^{d} \left(1 + \gamma_j \sum_{h \in \mathbb{Z} \backslash \{0\}} \frac{\exp(2\pi \mathrm{i} hk g_j/N)}{|h|^\alpha} \right) \left(\sum_{h \in \mathbb{Z} \backslash \{0\}} \frac{\exp(2\pi \mathrm{i} hk g/N}{|h|^\alpha} \right).
\end{aligned}
$$

Note that $\theta_{d+1,\alpha,\boldsymbol{\gamma}}(\boldsymbol{g}_d, g)$ is the only term in $E_{d+1}((\boldsymbol{g}_d, g))$ that depends on g. According to Algorithm 4.35, the element $g_{d+1} \in G_N$ is chosen such that $E_{d+1}((\boldsymbol{g}_d, g))$ is minimized and hence g_{d+1} is also a minimizer of $\theta_{d+1,\alpha,\boldsymbol{\gamma}}(\boldsymbol{g}_d, g)$. Hence we have $\theta_{d+1,\alpha,\boldsymbol{\gamma}}(\boldsymbol{g}_d, g_{d+1}) \leq \theta_{d+1,\alpha,\boldsymbol{\gamma}}(\boldsymbol{g}_d, g)$ for all $g \in G_N$. This implies that for any $\lambda \in (1/\alpha, 1]$ we have $[\theta_{d+1,\alpha,\boldsymbol{\gamma}}(\boldsymbol{g}_d, g_{d+1})]^\lambda \leq [\theta_{d+1,\alpha,\boldsymbol{\gamma}}(\boldsymbol{g}_d, g)]^\lambda$ for all $g \in G_N$ and further

$$
\theta_{d+1,\alpha,\boldsymbol{\gamma}}(\boldsymbol{g}_d, g_{d+1}) \leq \left(\frac{1}{N-1} \sum_{g \in G_N} [\theta_{d+1,\alpha,\boldsymbol{\gamma}}(\boldsymbol{g}_d, g)]^\lambda \right)^{1/\lambda}.
\tag{4.35}
$$

Now we show that

$$
[\theta_{d+1,\alpha,\boldsymbol{\gamma}}(\boldsymbol{g}_d, g)]^\lambda \leq \theta_{d+1,\alpha\lambda,\boldsymbol{\gamma}^\lambda}(\boldsymbol{g}_d, g).
\tag{4.36}
$$

To that end we use the function r_{α,γ_j}, as defined in Sect. 4.4, i.e., $r_{\alpha,\gamma_j}(0) := 1$ and $r_{\alpha,\gamma_j}(h) := \gamma_j^{-1} |h|^\alpha$ for $h \in \mathbb{Z} \backslash \{0\}$ to rewrite $\theta_{d+1,\alpha,\boldsymbol{\gamma}}(\boldsymbol{g}_d, g)$, for $g \in G_N$, in the form:

$$\theta_{d+1,\alpha,\boldsymbol{\gamma}}(\boldsymbol{g}_d, g)$$

$$= \frac{\gamma_{d+1}}{N} \sum_{k=0}^{N-1} \prod_{j=1}^{d} \left(\sum_{h\in\mathbb{Z}} \frac{\exp(2\pi\,\mathrm{i}hkg_j/N)}{r_{\alpha,\gamma_j}(h)} \right) \sum_{h\in\mathbb{Z}\backslash\{0\}} \frac{\exp(2\pi\,\mathrm{i}hkg/N)}{|h|^\alpha}$$

$$= \frac{\gamma_{d+1}}{N} \sum_{k=0}^{N-1} \sum_{\substack{\boldsymbol{h}\in\mathbb{Z}^{d+1}\\ h_{d+1}\neq 0}} \frac{\exp(2\pi\,\mathrm{i}k\boldsymbol{h}\cdot(\boldsymbol{g}_d,g)/N)}{|h_{d+1}|^\alpha \prod_{j=1}^{d} r_{\alpha,\gamma_j}(h_j)}.$$

Note that according to (4.1) the latter expression is zero if $\boldsymbol{h}\cdot(\boldsymbol{g}_d, g) \not\equiv 0 \pmod{N}$, so

$$\theta_{d+1,\alpha,\boldsymbol{\gamma}}(\boldsymbol{g}_d, g) = \frac{\gamma_{d+1}}{N} \sum_{k=0}^{N-1} \sum_{\substack{\boldsymbol{h}\in\mathbb{Z}^{d+1}\\ h_{d+1}\neq 0\\ \boldsymbol{h}\cdot(\boldsymbol{g}_d,g)\equiv 0 \pmod{N}}} \frac{1}{|h_{d+1}|^\alpha \prod_{j=1}^{d} r_{\alpha,\gamma_j}(h_j)}$$

$$= \sum_{\substack{\boldsymbol{h}\in\mathbb{Z}^{d+1}\\ h_{d+1}\neq 0\\ \boldsymbol{h}\cdot(\boldsymbol{g}_d,g)\equiv 0 \pmod{N}}} \frac{\gamma_{d+1}}{|h_{d+1}|^\alpha} \prod_{j=1}^{d} \frac{1}{r_{\alpha,\gamma_j}(h_j)}. \tag{4.37}$$

Note that for $\lambda \in (1/\alpha, 1]$ we have $(r_{\alpha,\gamma_j}(h_j))^\lambda = r_{\alpha\lambda,\gamma_j^\lambda}(h_j)$. Thus, applying (4.26) to (4.37) yields

$$[\theta_{d+1,\alpha,\boldsymbol{\gamma}}(\boldsymbol{g}_d, g)]^\lambda \leq \sum_{\substack{\boldsymbol{h}\in\mathbb{Z}^{d+1}\\ h_{d+1}\neq 0\\ \boldsymbol{h}\cdot(\boldsymbol{g}_d,g)\equiv 0 \pmod{N}}} \frac{\gamma_{d+1}^\lambda}{|h_{d+1}|^{\alpha\lambda}} \prod_{j=1}^{d} \frac{1}{r_{\alpha\lambda,\gamma_j^\lambda}(h_j)}$$

$$= \theta_{d+1,\alpha\lambda,\boldsymbol{\gamma}^\lambda}(\boldsymbol{g}_d, g),$$

and thus (4.36) is shown. Inserting into (4.35) yields

$$\theta_{d+1,\alpha,\boldsymbol{\gamma}}(\boldsymbol{g}_d, g_{d+1}) \leq \left(\frac{1}{N-1} \sum_{g\in G_N} \theta_{d+1,\alpha\lambda,\boldsymbol{\gamma}^\lambda}(\boldsymbol{g}_d, g) \right)^{1/\lambda}. \tag{4.38}$$

We now consider the average of $\theta_{d+1,\alpha\lambda,\boldsymbol{\gamma}^\lambda}(\boldsymbol{g}_d, g)$ over all values of $g \in G_N$,

$$\Theta_{d+1,\alpha\lambda,\boldsymbol{\gamma}^\lambda}(\boldsymbol{g}_d) := \frac{1}{N-1} \sum_{g\in G_N} \theta_{d+1,\alpha\lambda,\boldsymbol{\gamma}^\lambda}(\boldsymbol{g}_d, g)$$

$$= \frac{\gamma_{d+1}^\lambda}{N} \sum_{k=0}^{N-1} \frac{\mathcal{I}_{N,\alpha\lambda}(k)}{N-1} \prod_{j=1}^{d} \left(1 + \gamma_j^\lambda \sum_{h\in\mathbb{Z}\backslash\{0\}} \frac{\exp(2\pi\,\mathrm{i}hkg_j/N)}{|h|^{\alpha\lambda}} \right),$$

where $\mathcal{T}_{N,\alpha\lambda}(k)$ is from Lemma 4.37. Note that $\theta_{d+1,\alpha\lambda,\boldsymbol{\gamma}^\lambda}(\boldsymbol{g}_d, g)$ is necessarily a real number and so is the average $\Theta_{d+1,\alpha\lambda,\boldsymbol{\gamma}^\lambda}(\boldsymbol{g}_d)$. Thus we obtain

$$
\begin{aligned}
\Theta_{d+1,\alpha\lambda,\boldsymbol{\gamma}^\lambda}(\boldsymbol{g}_d) &\le \frac{\gamma_{d+1}^\lambda}{N} \sum_{k=0}^{N-1} \frac{|\mathcal{T}_{N,\alpha\lambda}(k)|}{N-1} \prod_{j=1}^{d} \left(1 + \gamma_j^\lambda \sum_{h \in \mathbb{Z}\setminus\{0\}} \frac{1}{|h|^{\alpha\lambda}} \right) \\
&= \frac{\gamma_{d+1}^\lambda}{N} \frac{1}{N-1} \sum_{k=0}^{N-1} |\mathcal{T}_{N,\alpha\lambda}(k)| \left(\prod_{j=1}^{d} (1 + 2\gamma_j^\lambda \zeta(\alpha\lambda)) \right) \\
&\le \frac{4\gamma_{d+1}^\lambda \zeta(\alpha\lambda)}{N} \prod_{j=1}^{d} (1 + 2\gamma_j^\lambda \zeta(\alpha\lambda)),
\end{aligned}
$$

where we used Lemma 4.37. Inserting into (4.38) yields

$$
\theta_{d+1,\alpha,\boldsymbol{\gamma}}(\boldsymbol{g}_d, g_{d+1}) \le \left(\frac{4\gamma_{d+1}^\lambda \zeta(\alpha\lambda)}{N} \right)^{1/\lambda} \prod_{j=1}^{d} \left(1 + 2\gamma_j^\lambda \zeta(\alpha\lambda) \right)^{1/\lambda}. \tag{4.39}
$$

Finally, we insert the induction Assumptions (4.32) and (4.39) into (4.34) in order to obtain for a minimizer g_{d+1} the estimate

$$
\begin{aligned}
E_{d+1}&((\boldsymbol{g}_d, g_{d+1})) \\
&\le E_d(\boldsymbol{g}_d) + \left(\frac{4\gamma_{d+1}^\lambda \zeta(\alpha\lambda)}{N} \right)^{1/\lambda} \prod_{j=1}^{d} \left(1 + 2\gamma_j^\lambda \zeta(\alpha\lambda) \right)^{1/\lambda} \\
&\le \frac{2^{1/\lambda}}{N^{1/\lambda}} \prod_{j=1}^{d} \left(1 + 2\gamma_j^\lambda \zeta(\alpha\lambda) \right)^{1/\lambda} + \left(\frac{4\gamma_{d+1}^\lambda \zeta(\alpha\lambda)}{N} \right)^{1/\lambda} \prod_{j=1}^{d} \left(1 + 2\gamma_j^\lambda \zeta(\alpha\lambda) \right)^{1/\lambda} \\
&= \frac{2^{1/\lambda}}{N^{1/\lambda}} \prod_{j=1}^{d} \left(1 + 2\gamma_j^\lambda \zeta(\alpha\lambda) \right)^{1/\lambda} \left(1 + (2\gamma_{d+1}^\lambda \zeta(\alpha\lambda))^{1/\lambda} \right) \\
&\le \frac{2^{1/\lambda}}{N^{1/\lambda}} \prod_{j=1}^{d+1} \left(1 + 2\gamma_j^\lambda \zeta(\alpha\lambda) \right)^{1/\lambda},
\end{aligned}
$$

where in the last step we again used (4.26). $\qquad\square$

A fast implementation of Algorithm 4.35 can be executed very similar to the fast implementation of Algorithm 4.4 as presented in Sect. 4.2.

4.8 An Outlook on Lattice Rules for Nonperiodic Functions

In Sects. 4.4–4.7 we saw that lattice rules perform outstandingly well when applied to integration of smooth and periodic functions. The advantages are excellent convergence rates and, at the same time, no or a rather weak influence of the dimension under suitable conditions on the implied product weights leading to (strong) polynomial tractability. However, for many applications periodicity is a strong assumption that we cannot expect to be satisfied. The good news is that lattice rules are also the basis for integration rules which work well for functions which are not necessarily periodic. A first example has already been discussed in Sect. 4.3, where we analyzed the weighted star discrepancy, which corresponds to the worst-case error in a suitable function space, as shown in Sect. 3.5.

Here we briefly discuss two methods that allow to obtain good convergence rates and tractability properties for smooth and not necessarily periodic functions using lattice rules, namely, *shifted lattice rules* and *folded lattice rules*. As reference space for these methods we use the weighted ANOVA Sobolev space.

For product weights $\gamma = \{\gamma_{\mathfrak{u}} = \prod_{j \in \mathfrak{u}} \gamma_j \ : \ \mathfrak{u} \subseteq [s]\}$, with coordinate weights $(\gamma_j)_{j \geq 1}$ define a reproducing kernel $K_{\mathrm{Sob},s,\gamma} : [0,1]^s \times [0,1]^s \to \mathbb{R}$ by

$$K_{\mathrm{Sob},s,\gamma}(x, y) := \prod_{j=1}^{s} \left(1 + \gamma_j \eta(x_j, y_j)\right), \tag{4.40}$$

where $x = (x_1, \ldots, x_s)$ and $y = (y_1, \ldots, y_s)$ are in $[0,1]^s$, and where η is given by

$$\eta(x, y) := B_1(x) B_1(y) + \frac{B_2(|x - y|)}{2}.$$

Here B_1 is the first Bernoulli polynomial which is defined as $B_1(x) := x - 1/2$ and B_2 is the second Bernoulli polynomial which is defined as $B_2(x) := x^2 - x + 1/6$.

Definition 4.38 The *weighted ANOVA Sobolev space* $\mathcal{H}_{\mathrm{Sob},s,\gamma}$ (of smoothness one) is the reproducing kernel Hilbert space with kernel $K_{\mathrm{Sob},s,\gamma}$ and the corresponding norm

$$\|f\|_{\mathrm{Sob},s,\gamma} = \left(\sum_{\mathfrak{u} \subseteq [s]} \frac{1}{\gamma_{\mathfrak{u}}} \int_{[0,1]^{|\mathfrak{u}|}} \left(\int_{[0,1]^{s-|\mathfrak{u}|}} \frac{\partial^{|\mathfrak{u}|} f}{\partial x_{\mathfrak{u}}}(x) \, dx_{[s] \setminus \mathfrak{u}} \right)^2 dx_{\mathfrak{u}} \right)^{1/2}.$$

The space $\mathcal{H}_{\mathrm{Sob},s,\gamma}$ is also referred to as the *weighted unanchored Sobolev space*.

For example, for dimension $s = 1$ and a generic weight $\gamma > 0$ the norm is

$$\|f\|_{\mathrm{Sob},1,\gamma} = \left(\left(\int_0^1 f(x) \, dx \right)^2 + \frac{1}{\gamma} \int_0^1 (f'(x))^2 \, dx \right)^{1/2}.$$

Shifted Lattice Rules

A shifted lattice rule is obtained when the points of a lattice point set are shifted modulo one by a shift $\boldsymbol{\Delta}$.

Definition 4.39 For a lattice point set $\mathcal{P}(\boldsymbol{g}, N)$ and a point $\boldsymbol{\Delta} \in [0, 1]^s$ the *shifted lattice point set* is defined as

$$\mathcal{P}_{\boldsymbol{\Delta}}(\boldsymbol{g}, N) := \left\{ \left\{ \frac{k}{N}\boldsymbol{g} + \boldsymbol{\Delta} \right\} : k \in \{0, 1, \dots, N-1\} \right\}.$$

Here $\{(k/N)\boldsymbol{g} + \boldsymbol{\Delta}\}$ means component-wise addition modulo one. In this context, the point $\boldsymbol{\Delta}$ is called the *shift*. A QMC rule that uses a shifted lattice point set as underlying sample nodes is called a *shifted lattice rule*.

It proves beneficial to choose the shift $\boldsymbol{\Delta}$ at random, where the random shift is a vector whose components are i.i.d. (independent and identically distributed) and uniformly distributed in the unit interval. Write $Q_{N,s}(f, \boldsymbol{\Delta}) := \frac{1}{N} \sum_{k=0}^{N-1} f(\{x_k + \boldsymbol{\Delta}\})$ for a shifted lattice rule, where $x_k = (k/N)\boldsymbol{g}$. Taking the expectation with respect to all shifts $\boldsymbol{\Delta} \in [0, 1]^s$ we have

$$\mathbb{E}\left[Q_{N,s}(f, \boldsymbol{\Delta}) \right] = \frac{1}{N} \sum_{k=0}^{N-1} \mathbb{E}[f(\{x_k + \boldsymbol{\Delta}\})]$$

$$= \frac{1}{N} \sum_{k=0}^{N-1} \int_{[0,1]^s} f(\{x_k + \boldsymbol{\delta}\}) \, \mathrm{d}\boldsymbol{\delta}$$

$$= \int_{[0,1]^s} f(\boldsymbol{\delta}) \, \mathrm{d}\boldsymbol{\delta}.$$

This means that the estimator $Q_{N,s}(f, \boldsymbol{\Delta})$ is an *unbiased* estimator for the integral of f over $[0, 1]^s$. For this unbiased estimator we study the root mean square worst-case error for the weighted ANOVA Sobolev space.

Definition 4.40 Let $e^2(\mathcal{H}_{\mathrm{Sob},s,\boldsymbol{\gamma}}, \mathcal{P}(\boldsymbol{g}, N))$ be the worst-case error of a lattice rule applied to functions from the weighted ANOVA Sobolev space $\mathcal{H}_{\mathrm{Sob},s,\boldsymbol{\gamma}}$. Assume that the components of the shift $\boldsymbol{\Delta}$ are i.i.d. uniformly distributed on $[0, 1]$. Then the root mean square worst-case error of the shifted lattice rule,

$$e_{\mathrm{sh}}(\mathcal{H}_{\mathrm{Sob},s,\boldsymbol{\gamma}}, \mathcal{P}(\boldsymbol{g}, N)) := \sqrt{\mathbb{E}\left[e^2(\mathcal{H}_{\mathrm{Sob},s,\boldsymbol{\gamma}}, \mathcal{P}_{\boldsymbol{\Delta}}(\boldsymbol{g}, N)) \right]}$$

where the expectation is taken with respect to the shift $\boldsymbol{\Delta}$, is referred to as the *shift-averaged worst-case error*.

Note that since the components of $\boldsymbol{\Delta}$ are i.i.d. uniformly distributed on $[0, 1]$, we have

$$\mathbb{E}\left[e^2(\mathcal{H}_{\mathrm{Sob},s,\boldsymbol{\gamma}}, \mathcal{P}_{\boldsymbol{\Delta}}(\boldsymbol{g}, N))\right] = \int_{[0,1]^s} e^2(\mathcal{H}_{\mathrm{Sob},s,\boldsymbol{\gamma}}, \mathcal{P}_{\boldsymbol{\delta}}(\boldsymbol{g}, N))\, \mathrm{d}\boldsymbol{\delta}.$$

Furthermore, since the expectation (the average) is always at least as large as the minimum there exists at least one shift $\boldsymbol{\Delta} \in [0, 1]^s$ such that

$$e(\mathcal{H}_{\mathrm{Sob},s,\boldsymbol{\gamma}}, \mathcal{P}_{\boldsymbol{\Delta}}(\boldsymbol{g}, N)) \leq e_{\mathrm{sh}}(\mathcal{H}_{\mathrm{Sob},s,\boldsymbol{\gamma}}, \mathcal{P}(\boldsymbol{g}, N)).$$

Using a general theory about so-called *shift invariant kernels* it can now be shown (see, e.g., [26, Sect. 7.1]) that

$$e_{\mathrm{sh}}(\mathcal{H}_{\mathrm{Sob},s,\boldsymbol{\gamma}}, \mathcal{P}(\boldsymbol{g}, N)) = e(\mathcal{H}_{s,2,\widetilde{\boldsymbol{\gamma}}}, \mathcal{P}(\boldsymbol{g}, N)), \tag{4.41}$$

where on the right-hand side of this identity we have the worst-case error of the lattice rule based on the (unshifted) lattice point set $\mathcal{P}(\boldsymbol{g}, N)$ in the weighted Korobov space with smoothness parameter $\alpha = 2$ and weights $\widetilde{\boldsymbol{\gamma}} = \{\widetilde{\gamma}_{\mathfrak{u}} : \mathfrak{u} \subseteq [s]\}$, where

$$\widetilde{\gamma}_{\mathfrak{u}} := \frac{\gamma_{\mathfrak{u}}}{(2\pi^2)^{|\mathfrak{u}|}} \quad \text{for all } \mathfrak{u} \subseteq [s]. \tag{4.42}$$

This means that the results for the weighted Korobov space with smoothness $\alpha = 2$ and slightly adapted weights directly apply to the shift-averaged worst-case error for the weighted ANOVA space. In particular, this is true for Theorems 4.30 and 4.36. However, the adaption of the weights does not affect the conditions for tractability (cf. Corollary 4.33). So shifted lattice rules are an excellent tool for numerical integration of functions from the weighted ANOVA space.

Folded Lattice Rules

For the construction of folded lattice rules one uses the *tent transformation* ϕ : $[0, 1] \to [0, 1]$,

$$\phi(x) := 1 - |2x - 1|,$$

and applies this transformation coordinate-wise to points $\boldsymbol{x} = (x_1, \ldots, x_s)$ in the s-dimensional unit cube $[0, 1]^s$, i.e.:

$$\phi(\boldsymbol{x}) := (\phi(x_1), \ldots, \phi(x_s)).$$

Definition 4.41 For a lattice point set $\mathcal{P}(\boldsymbol{g}, N)$ the *folded lattice point set* is defined as

$$\mathcal{P}_{\phi}(\boldsymbol{g}, N) := \left\{ \phi\left(\left\{\frac{k}{N}\, \boldsymbol{g}\right\}\right) : k \in \{0, 1, \ldots, N - 1\} \right\}.$$

A QMC rule that uses a folded lattice point set as underlying sample nodes is called a *folded lattice rule*.

As for shifted lattice rules the worst-case error of a folded lattice rule in the weighted ANOVA space can be related to the worst-case error of the unfolded version of the lattice rule in the weighted Korobov space of smoothness $\alpha = 2$ with slightly adapted weights. In particular, it is true that

$$e(\mathcal{H}_{\text{Sob},s,\gamma}, \mathcal{P}_\phi(\boldsymbol{g}, N)) \leq e(\mathcal{H}_{s,2,\tilde{\gamma}}, \mathcal{P}(\boldsymbol{g}, N)), \tag{4.43}$$

where the weights $\tilde{\gamma}$ are as in (4.42); see, e.g., [26, Sect. 7.3]. Again, the results for the weighted Korobov space, in particular, Theorems 4.30 and 4.36 and the sufficient condition in Corollary 4.33 apply. So also folded lattice rules are an excellent tool for numerical integration of functions from the weighted ANOVA space.

Further Reading

Standard references for the theory of lattice point sets and lattice rules are the books of Niederreiter [110] and of Sloan and Joe [135]. Both books can be warmly recommended for further reading. A comprehensive account of the development since the publication of these two standard works can be found in [26]. Further books dealing with lattice rules are the ones by Hua and Wang [71] and by Korobov [83], where the latter being available in Russian language only. For technical reasons we often restricted ourselves to prime N within this chapter. We remark that the most existence results also hold for arbitrary integers $N \geq 2$ and refer to [26,110] for further information. Theorem 4.5 was first shown by Joe in [76]. The proof of Theorem 4.6 can be found in [89]. The currently best existence result (4.6) for lattice point sets with low discrepancy was proved by Larcher [88] for $s = 2$ and by Bykovskii [15] for general s. The CBC construction of lattice point sets was introduced by Korobov [83] and later re-invented by Sloan and Reztsov [136]. In the statement and proofs of the error bounds we restricted ourselves to prime number N for simplicity. The case of general $N \geq 2$ is summarized in [26, Chap. 3]. The fast CBC algorithm was invented by Nuyens and Cools. Detailed information can be found in [123] and implementations of the fast CBC algorithm can be found in [121,124]; see also [26, Chaps. 3 and 4]. The fast implementation of Algorithm 4.35 is detailed in [26, Sect. 3.4]. Weighted star discrepancy of lattice point sets was first studied by Joe in [77]. Our exposition of the proof of Theorem 4.27 is based on the proof of [140, Lemma 3.1] by Temlyakov. A proof of Theorem 4.27 for linear algorithms can be found in [34], see also Bakhvalov [8,9] and Temlyakov [140]. More information on tractability for the weighted Korobov space can be found in the paper by Sloan and Woźniakowski [138]. The current state of the art concerning tractability is summarized in the three volumes [118–120] by Novak and Woźniakowski (in particular, [119, Chap. 16] is devoted to multivariate integration for Korobov spaces). The curse of dimensionality for the integration of monotone functions and of convex functions was shown by Hinrichs, Novak, and Woźniakowski [64] and for smooth functions by Hinrichs, Novak, Ullrich, and Woźniakowski [62,63]. The necessity of the conditions for (strong) polynomial tractability from Corollary 4.33 for arbitrary algorithms using N function evaluations was shown by Hickernell and

Woźniakowski [60]. Shifted lattice rules are a popular randomized variant of lattice rules that are considered in a multitude of papers and books. Shift-invariant kernels were introduced by Hickernell [58, Chap. 5]. For more information we refer to [26, Sect. 7.1] (where also the proof of (4.41) is to be found) and [29] and the references therein. The first to apply the tent transformation to lattice rules was Hickernell [59], who also required an additional shift, which could be removed successfully by Dick, Nuyens, and Pillichshammer [32]. The material presented here is taken from [26, Sect. 7.3], where also the proof of (4.43) can be found. A similar result like (4.43) for a weighted ANOVA Sobolev space of higher smoothness was shown by Goda, Suzuki, and Yoshiki [49].

An important application area of QMC methods, and in particular of lattice rules, is partial differential equations with random coefficients. There is a large and fast growing amount of literature about this area. For information we refer to [86,87] and [26, Appendix A] and the references therein.

A software tool called "LatNet Builder" (see [96]) for constructing lattice rules can be found under: https://github.com/umontreal-simul/latnetbuilder.

Exercises

4.1. Consider lattice points of the form $v_s(g) := (g_1, g_2, \ldots, g_s) \in \mathbb{Z}^s$, where $g_i \equiv g^{i-1} \pmod{N}$ with fixed $g \in G_N := \{1, \ldots, N-1\}$. Such a generating vector is often called a *Korobov vector*. Let R_N be defined as in (4.3) and let N be a prime number. Put $S_N := \sum_{h \in C_1^*(N)} |h|^{-1}$. Show that

$$\frac{1}{N-1} \sum_{g \in G_N} R_N(v_s(g)) \leq \frac{s-1}{N-1}(1+S_N)^s.$$

4.2. Let N be a prime number. Show that for all $\varepsilon \in [0, 1)$ there exist more than $\varepsilon|G_N|$ elements $g \in G_N$, such that

$$D_N(\mathcal{P}(v_s(g), N)) \leq \frac{s}{N} + \frac{1}{1-\varepsilon}\frac{s-1}{N-1}(1-S_N)^s.$$

Hint: Use Exercise 4.1 to show that for all $c \geq 1$ we have

$$1 \geq \frac{c}{N-1}\left|\left\{g \in G_N \,:\, R_N(v_s(g)) > c\frac{s-1}{N-1}(1+S_N)^s\right\}\right|.$$

4.3. Prove Theorem 4.14.

4.4. Let $g : [0, 1] \to \mathbb{R}$,

$$g(s) = \begin{cases} x^2 - \frac{x}{2} & \text{if } x \in [0, 1/2], \\ \frac{3x}{2} - x^2 - \frac{1}{2} & \text{if } x \in [1/2, 1]. \end{cases}$$

Show that $g \in \mathcal{H}_{2,\gamma}$ but $g \notin C^2([0, 1])$.

4.5. Show that the QMC mean square worst-case error in $\mathcal{H}_{s,\alpha,\gamma}$ is

$$\mathbb{E}[e^2(\mathcal{H}_{s,\alpha,\gamma}, \mathcal{P})] = \frac{1}{N}\left[-1 + \prod_{j=1}^{s}(1 + 2\gamma_j \zeta(\alpha))\right].$$

4.6. Prove the formula for the Fourier expansion of the Bernoulli polynomials given in Remark 4.24 for the second Bernoulli polynomial $B_2(x) = x^2 - x + \frac{1}{6}$.

4.7. Use the Fourier expansion of $B_\alpha(x)$ from Remark 4.24 to show the formula for $e^2(\mathcal{H}_{s,\alpha}, \mathcal{P}(\boldsymbol{g}, N))$ for all even $\alpha \geq 2$ as stated in the same remark.

4.8. Use Exercise 3.14 to show that the initial error $e(0, s)$ in $\mathcal{H}_{s,\alpha,\gamma}$ equals one.

4.9. Show that integration in $\mathcal{H}_{s,\alpha,\gamma}$ is polynomially tractable if (4.28) holds.

4.10. Besides polynomial and strong polynomial tractability there are many other notions of tractability such as *weak tractability*. We say that we have weak tractability if the information complexity $N_{\mathcal{H}(K)}(\varepsilon, s)$ satisfies

$$\lim_{\varepsilon^{-1}+s \to \infty} \frac{\log N_{\mathcal{H}(K)}(\varepsilon, s)}{\varepsilon^{-1} + s} = 0.$$

Show that weak tractability means that $N_{\mathcal{H}(K)}(\varepsilon, s)$ is asymptotically much smaller than $q^{\varepsilon^{-1}+s}$ for any $q > 1$. *Remark:* Problems that are not weakly tractable are said to be *intractable*. This means that the information complexity depends exponentially on ε^{-1} or s, whereas the curse of dimensionality means that the information complexity depends exponentially on s.

4.11. Consider the weighted Korobov space $\mathcal{H}_{s,\alpha,\gamma}$. Show that a sufficient condition for weak tractability is $\lim_{s \to \infty} \frac{1}{s} \sum_{j=1}^{s} \gamma_j = 0$. *Remark:* This condition is also necessary (see [26]).

5.1 Definition and Existence

We are interested in point sets with very low star discrepancy. This means that we aim on finding point sets $\mathcal{P}$ for which the absolute local discrepancy,

$$
|\Delta_{\mathcal{P},N}(\boldsymbol{y})| = \left| \frac{A([\boldsymbol{0}, \boldsymbol{y}), \mathcal{P}, N)}{N} - \lambda_s([\boldsymbol{0}, \boldsymbol{y})) \right|,
$$

is as small as possible for all $\boldsymbol{y} \in (0, 1]^s$. Our strategy for achieving this goal is to discretize the problem and to investigate point sets $\mathcal{P}$ for which we have

$$
A(J, \mathcal{P}, N) = N\lambda_s(J) \tag{5.1}
$$

for all J from a sufficiently large class of intervals J and hope that this already implies a low star discrepancy of $\mathcal{P}$. If (5.1) holds then we say that the point set $\mathcal{P}$ is *fair* with respect to J. It is obvious that a given point set $\mathcal{P}$ cannot be fair with respect to all intervals since, for example, for any N-element point set $\mathcal{P}$ in $[0, 1)^s$ we have $D_N^*(\mathcal{P}) \geq (2N)^{-1}$.

Now we try to find classes of intervals with the property that fairness of a point set $\mathcal{P}$ with respect to all intervals from this class implies low star discrepancy of $\mathcal{P}$.

1. Attempt: On first sight it is near at hand to consider the class of intervals of the form:

$$
J = \prod_{j=1}^{s} \left[\frac{a_j}{b}, \frac{a_j + 1}{b} \right), \tag{5.2}
$$

where $a_j \in \{0, 1, \ldots, b - 1\}$ for $j = 1, \ldots, s$. Assume that we are given a fair point set $\mathcal{P}$ with respect to these intervals. This means that each of the intervals of the form (5.2) contains the same number of elements from $\mathcal{P}$, and hence the cardinality

© The Author(s), under exclusive license to Springer Nature Switzerland AG 2026
G. Leobacher and F. Pillichshammer, *Introduction to Quasi-Monte Carlo Integration and Applications*, Compact Textbooks in Mathematics,
https://doi.org/10.1007/978-3-032-05446-3_5

Fig. 5.1 The point set $\Gamma_{3,2}$
with fair intervals

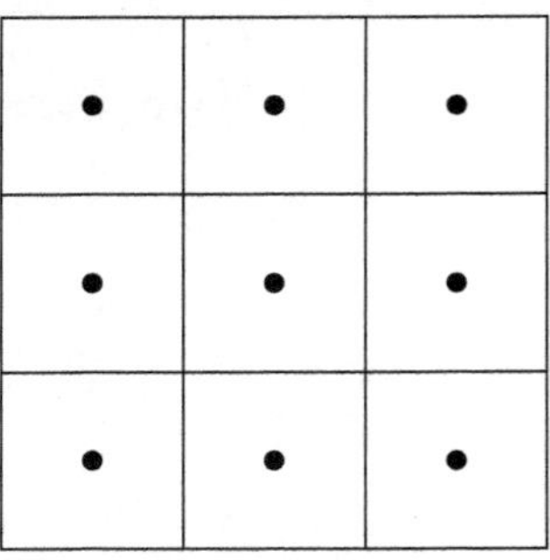

of $\mathcal{P}$ is a multiple of b^s. Here we observe that the regular lattice $\Gamma_{b,s}$ also satisfies these requirements (see Fig. 5.1). But the discrepancy of $\Gamma_{b,s}$ is of order $1/b$ which is rather large. Hence fairness with respect to the class of intervals of the form (5.2) does not necessarily lead to small discrepancy. This means that this class is in some sense too small to achieve our goal.

2. Attempt: We have seen that we have to demand fairness for a larger and, in some sense, finer class of intervals to find point sets with low discrepancy. This class will consist of intervals of a special shape which we define now:

Definition 5.1 Let $b \in \mathbb{N}, b \geq 2$. An *elementary interval* in base b is an interval of the form

$$J = \prod_{j=1}^{s} \left[\frac{a_j}{b^{d_j}}, \frac{a_j + 1}{b^{d_j}} \right)$$

with $d_j \in \mathbb{N}_0$ and $a_j \in \{0, 1, \ldots, b^{d_j} - 1\}$ for all $j = 1, 2, \ldots, s$.

Now we consider finite point sets which are fair with respect to all elementary intervals of prescribed volume. This leads to the following definition:

Definition 5.2 Let $m, s, b \in \mathbb{N}, b \geq 2$. A $(0, m, s)$-*net in base b* is a b^m-element point set in $[0, 1)^s$ which is fair with respect to all s-dimensional elementary intervals in base b having volume b^{-m}.

Let us provide an example.

Example 5.3 A $(0, 4, 2)$-net in base 2 is a $2^4 = 16$-element point set $\mathcal{P}$ in $[0, 1)^2$ for which every two-dimensional elementary interval in base 2 of area $2^{-4} = 1/16$ contains exactly one element of $\mathcal{P}$ (see Fig. 5.2).

We will show later that $(0, m, s)$-nets are in fact point sets with low discrepancy. Before we do so, we should be concerned with the existence of $(0, m, s)$-nets, i.e., with the question for which parameters $m, s, b \in \mathbb{N}$, where $b \geq 2$, a $(0, m, s)$-net

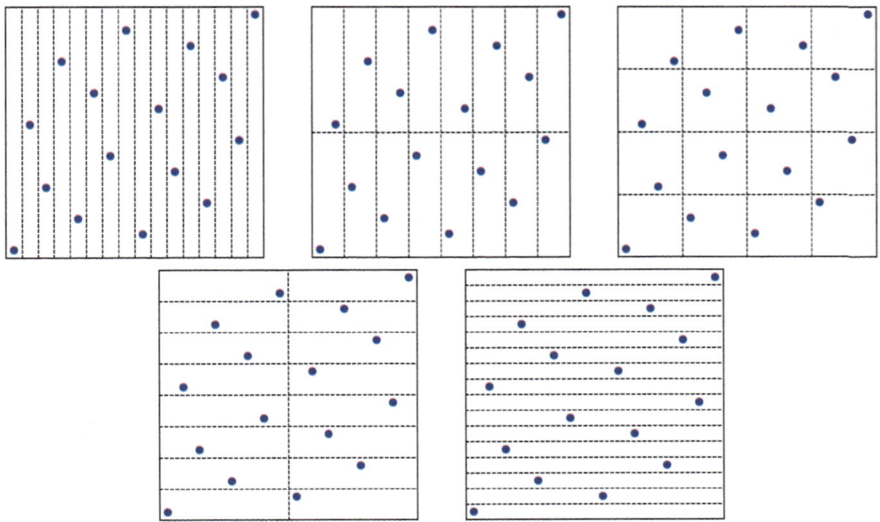

Fig. 5.2 A $(0, 4, 2)$-net in base 2; every two-dimensional elementary interval in base 2 of area 2^{-4} contains exactly one point

in base b exists. For the moment we will only show a necessary condition. As we will see, this condition is independent of the choice of m, but depends on a relation between s and b. A sufficient condition, and even an explicit construction, for the case of prime power bases b will be provided in Sect. 5.4.

We prove a series of lemmas which finally lead to the necessary condition for the existence of $(0, m, s)$-nets in base b. In the first lemma we show that from a given $(0, m, s)$-net in base b one can always construct a $(0, m, r)$-net in base b of dimension $r \leq s$. For $\boldsymbol{x} = (x_1, \ldots, x_s)$ and for $r \in \{1, \ldots, s\}$ let $\pi_r(\boldsymbol{x}) := (x_1, \ldots, x_r)$ be the projection of $\boldsymbol{x}$ onto the first r components of $\boldsymbol{x}$. For a point set $\mathcal{P}$ the projection $\pi_r(\mathcal{P})$ is explained element-wise.

Lemma 5.4 *Let $\mathcal{P}$ be a $(0, m, s)$-net in base b. Then for every $r \in \{1, \ldots, s\}$ the point set $\pi_r(\mathcal{P})$ is a $(0, m, r)$-net in base b.*

Proof Let J' be an r-dimensional elementary interval in base b of volume b^{-m}. Then the interval $J = J' \times [0, 1)^{s-r}$ is an s-dimensional elementary interval in base b which also has volume b^{-m}. By the $(0, m, s)$-net property the interval J contains exactly one element of $\mathcal{P}$ and therefore J' contains exactly one element of $\mathcal{P}'$ as well. $\qquad\square$

Lemma 5.5 *If a $(0, m, s)$-net in base b with $m \geq 2$ exists, then there exists a $(0, 2, s)$-net in base b.*

Proof Let $\mathcal{P}$ be a $(0, m, s)$-net in base b. Then the interval

$$
J = \left[0, \frac{1}{b^{m-2}}\right) \times \prod_{i=2}^{s} [0, 1) = \bigcup_{k=0}^{b^2-1} \left(\left[\frac{k}{b^m}, \frac{k+1}{b^m}\right) \times [0, 1)^{s-1}\right)
$$

contains exactly b^2 elements of $\mathcal{P}$. We denote these elements by $z_0, \ldots, z_{b^2-1}$. Multiplying the first component of each of these b^2 points with b^{m-2} gives a point set $\{y_0, \ldots, y_{b^2-1}\}$ in $[0, 1)^s$. We show that this point set is a $(0, 2, s)$-net in base b. To this end we have to prove that every elementary interval of the form

$$
E = \prod_{j=1}^{s} \left[\frac{a_j}{b^{d_j}}, \frac{a_j+1}{b^{d_j}}\right)
$$

of volume b^{-2} contains exactly one element of the set $\{y_0, \ldots, y_{b^2-1}\}$. The number of indices $n \in \{0, 1, \ldots, b^2 - 1\}$ satisfying $y_n \in E$ equals the number of indices $n \in \{0, 1, \ldots, b^2 - 1\}$, for which $z_n \in E'$, where

$$
E' = \underbrace{\left[\frac{a_1}{b^{d_1} b^{m-2}}, \frac{a_1+1}{b^{d_1} b^{m-2}}\right)}_{\subseteq [0, b^{-m+2})} \times \prod_{j=2}^{s} \left[\frac{a_j}{b^{d_j}}, \frac{a_j+1}{b^{d_j}}\right).
$$

This, however, is exactly the number of points of $\mathcal{P}$ which belong to the interval E'. Since $\mathcal{P}$ is a $(0, m, s)$-net in base b and since E' is an elementary interval of volume b^{-m}, this number is exactly one and the proof is finished. $\qquad\square$

Lemma 5.6 *A $(0, 2, b + 2)$-net in base b cannot exist.*

Proof Assume that there exists a $(0, 2, b + 2)$-net $\mathcal{P} = \{x_0, \ldots, x_{b^2-1}\}$ in base b.

Every point $x_n = (x_{n,1}, \ldots, x_{n,b+2})$ in $\mathcal{P}$ corresponds to a $b + 2$-dimensional vector

$$
x_n \leftrightarrow \begin{pmatrix} a_n^{(1)} \\ \vdots \\ a_n^{(b+2)} \end{pmatrix},
$$

where $a_n^{(j)} = \lfloor bx_{n,j} \rfloor \in \{0, 1, \ldots, b-1\}$ is chosen in such a way that x_n belongs to the interval $\prod_{j=1}^{b+2} [a_n^{(j)}/b, (a_n^{(j)} + 1)/b)$. Hence the b^2 elements of $\mathcal{P}$ correspond to the following $(b+2) \times b^2$ array:

$$
\begin{array}{cccc}
\boldsymbol{x}_0 & \boldsymbol{x}_1 & \ldots & \boldsymbol{x}_{b^2-1} \\
\updownarrow & \updownarrow & & \updownarrow \\
a_0^{(1)} & a_1^{(1)} & \ldots & a_{b^2-1}^{(1)} \\
\vdots & \vdots & & \vdots \\
a_0^{(b+2)} & a_1^{(b+2)} & \ldots & a_{b^2-1}^{(b+2)}
\end{array}
\tag{5.3}
$$

We prove that this array satisfies the following "orthogonality property": For any two rows of this array, say

$$
a_0^{(i)} \; a_1^{(i)} \; \ldots \; a_{b^2-1}^{(i)},
$$
$$
a_0^{(j)} \; a_1^{(j)} \; \ldots \; a_{b^2-1}^{(j)},
$$

where $i \neq j$, the b^2 two-dimensional columns

$$
\left(\begin{array}{c} a_k^{(i)} \\ a_k^{(j)} \end{array} \right)_{k=0,\ldots,b^2-1}
$$

attain any possible value $\left(\begin{array}{c} A \\ B \end{array} \right)$, with $A, B \in \{0, \ldots, b-1\}$, exactly once.

Assume to the contrary that some $\left(\begin{array}{c} A \\ B \end{array} \right)$ appears not exactly once. Then the elementary interval

$$
[0, 1)^{i-1} \times \left[\frac{A}{b}, \frac{A+1}{b} \right) \times [0, 1)^{j-i-1} \times \left[\frac{B}{b}, \frac{B+1}{b} \right) \times [0, 1)^{b+2-j}
$$

of volume b^{-2} does not contain exactly one element of $\mathcal{P}$, which contradicts the $(0, 2, b+2)$-net property in base b of $\mathcal{P}$. Hence the orthogonality property is proven.

In particular, any possible value $A \in \{0, \ldots, b-1\}$ must occur in any row of (5.3) exactly b-times. This, however, cannot be satisfied for all possible pairs of rows of (5.3) as we show now:

Assume that any two of the $b+2$ rows of (5.3) satisfy the orthogonality property. We may assume that in array (5.3) the values of the first column all equal 1, since a permutation of the values $A \in \{0, \ldots, b-1\}$ in a single row of the array does not affect the orthogonality property.

Then, by the orthogonality property, in any of the remaining $b^2 - 1$ columns, 1 can occur at most once. But in each row, 1 must occur b-times, and hence we would require a place for the $(b-1)(b+2)$ remaining 1's in these $b^2 - 1$ columns. Since $(b-1)(b+2) = b^2 + b - 2 > b^2 - 1$, we have a contradiction. $\qquad \square$

Now we come to the desired necessary condition for the existence of $(0, m, s)$-nets in base b.

Theorem 5.7 (Niederreiter) *Let $s, b \in \mathbb{N}$, $b \geq 2$. Assume that for $m \in \mathbb{N}$ with $m \geq 2$ there exists a $(0, m, s)$-net in base b. Then $s \leq b + 1$.*

Proof Assume that a $(0, m, s)$-net in base b where $m \geq 2$ and $s \geq b + 2$ exist. Then, according to Lemmas 5.4 and 5.5, there exists a $(0, 2, b + 2)$-net in base b, which contradicts Lemma 5.6. □

In other words, a $(0, m, s)$-net in base b with $m \geq 2$ cannot exist as long as $s \geq b + 2$. For example, there is no $(0, m, s)$-net in base 2 for $m \geq 2$ and $s \geq 4$.

As $(0, m, s)$-nets in base b do not exist for all parameters $s, b \in \mathbb{N}$, $b \geq 2$, we now weaken the condition from their definition.

Definition 5.8 Let $m, s, b \in \mathbb{N}$, $b \geq 2$, and let $t \in \{0, \ldots, m\}$. A (t, m, s)-*net in base b* is a b^m-element point set $\mathcal{P}$ in $[0, 1)^s$ which is fair with respect to every s-dimensional elementary interval in base b having volume b^{-m+t}. The parameter t is called the *quality parameter* of the net. Moreover, $\mathcal{P}$ is called a *strict (t, m, s)-net in base b*, if t is the smallest number $u \in \{0, \ldots, m\}$, such that $\mathcal{P}$ is a (u, m, s)-net in base b.

Some remarks on the definition of (t, m, s)-nets in base b are appropriate.

Remark 5.9 1. An s-dimensional elementary interval in base b of volume b^{-m+t} is an interval of the form

$$E = \prod_{i=1}^{s} \left[\frac{a_i}{b^{d_i}}, \frac{a_i + 1}{b^{d_i}} \right)$$

where $d_1 + \cdots + d_s = m - t$. Every such interval contains exactly b^t elements of a (t, m, s)-net in base b. The larger t is, the larger are the intervals under consideration and the smaller is the class of considered intervals. For such intervals we have $A(E, \mathcal{P}, b^m) = b^t = b^m \lambda_s(E)$.
2. Every b^m-element point set in $[0, 1)^s$ is a (t, m, s)-net in base b for some $t \in \{0, \ldots, m\}$. In the worst case it is a (m, m, s)-net in base b.
3. Every $(0, m, s)$-net in base b is strict by definition.
4. Every elementary interval in base b of volume b^{-m+t} is a disjoint union of b^t elementary intervals in base b of volume b^{-m}. Hence, every $(0, m, s)$-net in base b is also a (t, m, s)-net in base b for every $t \in \{0, \ldots, m\}$.
5. Similarly, every (t, m, s)-net in base b is also a (u, m, s)-net in base b for every $u \in \{t, \ldots, m\}$.
6. From Lemma 5.5 we know that from a $(0, m, s)$-net in base b with $m \geq 2$ one can construct a $(0, 2, s)$-net in base b. More generally, but in the same way, one

can construct from a (t, m, s)-net in base b a (t, u, s)-net in base b for every $u \in \{t, \ldots, m\}$.

7. From Lemma 5.4 we know that from a $(0, m, s)$-net in base b a $(0, m, r)$-net in base b with $r \in \{1, \ldots, s\}$ can be constructed. More generally, but in the same way, one can construct from a (t, m, s)-net in base b a (t, m, r)-net in base b for any $r \in \{1, \ldots, s\}$.

8. It can be shown that, for all $s, b \in \mathbb{N}$, $b \geq 2$, there exists a number $t_b(s) \in \mathbb{N}_0$ such that for every $m \geq t_b(s)$ there exists a $(t_b(s), m, s)$-net in base b. It is a remarkable result by H. Niederreiter and C.P. Xing that $t_b(s) \asymp_b s$ asymptotically for $s \to \infty$.

Example 5.10 The regular lattice $\Gamma_{n,s}$ as defined in (2.7) with $n = b^L$ (i.e., with b^{sL} elements) is a strict $(sL - L, sL, s)$-net in base b.

Proof The structure of the regular grid $\Gamma_{b^L,s}$ is such that every interval of the form,

$$M_{\boldsymbol{k}} = \prod_{j=1}^{s} \left[\frac{k_j}{b^L}, \frac{k_j + 1}{b^L} \right)$$

where $\boldsymbol{k} = (k_1, \ldots, k_s) \in \{0, \ldots, b^L - 1\}^s$, contains exactly one element. Let

$$J = \prod_{j=1}^{s} \left[\frac{a_j}{b^{d_j}}, \frac{a_j + 1}{b^{d_j}} \right)$$

be an elementary interval of volume b^{-L}. Then we can write J as disjoint union of intervals $M_{\boldsymbol{k}}$ in the following way:

$$J = \bigcup_{k_1 = b^{L-d_1} a_1}^{b^{L-d_1}(a_1+1)-1} \cdots \bigcup_{k_s = b^{L-d_s} a_s}^{b^{L-d_s}(a_s+1)-1} M_{\boldsymbol{k}} \quad \text{with } d_1 + \cdots + d_s = L.$$

Since each of the intervals $M_{\boldsymbol{k}}$ contains exactly one element of the regular lattice, it follows that J contains exactly b^{sL-L} elements of the regular lattice. This shows that the regular lattice with b^{sL} elements is an $(sL - L, sL, s)$-net in base b.

It remains to show strictness. To this end we consider the elementary interval

$$J' = \left[0, \frac{1}{b^{L+1}} \right) \times [0, 1)^{s-1}.$$

This interval has volume $b^{-(L+1)}$ and contains no elements of the regular lattice. Hence we have

$$A(J', \Gamma_{b^L,s}, b^{sL}) - b^{sL} \lambda_s(B) = -b^{(s-1)L-1} \neq 0.$$

$\square$

A disadvantage of nets in base b is that the number of points is restricted to a power of b. At first glance one could argue that we can always choose b arbitrarily large and $m = 1$, which would mean that there is no restriction at all. However, it is intuitively obvious (and is supported by the discrepancy estimates in Sect. 5.2) that the structure of a (t, m, s)-net in base b becomes strong only if m is large compared to b. Hence, for a given number N of points, it is sometimes better to realize the point set with a small base b, i.e., a larger value for m, and with a suboptimal quality parameter t, instead of choosing a large base b (e.g., $b = N$) (and therefore a small m, for instance, $m = 1$), with optimal quality parameter.

To overcome this problem one considers infinite sequences which have a certain "net structure" in the sense that the elements of certain finite subsequences form (t, m, s)-nets in base b. This leads to the definition of (t, s)-sequences.

Definition 5.11 Let $s, b \in \mathbb{N}$, $b \geq 2$, and let $t \in \mathbb{N}_0$. An infinite sequence $(\boldsymbol{x}_n)_{n \in \mathbb{N}_0}$ of points in $[0, 1)^s$ is called a (t, s)-*sequence in base* b if for all integers $m > t$ and $k \geq 0$ the point set consisting of the points $\boldsymbol{x}_{kb^m}, \boldsymbol{x}_{kb^m+1}, \ldots, \boldsymbol{x}_{kb^m+b^m-1}$ forms a (t, m, s)-net in base b. The parameter t is called the *quality parameter* of the (t, s)-sequence.

Definition 5.12 A (t, s)-sequence in base b with $t \geq 1$ is called a *strict* (t, s)-*sequence in base* b if it is not a $(t - 1, s)$-sequence in base b. A $(0, s)$-sequence is called strict by definition.

We give a first example.

Example 5.13 The van der Corput sequence in base b is a $(0, 1)$-sequence in base b. This follows from the proof of Proposition 2.10, where we have shown that for every $m \in \mathbb{N}_0$ every elementary interval in base b of length b^{-m} contains exactly one of b^m consecutive elements of the van der Corput sequence in base b.

We have shown in Proposition 2.10 that the van der Corput sequence is uniformly distributed modulo one. This result can be generalized in the following sense.

Theorem 5.14 A (t, s)-*sequence in base* b *is uniformly distributed modulo one.*

For the proof of this result we require the following lemma.

Lemma 5.15 *Let* $\mathcal{P}$ *be a* (t, m, s)-*net in base* b. *Let* $J_1, \ldots, J_r$ *be disjoint s-dimensional elementary intervals in base b of volume* b^{t-m} *and let* $J = J_1 \cup \ldots \cup J_r$. *Then we have*

$$A(J, \mathcal{P}, b^m) = b^m \lambda_s(J).$$

Proof For every J_l we have $A(J_l, \mathcal{P}, b^m) = b^t = b^m \lambda_s(J_l)$, and hence

$$A(J, \mathcal{P}, b^m) = \sum_{l=1}^{r} A(J_l, \mathcal{P}, b^m) = b^m \sum_{l=1}^{r} \lambda_s(J_l) = b^m \lambda_s(J). \qquad \square$$

Proof of Theorem 5.14. Let $\mathcal{S} = (x_n)_{n \in \mathbb{N}_0}$ be a (t, s)-sequence in base b. Further let $J := \prod_{j=1}^{s} [\alpha_j, \beta_j)$ with $0 \le \alpha_j < \beta_j \le 1$, be an arbitrary sub-interval of $[0, 1)^s$, and let $\varepsilon > 0$ be given.

Let $r \in \mathbb{N}$ be fixed and such that $2sb^{-r} < \varepsilon/2$, let $l = rs$ and let m be fixed such that $m - t \ge l$. Let $A_j, B_j \in \{0, 1, \ldots, b^r - 1\}$ be such that

$$\frac{A_j}{b^r} \le \alpha_j < \frac{A_j+1}{b^r} \quad \text{and} \quad \frac{B_j}{b^r} \le \beta_j < \frac{B_j+1}{b^r} \quad \text{for } j = 1, \ldots, s.$$

Then for

$$J_1 := \prod_{j=1}^{s} \left[\frac{A_j + 1}{b^r}, \frac{B_j}{b^r} \right) \quad \text{and} \quad J_2 := \prod_{j=1}^{s} \left[\frac{A_j}{b^r}, \frac{B_j + 1}{b^r} \right)$$

we have

$$J_1 \subseteq J \subseteq J_2 \subseteq [0, 1)^s,$$

and, by Lemma 2.16, $\lambda_s(J_2 \setminus J_1) \le 2sb^{-r}$. Furthermore,

$$J_1 = \bigcup_{a_1=A_1+1}^{B_1-1} \cdots \bigcup_{a_s=A_s+1}^{B_s-1} \prod_{j=1}^{s} \left[\frac{a_j}{b^r}, \frac{a_j + 1}{b^r} \right),$$

i.e., J_1 is a union of at most b^l elementary intervals of volume $b^{-rs} = b^{-l}$. In the same way, J_2 is a union of at most b^l elementary intervals of volume $b^{-rs} = b^{-l}$. Hence, by Lemma 5.15 and since $m - t \ge l$, the point sets $\{x_{kb^m}, x_{kb^m+1}, \ldots, x_{kb^m+b^m-1}\}$ are fair with respect to the intervals J_1 and J_2. Therefore, for all $N \in \mathbb{N}$, we have

$$A(J, \mathcal{S}, N) - N\lambda_s(J) \le A(J_2, \mathcal{S}, N) - N\lambda_s(J_2) + N\frac{2s}{b^r}$$

$$\le A(J_2, \mathcal{S}, \lfloor Nb^{-m} \rfloor b^m) - \lfloor Nb^{-m} \rfloor b^m \lambda_s(J_2) + b^m + N\frac{2s}{b^r}$$

$$= b^m + N\frac{2s}{b^r}$$

and

$$A(J, \mathcal{S}, N) - N\lambda_s(J) \ge A(J_1, \mathcal{S}, N) - N\lambda_s(J_1) - N\frac{2s}{b^r}$$

$$\ge A(J_1, \mathcal{S}, \lceil Nb^{-m} \rceil b^m) - \lceil Nb^{-m} \rceil b^m \lambda_s(J_1) - b^m - N\frac{2s}{b^r}$$

$$= -b^m - N\frac{2s}{b^r},$$

so that

$$\left| \frac{A(J, \mathcal{S}, N)}{N} - \lambda_s(J) \right| \leq \frac{b^m}{N} + \frac{2s}{b^r} < \varepsilon$$

for N large enough. Hence the result follows. $\qquad\square$

As for $(0, m, s)$-nets in base b, it is clear that a $(0, s)$-sequence in base b cannot exist for all dimensions s. A necessary condition for the existence of $(0, s)$-sequences in base b can be obtained from Theorem 5.7.

Theorem 5.16 (Niederreiter) *Let $s, b \in \mathbb{N}, b \geq 2$. The existence of a $(0, s)$-sequence in base b implies $s \leq b$.*

For the proof of Theorem 5.16 we need the following lemma which shows that if a (t, s)-sequence in base b exists, then, for every $m \geq t$, also a $(t, m, s + 1)$-net in base b exists.

Lemma 5.17 *Let $(x_n)_{n \in \mathbb{N}_0}$ be a (t, s)-sequence in base b. Then, for every m, the point set $\{y_0, y_1, \ldots, y_{b^m - 1}\}$ with $y_k := \left(kb^{-m}, x_k\right)$ is a $(t, m, s + 1)$-net in base b.*

Proof Let $J = \prod_{j=1}^{s+1} \left[\frac{A_j}{b^{d_j}}, \frac{A_j+1}{b^{d_j}} \right)$ be an elementary interval of volume b^{-m+t}. Then $y_k \in J$ if and only if

$$\frac{k}{b^m} \in \left[\frac{A_1}{b^{d_1}}, \frac{A_1 + 1}{b^{d_1}} \right) \quad \text{and} \quad x_k \in \prod_{j=2}^{s+1} \left[\frac{A_j}{b^{d_j}}, \frac{A_j + 1}{b^{d_j}} \right).$$

The first condition implies

$$A_1 b^{m-d_1} \leq k < A_1 b^{m-d_1} + b^{m-d_1}.$$

Since $(x_n)_{n \in \mathbb{N}_0}$ is a (t, s)-sequence in base b, the points $x_{A_1 b^{m-d_1}+l}$ for $l = 0, \ldots, b^{m-d_1} - 1$, form a $(t, m - d_1, s)$-net in base b. The interval

$$\prod_{j=2}^{s+1} \left[\frac{A_j}{b^{d_j}}, \frac{A_j + 1}{b^{d_j}} \right)$$

has volume $b^{-d_2 - \cdots - d_{s+1}} = b^{-m+d_1+t}$ and therefore contains exactly b^t of the points $x_{A_1 b^{m-d_1}+l}$ for $l = 0, \ldots, b^{m-d_1} - 1$. Consequently, J contains exactly b^t of the y_k's and the result follows. $\qquad\square$

Proof of Theorem   5.16. Assume that there exists a $(0, s)$-sequence in base b. Then, according to Lemma 5.17, there also exists a $(0, m, s + 1)$-net in base b. Then it follows from Theorem 5.7 that $s + 1 \leq b + 1$, and hence $s \leq b$. $\qquad\square$

5.2 The Star Discrepancy of (t, m, s)-Nets and (t, s)-Sequences

In this section we will show that (t, m, s)-nets and (t, s)-sequences defined above have in fact the desired property of having low discrepancy.

First we consider (t, m, s)-nets in base b for which we have the following fundamental discrepancy estimate.

Theorem 5.18 *For the star discrepancy of a (t, m, s)-net $\mathcal{P}$ in base b we have*

$$D^*_{b^m}(\mathcal{P}) \leq \frac{1}{b^{m-t}} \sum_{k=0}^{s-1} \binom{m-t}{k} (b-1)^k.$$

For the proof of this estimate we require the following lemma.

Lemma 5.19 *For $s, m \in \mathbb{N}$ we have*

$$\sum_{\substack{k_1,\ldots,k_s \in \mathbb{N} \\ k_1 + \cdots + k_s \leq m}} 1 = \binom{m}{s}.$$

Proof We have

$$\sum_{\substack{k_1,\ldots,k_s \in \mathbb{N} \\ k_1 + \cdots + k_s \leq m}} 1 = \sum_{l=s}^{m} \sum_{\substack{k_1,\ldots,k_s \in \mathbb{N} \\ k_1 + \cdots + k_s = l}} 1.$$

For $l \geq s$ the number of solutions of the equation $k_1 + \cdots + k_s = l$ with $k_i \in \mathbb{N}$ equals the number of solutions of the equation $k_1 + \cdots + k_s = l - s$ with $k_i \in \mathbb{N}_0$, which in turn equals $\binom{l-1}{s-1}$. Consequently,

$$\sum_{\substack{k_1,\ldots,k_s \in \mathbb{N} \\ k_1 + \cdots + k_s \leq m}} 1 = \sum_{l=s}^{m} \binom{l-1}{s-1} = \binom{m}{s},$$

where the last equality can be shown by induction on m. $\qquad\square$

Proof of Theorem   5.18. We only consider the case $t = 0$; the general case can be proven along the same lines by considering the same b-adic expansions, only up to $m - t$ instead of m.

Thus, let $\mathcal{P}$ be a $(0, m, s)$-net in base b. Consider an interval $B = [0, \alpha^{(1)}) \times \cdots \times [0, \alpha^{(s)})$ with

$$\alpha^{(j)} = \frac{a_1^{(j)}}{b} + \frac{a_2^{(j)}}{b^2} + \cdots + \frac{a_m^{(j)}}{b^m} + \cdots$$

for $j = 1, 2, \ldots, s$. The fundamental idea of the proof is to approximate the interval B from the interior and exterior by sets B_s and B'_s, respectively, which can be written as disjoint unions of elementary intervals of volume b^{-m}. Then we have $B_s \subseteq B \subseteq B'_s$ such that

$$A(B_s, \mathcal{P}, b^m) \leq A(B, \mathcal{P}, b^m) \leq A(B'_s, \mathcal{P}, b^m)$$

and, from Lemma 5.15,

$$\frac{A(B_s, \mathcal{P}, b^m)}{b^m} = \lambda_s(B_s) \leq \lambda_s(B) \leq \lambda_s(B'_s) = \frac{A(B'_s, \mathcal{P}, b^m)}{b^m}.$$

Thus we get

$$\left| \frac{A(B, \mathcal{P}, b^m)}{b^m} - \lambda_s(B) \right| \leq \lambda_s(B'_s) - \lambda_s(B_s) = \lambda_s(B'_s \setminus B_s),$$

and the theorem will be proven once we succeed in showing that B_s and B'_s can always be constructed in a way that guarantees

$$\lambda_s(B'_s \setminus B_s) \leq \frac{1}{b^m} \sum_{k=0}^{s-1} \binom{m}{k} (b-1)^k.$$

Of course, this is the hard part of the proof. We use induction on s.

- $s = 1$: Let

$$B_1 := \left[0, \frac{a_1^{(1)}}{b} + \cdots + \frac{a_m^{(1)}}{b^m} \right) \quad \text{and} \quad B'_1 := \left[0, \frac{a_1^{(1)}}{b} + \cdots + \frac{a_m^{(1)}}{b^m} + \frac{1}{b^m} \right).$$

  Then

$$B_1 = \bigcup_{k=0}^{b^{m-1}a_1^{(1)} + \cdots + a_m^{(1)} - 1} \left[\frac{k}{b^m}, \frac{k+1}{b^m} \right)$$

  is a disjoint union of elementary intervals of length b^{-m} and the same is true for B'_1. Furthermore, we have $B_1 \subseteq B \subseteq B'_1$ and $\lambda_1(B'_1 \setminus B_1) = \frac{1}{b^m} = \frac{1}{b^m} \sum_{k=0}^{0} \binom{m}{k} (b-1)^k$.

- For illustration purposes we give the step $1 \to 2$ separately before showing the general induction step: Let

$$I_1 := \left[0, \frac{a_1^{(1)}}{b} \right) \times \left[0, \frac{a_1^{(2)}}{b} + \cdots + \frac{a_{m-1}^{(2)}}{b^{m-1}} \right)$$

$$J_1 := \left[0, \frac{a_1^{(1)}}{b} \right) \times \left[0, \frac{a_1^{(2)}}{b} + \cdots + \frac{a_{m-1}^{(2)}}{b^{m-1}} + \frac{1}{b^{m-1}} \right).$$

Then $I_1 \subseteq B$ and

$$I_1 = \bigcup_{k=0}^{b^{m-2}a_1^{(2)}+\cdots+a_{m-1}^{(2)}-1} \left[0, \frac{a_1^{(1)}}{b}\right) \times \left[\frac{k}{b^{m-1}}, \frac{k+1}{b^{m-1}}\right).$$

If $a_1^{(1)} \neq 0$, then I_1 is a disjoint union of two-dimensional elementary intervals of area b^{-m}. If, however, $a_1^{(1)} = 0$, then $I_1 = \emptyset$ and is therefore trivially a disjoint union of elementary intervals of area b^{-m}. For the same reasons J_1 is a disjoint union of elementary intervals of area b^{-m}.
For $k \in \{2, \ldots, m-1\}$ put

$$I_k := \left[\frac{a_1^{(1)}}{b} + \cdots + \frac{a_{k-1}^{(1)}}{b^{k-1}}, \frac{a_1^{(1)}}{b} + \cdots + \frac{a_k^{(1)}}{b^k}\right) \times \left[0, \frac{a_1^{(2)}}{b} + \cdots + \frac{a_{m-k}^{(2)}}{b^{m-k}}\right),$$

$$J_k := \left[\frac{a_1^{(1)}}{b} + \cdots + \frac{a_{k-1}^{(1)}}{b^{k-1}}, \frac{a_1^{(1)}}{b} + \cdots + \frac{a_k^{(1)}}{b^k}\right) \times \left[0, \frac{a_1^{(2)}}{b} + \cdots + \frac{a_{m-k}^{(2)}}{b^{m-k}} + \frac{1}{b^{m-k}}\right)$$

and

$$I_m := \left[\frac{a_1^{(1)}}{b} + \cdots + \frac{a_{m-1}^{(1)}}{b^{m-1}}, \frac{a_1^{(1)}}{b} + \cdots + \frac{a_m^{(1)}}{b^m}\right) \times [0, 0) = \emptyset,$$

$$J_m := \left[\frac{a_1^{(1)}}{b} + \cdots + \frac{a_{m-1}^{(1)}}{b^{m-1}}, \frac{a_1^{(1)}}{b} + \cdots + \frac{a_m^{(1)}}{b^m}\right) \times [0, 1).$$

Again it can be checked that I_k, J_k are disjoint unions of elementary intervals of area b^{-m}.
Now let

$$B_2 := \bigcup_{k=1}^{m} I_k \quad \text{and} \quad B_2' := \bigcup_{k=1}^{m} J_k \cup (B_1' \setminus B_1 \times [0, 1)),$$

where B_1 and B_1' are defined as in the case for $s = 1$ to approximate the interval $[0, \alpha^{(1)})$, such that

$$B_1' \setminus B_1 = \left[\frac{a_1^{(1)}}{b} + \cdots + \frac{a_m^{(1)}}{b^m}, \frac{a_1^{(1)}}{b} + \cdots + \frac{a_m^{(1)}}{b^m} + \frac{1}{b^m}\right)$$

and $\lambda_1(B'_1 \setminus B_1) = \frac{1}{b^m}$. Then we have $B_2 \subseteq B \subseteq B'_2$ and

$$\begin{aligned}
\lambda_2(B'_2 \setminus B_2) &= \sum_{k=1}^{m} \lambda_2(J_k \setminus I_k) + \lambda_1(B'_1 \setminus B_1) \\
&= \sum_{k=1}^{m} \frac{a_k^{(1)}}{b^k} \frac{1}{b^{m-k}} + \frac{1}{b^m} \\
&\leq \frac{m}{b^m}(b-1) + \frac{1}{b^m} \\
&= \frac{1}{b^m} \sum_{k=0}^{1} \binom{m}{k} (b-1)^k.
\end{aligned}$$

- $s - 1 \to s$: For $k_1, \ldots, k_{s-1} \in \mathbb{N}$ with $k_1 + \cdots + k_{s-1} \leq m$ put

$$I_{k_1,\ldots,k_{s-1}} := \prod_{l=1}^{s-1} \left[\frac{a_1^{(l)}}{b} + \cdots + \frac{a_{k_l-1}^{(l)}}{b^{k_l-1}}, \frac{a_1^{(l)}}{b} + \cdots + \frac{a_{k_l-1}^{(l)}}{b^{k_l-1}} + \frac{a_{k_l}^{(l)}}{b^{k_l}} \right).$$

Furthermore, let

$$I'_{k_1,\ldots,k_{s-1}} := I_{k_1,\ldots,k_{s-1}} \times \left[0, \frac{a_1^{(s)}}{b} + \cdots + \frac{a_{m-k_1-\cdots-k_{s-1}}^{(s)}}{b^{m-k_1-\cdots-k_{s-1}}} \right) \quad \text{and}$$

$$J'_{k_1,\ldots,k_{s-1}} := I_{k_1,\ldots,k_{s-1}} \times \left[0, \frac{a_1^{(s)}}{b} + \cdots + \frac{a_{m-k_1-\cdots-k_{s-1}}^{(s)}}{b^{m-k_1-\cdots-k_{s-1}}} + \frac{1}{b^{m-k_1-\cdots-k_{s-1}}} \right)$$

for $k_1 + \cdots + k_{s-1} \leq m - 1$, and

$$I'_{k_1,\ldots,k_{s-1}} := I_{k_1,\ldots,k_{s-1}} \times [0,0) = \emptyset \quad \text{and}$$

$$J'_{k_1,\ldots,k_{s-1}} := I_{k_1,\ldots,k_{s-1}} \times [0,1)$$

for $k_1 + \cdots + k_{s-1} = m$.

It is readily checked that the intervals $I'_{k_1,\ldots,k_{s-1}}$ are again disjoint unions of elementary intervals of volume b^{-m}. Furthermore,

$$I'_{k_1,\ldots,k_{s-1}} \cap I'_{j_1,\ldots,j_{s-1}} = \emptyset = J'_{k_1,\ldots,k_{s-1}} \cap J'_{j_1,\ldots,j_{s-1}}$$

for $(k_1, \ldots, k_{s-1}) \neq (j_1, \ldots, j_{s-1})$. Put

$$B_s := \bigcup_{\substack{k_1,\ldots,k_{s-1} \in \mathbb{N} \\ k_1+\cdots+k_{s-1} \leq m}} I'_{k_1,\ldots,k_{s-1}},$$

$$B'_s := \bigcup_{\substack{k_1,\ldots,k_{s-1} \in \mathbb{N} \\ k_1+\cdots+k_{s-1} \leq m}} J'_{k_1,\ldots,k_{s-1}} \cup (B'_{s-1} \setminus B_{s-1}) \times [0,1),$$

where B_{s-1} and B'_{s-1} are the sets from the $(s-1)$-dimensional case approximating $[0, \alpha^{(1)}) \times \cdots \times [0, \alpha^{(s-1)})$ with

$$\lambda_{s-1}(B'_{s-1} \setminus B_{s-1}) \leq \frac{1}{b^m} \sum_{k=0}^{s-2} \binom{m}{k} (b-1)^k.$$

Then we have $B_s \subseteq B \subseteq B'_s$ and, like in the case $s = 2$, we obtain

$$
\begin{aligned}
\lambda_s(B'_s \setminus B_s) &= \sum_{\substack{k_1,\ldots,k_{s-1}=1 \\ k_1+\cdots+k_{s-1}\leq m}}^{m} \lambda_s(J'_{k_1,\ldots,k_{s-1}} \setminus I'_{k_1,\ldots,k_{s-1}}) + \lambda_{s-1}(B'_{s-1} \setminus B_{s-1}) \\
&= \sum_{\substack{k_1,\ldots,k_{s-1}=1 \\ k_1+\cdots+k_{s-1}\leq m}}^{m} \frac{1}{b^{m-k_1-\cdots-k_{s-1}}} \prod_{l=1}^{s-1} \frac{a_{k_l}^{(l)}}{b^{k_l}} + \frac{1}{b^m} \sum_{k=0}^{s-2} \binom{m}{k} (b-1)^k \\
&\leq \frac{(b-1)^{s-1}}{b^m} \sum_{\substack{k_1,\ldots,k_{s-1}=1 \\ k_1+\cdots+k_{s-1}\leq m}}^{m} 1 + \frac{1}{b^m} \sum_{k=0}^{s-2} \binom{m}{k} (b-1)^k \\
&= \frac{1}{b^m} \sum_{k=0}^{s-1} \binom{m}{k} (b-1)^k,
\end{aligned}
$$

where we have used Lemma 5.19.

Thus the proof is finished. $\qquad\square$

Corollary 5.20 *For the star discrepancy of a (t, m, s)-net $\mathcal{P}$ in base b with $m \geq 2s - 2 + t$ we have*

$$D_{b^m}^*(\mathcal{P}) \leq \frac{b^t}{b^m} \left(1 + \frac{(b-1)^{s-1}}{(s-2)!} (m-t)^{s-1}\right).$$

Proof For $m \geq 2s - 2 + t$ and $k \in \{1, \ldots, s-1\}$ we have

$$\binom{m-t}{k} \leq \binom{m-t}{s-1} = \frac{(m-t)(m-t-1)\cdots(m-t-s+2)}{(s-1)!} \leq \frac{(m-t)^{s-1}}{(s-1)!}.$$

Using this estimate in Theorem 5.18 gives the desired result. $\qquad\square$

Since $m = (\log b^m)/\log b$, it follows that a $(0, m, s)$-net in base b has star discrepancy of order of magnitude $O((\log N)^{s-1}/N)$, where $N = b^m$ with an implied constant only depending on s and b. Furthermore, recall from Remark 5.9 that for all $s, b \in \mathbb{N}$, $b \geq 2$, there exists a $t_b(s)$ such that for all $m \geq t_b(s)$ a $(t_b(s), m, s)$-net in base b exists and $t_b(s) \asymp_b s$. Let $t_b(s) \leq cs$ for $c = c(b) > 0$ and let $(\mathcal{P}_m)_{m\geq t_b(s)}$

be a sequence of $(t_b(s), m, s)$-nets in base b. Using Stirling's formula we find that $n! \geq \sqrt{2\pi n}(n/e)^n$ for $n \in \mathbb{N}$, where $e = \exp(1)$. Then we have

$$\limsup_{m \to \infty} \frac{b^m D^*_{b^m}(\mathcal{P}_m)}{(\log b^m)^{s-1}} \leq \frac{b^{cs}(b-1)^{s-1}}{(s-2)!(\log b)^{s-1}} \leq \frac{b^{2c}(b-1)}{\sqrt{2\pi(s-2)}(\log b)^{s-1}} \left(\frac{eb^c(b-1)}{s-2}\right)^{s-2}.$$

This expression tends to zero at a superexponential rate as $s \to \infty$.

Now we turn to discrepancy estimates for (t, s)-sequences in base b. To this end, assume that $\Delta_b(t, m, s)$ is a number such that

$$b^m D^*_{b^m}(\mathcal{P}) \leq \Delta_b(t, m, s)$$

holds for the star discrepancy of any (t, m, s)-net $\mathcal{P}$ in base b. The star discrepancy D^*_N of the infinite (t, s)-sequence $\mathcal{S} = (\boldsymbol{x}_n)_{n \in \mathbb{N}_0}$ in base b is by definition the star discrepancy of the point set $\mathcal{P}_N = \{\boldsymbol{x}_0, \ldots, \boldsymbol{x}_{N-1}\}$ consisting of the first N elements of $\mathcal{S}$. Let N have b-adic representation $N = a_r b^r + \cdots + a_1 b + a_0$. Then we partition the point set $\mathcal{P}_N$ into the disjoint point sets

$$\mathcal{P}_{m,a} = \{\boldsymbol{x}_{a_r b^r + \cdots + a_{m+1} b^{m+1} + a b^m + k} : k = 0, \ldots, b^m - 1\},$$

for $m = 0, \ldots, r$ and $a = 1, \ldots, a_m$. Hence,

$$\mathcal{P}_N = \bigcup_{m=0}^{r} \bigcup_{a=0}^{a_m - 1} \mathcal{P}_{m,a}.$$

Every $\mathcal{P}_{m,a}$ is a (t, m, s)-net in base b and b^m-times its star discrepancy is at most $\Delta_b(t, m, s)$. Using the triangle inequality for the star discrepancy (see Exercise 2.4), we obtain the following result.

Lemma 5.21 *Let $\mathcal{S}$ be a (t, s)-sequence in base b. Let $N \in \mathbb{N}$ with b-adic expansion $N = a_r b^r + \cdots + a_1 b + a_0$. Then*

$$N D^*_N(\mathcal{S}) \leq \sum_{m=0}^{r} a_m \Delta_b(t, m, s).$$

Theorem 5.18 shows that we can choose $\Delta_b(t, m, s) = b^t \sum_{k=0}^{s-1} \binom{m-t}{k}(b-1)^k$. Using this estimate in conjunction with Lemma 5.21 we obtain the following estimate.

Theorem 5.22 *For the star discrepancy of a (t, s)-sequence $\mathcal{S}$ in base b we have*

$$D^*_N(\mathcal{S}) \leq \frac{b^t(b-1)}{N} \sum_{m=0}^{r} \sum_{k=0}^{s-1} \binom{m-t}{k}(b-1)^k,$$

where $r := \lfloor (\log N)/(\log b) \rfloor$.

For $s \geq 2$ and $r \to \infty$ we have

$$\sum_{\substack{m=0 \\ m \geq 2s-2+t}}^{r} \sum_{k=0}^{s-1} \binom{m-t}{k}(b-1)^k \leq \sum_{\substack{m=0 \\ m \geq 2s-2+t}}^{r} \left(1 + \frac{m^{s-1}}{(s-2)!}(b-1)^{s-1}\right)$$

$$\leq r+1+ \frac{(b-1)^{s-1}}{(s-2)!} \sum_{m=0}^{r} m^{s-1}$$

$$= \frac{(b-1)^{s-1}(s-1)}{s!} r^s + O(r^{s-1})$$

and

$$\sum_{\substack{m=0 \\ m < 2s-2+t}}^{r} \sum_{k=0}^{s-1} \binom{m-t}{k}(b-1)^k = O(1).$$

Hence we conclude from Theorem 5.22 that

$$D_N^*(\mathcal{S}) \leq \frac{b^t(b-1)^s(s-1)}{s!(\log b)^s} \frac{(\log N)^s}{N} + O\left(\frac{(\log N)^{s-1}}{N}\right).$$

To compare this estimate with the Halton sequences we study the quantity d^* which was introduced in (2.17). The above bound on the star discrepancy implies that

$$d^*(\mathcal{S}) := \limsup_{N \to \infty} \frac{N D_N^*(\mathcal{S})}{(\log N)^s} \leq b^t \frac{(b-1)^s(s-1)}{s!(\log b)^s} \leq b^t \frac{(b-1)^s(s-1)e^s}{\sqrt{2\pi s}\, s^s(\log b)^s}.$$

It is known that for any $s, b \in \mathbb{N}$, $b \geq 2$, one can construct a $(t_b(s), s)$-sequence $\mathcal{S}_s^*$ in base b with $t_b(s) \asymp_b s$. For such sequences we obtain

$$\limsup_{s \to \infty} \frac{\log d^*(\mathcal{S}_s^*)}{s \log s} \leq -1$$

and hence $d^*(\mathcal{S}_s^*)$ tends to zero at a superexponential rate as $s \to \infty$. This result should be compared with the analog result for Halton sequences presented in Sect. 2.4, Eq. (2.18).

5.3 Digital Nets and Sequences

In the last section, we have shown that (t, m, s)-nets and (t, s)-sequences in base b have the desired property of having a low star discrepancy (at least in an asymptotic sense). Now the question arises how to construct concrete examples of (t, m, s)-nets and (t, s)-sequences in base b with good equidistribution properties. The concept of *digital* nets and sequences is a general framework for the construction of point sets and sequences, and it is the basis for virtually all concrete constructions of (t, m, s)-nets and (t, s)-sequences.

Although one can introduce digital nets and sequences in arbitrary integer bases $b \geq 2$, we restrict our treatment to prime power bases b only. The reason for this restriction is that exactly for such b there exists a finite field $\mathbb{F}_b$ of order b and this will be helpful in the construction. We denote the elements of $\mathbb{F}_b$ by $\overline{0}, \overline{1}, \ldots, \overline{b-1}$, respectively, where $\overline{0}$ denotes the neutral element of addition in $\mathbb{F}_b$. If b is a prime number, then we identify $\mathbb{F}_b$ with $\mathbb{Z}_b$, the set of residue classes modulo b equipped with arithmetic operations modulo b, which in turn we identify with the elements of $\{0, \ldots, b-1\}$.

Remark 5.23 (Some background on finite fields) Recall from Algebra that a finite field of order b exists if and only if b is a prime power. Furthermore, a finite field of prime power order b is uniquely determined up to isomorphisms. The construction of a finite field uses irreducible polynomials and is based on the following result: If K is a field and if $f \in K[x]$ is irreducible (i.e., it cannot be factored into the product of two or more non-trivial polynomials over K) then the residue class ring $K[x]/(f)$ is also a field.

Hence, to construct a finite field with $b = q^m$ elements, where q is a prime number and $m \in \mathbb{N}$, one can start with the finite field $\mathbb{F}_q$ (which we identify with the set of residue classes modulo q equipped with arithmetic operations modulo q) and an irreducible polynomial $f \in \mathbb{F}_q[x]$ of degree m (whose existence is also guaranteed by a result from algebra). Then $\mathbb{F}_q[x]/(f)$ is a field that contains exactly $b = q^m$ elements which is usually denoted by $\mathbb{F}_b$.

For example, to construct $\mathbb{F}_4$, the finite field with four elements, we start with $\mathbb{F}_2 = \{0, 1\}$ and choose $f(x) = x^2 + x + 1 \in \mathbb{F}_2[x]$, which is irreducible over $\mathbb{F}_2$. Then $\mathbb{F}_2[x]/(f)$ has the four elements $0, 1, [x], [x + 1]$ and we obtain the following operation tables for addition and multiplication, where we write $A = [x]$ and $B = [x + 1]$ and where we use arithmetic operations modulo (f), i.e., $x^2 = x + 1$:

$$
\begin{array}{c|cccc}
+ & 0 & 1 & A & B \\
\hline
0 & 0 & 1 & A & B \\
1 & 1 & 0 & B & A \\
A & A & B & 0 & 1 \\
B & B & A & 1 & 0
\end{array}
\qquad \text{and} \qquad
\begin{array}{c|cccc}
\cdot & 0 & 1 & A & B \\
\hline
0 & 0 & 0 & 0 & 0 \\
1 & 0 & 1 & A & B \\
A & 0 & A & B & 1 \\
B & 0 & B & 1 & A
\end{array}
$$

Digital (t, m, s)-Nets Over $\mathbb{F}_b$

To construct a digital (t, m, s)-net in a prime power base b, we use the finite field $\mathbb{F}_b$ with b elements and a bijection $\varphi : \{0, \ldots, b-1\} \to \mathbb{F}_b$ which maps the set of b-adic digits onto $\mathbb{F}_b$. We assume in the following that the elements of $\mathbb{F}_b$ are ordered in such a way that $\varphi(a) = \overline{a}$ for $a \in \{0, \ldots, b-1\}$. If b is a prime number we identify $\mathbb{F}_b$ with the elements of $\{0, \ldots, b-1\}$ and therefore we shall omit the bijection φ and the bar in this case.

Having set the stage, we now explain the digital method. Let $s, m \in \mathbb{N}$ and let b be a prime power. We aim at constructing a b^m-element point set $\{x_0, \ldots, x_{b^m-1}\}$ in $[0, 1)^s$. To generate such a point set we first choose $m \times m$ matrices $C_1, \ldots, C_s$ (one for each dimension) over $\mathbb{F}_b$, that is, with entries from $\mathbb{F}_b$.

Digital method. To generate the point $x_n = (x_{n,1}, \ldots, x_{n,s})$ for $n \in \{0, \ldots, b^m - 1\}$ we write n in its base b expansion

$$n = n_0 + n_1 b + \cdots + n_{m-1} b^{m-1}$$

with digits $n_j \in \{0, \ldots, b - 1\}$. Then take the m-dimensional column vector

$$\vec{n} := \begin{pmatrix} \varphi(n_0) \\ \varphi(n_1) \\ \vdots \\ \varphi(n_{m-1}) \end{pmatrix}$$

with entries from $\mathbb{F}_b$ and multiply the matrix C_j, $j = 1, 2, \ldots, s$, with this vector

$$C_j \vec{n} =: \begin{pmatrix} \overline{y}_{n,j,1} \\ \overline{y}_{n,j,2} \\ \vdots \\ \overline{y}_{n,j,m} \end{pmatrix}.$$

Transforming the entries of this vector with elements in $\mathbb{F}_b$ back into the set of base b digits $\{0, 1, \ldots, b - 1\}$ by applying the inverse map φ^{-1} we obtain $y_{n,j,l} = \varphi^{-1}(\overline{y}_{n,j,l})$ for $l = 1, 2, \ldots, m$, which are taken as the base b digits of the jth component of the point x_n, i.e.,

$$x_{n,j} := \frac{y_{n,j,1}}{b} + \frac{y_{n,j,2}}{b^2} + \cdots + \frac{y_{n,j,m}}{b^m}.$$

Finally, we put

$$x_n := (x_{n,1}, \ldots, x_{n,s}).$$

Every point set constructed with the digital method is a (t, m, s)-net in base b for some quality parameter $t \in \{0, \ldots, m\}$ since it is at least an (m, m, s)-net in base b. This leads to the following definition:

Definition 5.24 (*Digital net*) If the point set $\{x_0, \ldots, x_{b^m-1}\}$ constructed by the digital method is for some $t \in \{0, \ldots, m\}$ a (t, m, s)-net in base b, then it is called a *digital* (t, m, s)-*net over the field* $\mathbb{F}_b$ or, for short, a *digital net* with generating matrices $C_1, \ldots, C_s$. It is called a *strict* digital (t, m, s)-net over $\mathbb{F}_b$ if $\{x_0, \ldots, x_{b^m-1}\}$ is a strict (t, m, s)-net in base b.

In most cases, the finite field $\mathbb{Z}_b$ with b prime is chosen for practical applications, and indeed $\mathbb{Z}_2$ is the most frequent choice.

Example 5.25 We want to construct a digital $(t, 4, 2)$-net over $\mathbb{Z}_2$. First we have to choose two 4×4 matrices over $\mathbb{Z}_2$, for instance:

$$C_1 = \begin{pmatrix} 1\,0\,0\,0 \\ 0\,1\,0\,0 \\ 0\,0\,1\,0 \\ 0\,0\,0\,1 \end{pmatrix} \quad \text{and} \quad C_2 = \begin{pmatrix} 0\,0\,0\,1 \\ 0\,0\,1\,0 \\ 0\,1\,0\,0 \\ 1\,0\,0\,0 \end{pmatrix}. \tag{5.4}$$

As an example we show how to construct the point x_{10}. We have $n = 10 = 0 + 1 \cdot 2 + 0 \cdot 2^2 + 1 \cdot 2^3$ and therefore

$$C_1 \vec{n} = \begin{pmatrix} 1\,0\,0\,0 \\ 0\,1\,0\,0 \\ 0\,0\,1\,0 \\ 0\,0\,0\,1 \end{pmatrix} \begin{pmatrix} 0 \\ 1 \\ 0 \\ 1 \end{pmatrix} = \begin{pmatrix} 0 \\ 1 \\ 0 \\ 1 \end{pmatrix},$$

$$C_2 \vec{n} = \begin{pmatrix} 0\,0\,0\,1 \\ 0\,0\,1\,0 \\ 0\,1\,0\,0 \\ 1\,0\,0\,0 \end{pmatrix} \begin{pmatrix} 0 \\ 1 \\ 0 \\ 1 \end{pmatrix} = \begin{pmatrix} 1 \\ 0 \\ 1 \\ 0 \end{pmatrix}.$$

Hence, $x_{10,1} = \frac{1}{4} + \frac{1}{16} = \frac{5}{16}$ and $x_{10,2} = \frac{1}{2} + \frac{1}{8} = \frac{5}{8}$, and thus $x_{10} = \left(\frac{5}{16}, \frac{5}{8} \right)$.

We will show now how the (strict) quality parameter t is connected with the choice of the generating matrices $C_1, \ldots, C_s$. This can be answered with the help of the following quantity ρ which, in some sense, "measures" the "linear independence of the s matrices $C_1, \ldots, C_s$".

Definition 5.26 Let b be a prime power and let $C_1, \ldots, C_s$ be $m \times m$ matrices over $\mathbb{F}_b$. Let $\rho = \rho(C_1, \ldots, C_s)$ be the largest integer such that for any choice of $d_1, \ldots, d_s \in \mathbb{N}_0$ with

$$d_1 + \cdots + d_s = \rho$$

the following holds:

The first d_1 row vectors of C_1 together with
the first d_2 row vectors of C_2 together with

$\vdots$

the first d_s row vectors of C_s

(these are together ρ vectors in $\mathbb{F}_b^m$) are linearly independent over the finite field $\mathbb{F}_b$. We call ρ the *linear independence parameter* of the matrices $C_1, \ldots, C_s$.

Example 5.27 Consider C_1 and C_2 over $\mathbb{Z}_2$ from Example 5.25. Clearly, for these 4×4 matrices $\rho(C_1, C_2)$ is at most 4. It is easy to see that in the present case $\rho(C_1, C_2)$ is indeed 4, since for any choice of $d_1, d_2 \in \mathbb{N}_0$ with $d_1 + d_2 = 4$, the first d_1 rows of C_1 together with the first d_2 rows of C_2 provide the four canonical row vectors $(1, 0, 0, 0)$, $(0, 1, 0, 0)$, $(0, 0, 1, 0)$, $(0, 0, 0, 1)$, which are linearly independent over $\mathbb{Z}_2$. Hence, $\rho(C_1, C_2) = 4$.

Now we can determine the strict quality parameter t of a digital net generated by matrices $C_1, \ldots, C_s$ over $\mathbb{Z}_b$.

Theorem 5.28 (Niederreiter) *Let $s, m \in \mathbb{N}$ and let b be a prime power. The point set constructed by the digital method with the $m \times m$ matrices $C_1, \ldots, C_s$ over a finite field $\mathbb{F}_b$ is a strict $(m - \rho, m, s)$-net in base b, where $\rho = \rho(C_1, \ldots, C_s)$ is the linear independence parameter introduced in Definition 5.26.*

Proof First we have to show that every s-dimensional elementary b-adic interval of volume $b^{-\rho}$ contains exactly $b^{m-\rho}$ of the generated points. Let

$$
J = \prod_{j=1}^{s} \left[\frac{a_j}{b^{d_j}}, \frac{a_j + 1}{b^{d_j}} \right)
$$

with $d_1, \ldots, d_s \in \mathbb{N}_0$ such that $d_1 + \cdots + d_s = \rho$, and $a_j \in \{0, \ldots, b^{d_j} - 1\}$ for $j = 1, 2, \ldots, s$. We have to count the number of indices $n \in \{0, 1, \ldots, b^m - 1\}$ for which x_n belongs to J. We have $x_n \in J$ if and only if

$$
x_{n,j} \in \left[\frac{a_j}{b^{d_j}}, \frac{a_j + 1}{b^{d_j}} \right) \tag{5.5}
$$

for all $j = 1, 2, \ldots, s$. Let $a_j = e_{d_j}^{(j)} + \cdots + e_1^{(j)} b^{d_j - 1}$ and hence

$$
\frac{a_j}{b^{d_j}} = \frac{e_1^{(j)}}{b} + \cdots + \frac{e_{d_j}^{(j)}}{b^{d_j}}.
$$

Then (5.5) is equivalent to

$$
\frac{e_1^{(j)}}{b} + \cdots + \frac{e_{d_j}^{(j)}}{b^{d_j}} \leq x_{n,j} < \frac{e_1^{(j)}}{b} + \cdots + \frac{e_{d_j}^{(j)}}{b^{d_j}} + \frac{1}{b^{d_j}},
$$

which, in turn, is equivalent to

$$
x_{n,j} = \frac{e_1^{(j)}}{b} + \cdots + \frac{e_{d_j}^{(j)}}{b^{d_j}} + \cdots .
$$

Hence the condition (5.5) uniquely determines the first d_j base b digits of $x_{n,j}$.

According to the digital construction scheme, the ith base b digit of $x_{n,j}$ is given by φ^{-1} applied to the product $\vec{c}_i^{\,(j)}\vec{n}$ of the ith row vector $\vec{c}_i^{\,(j)}$ of C_j with the column vector $\vec{n}$, where φ is the bijection used in the construction. Hence $\boldsymbol{x}_n \in J$ if and only if the following system of equations over $\mathbb{F}_b$ is satisfied:

$$\vec{c}_1^{\,(1)}\vec{n} = \varphi(e_1^{(1)}),$$

$$\vdots$$

$$\vec{c}_{d_1}^{\,(1)}\vec{n} = \varphi(e_{d_1}^{(1)}),$$

$$\vec{c}_1^{\,(2)}\vec{n} = \varphi(e_1^{(2)}),$$

$$\vdots$$

$$\vec{c}_{d_2}^{\,(2)}\vec{n} = \varphi(e_{d_2}^{(2)}),$$

$$\vdots$$

$$\vec{c}_1^{\,(s)}\vec{n} = \varphi(e_1^{(s)}),$$

$$\vdots$$

$$\vec{c}_{d_s}^{\,(s)}\vec{n} = \varphi(e_{d_s}^{(s)}). \tag{5.6}$$

This system consists of exactly $d_1 + d_2 + \cdots + d_s = \rho$ equations. Let A be the $\rho \times m$ matrix over $\mathbb{F}_b$ with row vectors

$$\vec{c}_1^{\,(1)}, \ldots, \vec{c}_{d_1}^{\,(1)}, \vec{c}_1^{\,(2)}, \ldots, \vec{c}_{d_2}^{\,(2)}, \ldots, \vec{c}_1^{\,(s)}, \ldots, \vec{c}_{d_s}^{\,(s)} \tag{5.7}$$

and let

$$\vec{f}^{\,\top} := (\varphi(e_1^{(1)}), \ldots, \varphi(e_{d_1}^{(1)}), \varphi(e_1^{(2)}), \ldots, \varphi(e_{d_2}^{(2)}), \ldots, \varphi(e_1^{(s)}), \ldots, \varphi(e_{d_s}^{(s)})) \in \mathbb{F}_b^{\rho}.$$

Then the linear system (5.6) can be rewritten into the system

$$A\vec{n} = \vec{f}. \tag{5.8}$$

Since the system of row vectors (5.7) by the definition of ρ is linearly independent over $\mathbb{F}_b$, the $\rho \times m$ matrix A has rank ρ. This implies that (5.8) is resolvable and that the nullspace of A has dimension $m - \rho$. Hence the linear system (5.8) has exactly $b^{m-\rho}$ solutions and thus J contains $b^{m-\rho}$ elements from the digital net. This means that the matrices $C_1, \ldots, C_s$ generate a digital $(m - \rho, m, s)$-net over $\mathbb{F}_b$.

It remains to show the strictness of the quality parameter. If $\rho = m$, then there is nothing to prove. If $\rho \leq m - 1$, then there are $d_1, \ldots, d_s \in \mathbb{N}_0$ with $d_1 + \cdots + d_s = \rho + 1$, and such that the vectors

$$\vec{c}_1^{\,(1)}, \ldots, \vec{c}_{d_1}^{\,(1)}, \ldots, \vec{c}_1^{\,(s)}, \ldots, \vec{c}_{d_s}^{\,(s)}$$

are linearly dependent over $\mathbb{F}_b$. But then the system (5.6) with $\varphi(e_i^{(j)}) = 0$ for $i = 1, \ldots, d_j$ and $j = 1, \ldots, s$, also has $b^{m-\rho}$ solutions $\vec{n}$ (although it consists of $\rho + 1$ equations in m variables). This means that the corresponding elementary interval

$$\tilde{J} := \prod_{j=1}^{s} \left[\frac{a_j}{b^{d_j}}, \frac{a_j + 1}{b^{d_j}} \right)$$

with $a_j = e_{d_j}^{(j)} + \cdots + e_1^{(j)} b^{d_j - 1}$ for $j = 1, 2, \ldots, s$ and of volume $b^{-(\rho+1)}$ contains $b^{m-\rho}$ points of the digital net. Consequently,

$$A(\tilde{J}, b^m) - b^m \lambda_s(\tilde{J}) = b^{m-\rho} - b^{m-\rho-1} = (b-1)b^{m-\rho-1} \neq 0,$$

and so the matrices $C_1, \ldots, C_s$ generate a strict digital $(m - \rho, m, s)$-net over $\mathbb{F}_b$. $\qquad\square$

Digital (t, s)-Sequences Over $\mathbb{F}_b$

To construct a (t, s)-sequence in prime power base b by the digital method, we again use a finite field $\mathbb{F}_b$ and a bijection $\varphi : \{0, \ldots, b-1\} \to \mathbb{F}_b$ for which we demand $\varphi(0) = \overline{0}$.

First we have to choose $\mathbb{N} \times \mathbb{N}$ matrices $C_1, \ldots, C_s$ (one for each component) over $\mathbb{F}_b$. That is, matrices of the form:

$$C = \begin{pmatrix} c_{11} & c_{12} & c_{13} & \cdots \\ c_{21} & c_{22} & c_{23} & \cdots \\ c_{31} & c_{32} & c_{33} & \cdots \\ \vdots & \vdots & \vdots & \ddots \end{pmatrix} \in \mathbb{F}_b^{\mathbb{N} \times \mathbb{N}}.$$

Digital method. To generate one of the points $x_n = (x_{n,1}, \ldots, x_{n,s})$ for $n \in \mathbb{N}_0$ of the (t, s)-sequence, we again write n in its base b expansion $n = \sum_{i=0}^{\infty} a_i b^i$ with $a_i \in \{0, \ldots, b-1\}$ and $a_i = 0$ for all i large enough. Then take the column vector

$$\vec{n} := \begin{pmatrix} \varphi(a_0) \\ \varphi(a_1) \\ \varphi(a_2) \\ \vdots \end{pmatrix}$$

with entries in $\mathbb{F}_b$ and multiply the matrix C_j, $j = 1, 2, \ldots, s$, with this vector

$$C_j \vec{n} =: \begin{pmatrix} \overline{y}_{n,j,1} \\ \overline{y}_{n,j,2} \\ \vdots \end{pmatrix}.$$

Note that for the multiplication only finitely many of the entries of $\vec{n}$ are different from zero, as we assumed that $\varphi(0) = \overline{0}$. Transforming the entries of $C_j\vec{n}$ back into the set of base b digits $\{0, 1, \ldots, b - 1\}$ by applying the inverse map φ^{-1} we obtain

$$x_{n,j} := \frac{\varphi^{-1}(\overline{y}_{n,j,1})}{b} + \frac{\varphi^{-1}(\overline{y}_{n,j,2})}{b^2} + \cdots .$$

Finally, we put

$$\boldsymbol{x}_n := (x_{n,1}, \ldots, x_{n,s}).$$

Definition 5.29 We call the sequence $(\boldsymbol{x}_n)_{n \in \mathbb{N}_0}$ constructed in this way a *digital sequence over $\mathbb{F}_b$ with generating matrices $C_1, \ldots, C_s$*, or short, a *digital sequence*.

Remark 5.30 Depending on the matrices $C_1, \ldots, C_s$, it may happen that the vector $C_j\vec{n} =: \vec{y}_{n,j}$ contains infinitely many entries different from zero. For practical purposes this requires an adaptation of the point generation. Usually the vector $\vec{y}_{n,j}$ is truncated at a suitable place.

Furthermore, another theoretical problem may arise. It should be avoided that the vector $\vec{y}_{n,j}$ contains only finitely many elements different from $\varphi(b - 1)$. Because of the nonuniqueness of representation of the b-adic real numbers represented by such "digit vectors," the net structure of the sequence under consideration would be destroyed. This is the reason for the following additional "finiteness" condition on the matrices $C_1, \ldots, C_s$. Let

$$C_j = (c_{i,r}^{(j)})_{i,r \in \mathbb{N}} \in \mathbb{F}_b^{\mathbb{N} \times \mathbb{N}}$$

for $j = 1, \ldots, s$.

F : We demand that for all j and r we have $c_{i,r}^{(j)} = 0$ for all i large enough.

To eliminate the problems with ambiguous digit expansions without insisting on the finiteness condition **F** one can use a slightly revised version of the definition of digital sequences which is based on a certain truncation operator. Such sequences are then occasionally called *digital sequences in the broad sense*.

We will show in the following that under certain conditions on the generating matrices a digital sequence is a (t, s)-sequence in base b for some $t \in \mathbb{N}_0$ which we then call a *digital (t, s)-sequence over $\mathbb{F}_b$*. Before we do so we again have to introduce the quantities ρ_m, which in some sense "measure" the "linear independence" of the s infinite matrices $C_1, \ldots, C_s$.

Definition 5.31 Let $C_1, \ldots, C_s$ be $\mathbb{N} \times \mathbb{N}$ matrices over the finite field $\mathbb{F}_b$. For any $j = 1, 2, \ldots, s$ and $m \in \mathbb{N}$ we denote by $C_j^{(m)}$ the left upper $m \times m$ sub-matrix of C_j. Then

$$\rho_m = \rho_m(C_1, \ldots, C_s) := \rho(C_1^{(m)}, \ldots, C_s^{(m)}),$$

where ρ is the linear independence parameter defined for s-tuples of $m \times m$ matrices over $\mathbb{F}_b$ in Definition 5.26.

Now we can determine the quality parameter t of a digital sequence.

Theorem 5.32 (Niederreiter) *Let $C_1, \ldots, C_s$ be $\mathbb{N} \times \mathbb{N}$ matrices over $\mathbb{F}_b$. If there exists a $t \in \mathbb{N}_0$ such that for each $m > t$ we have $\rho_m(C_1, \ldots, C_s) \geq m - t$, then the digital sequence generated by $C_1, \ldots, C_s$ is a digital (t, s)-sequence over $\mathbb{F}_b$.*

Proof Let $(x_n)_{n \in \mathbb{N}_0}$ denote the digital sequence generated by $C_1, \ldots, C_s$. By the definition of a (t, s)-sequence we have to show that for any $m \in \mathbb{N}$ and any $k \in \mathbb{N}_0$ the point set

$$\mathcal{P}_{k,m} = \{x_{kb^m}, \ldots, x_{kb^m + b^m - 1}\}$$

is a (t, m, s)-net in base b. Indeed, for given k and m, and any $l \in \{0, \ldots, b^m - 1\}$ let $k = \kappa_{r+1}b^r + \cdots + \kappa_1$ and $l = \lambda_{m-1}b^{m-1} + \cdots + \lambda_0$ be the base b representations of k and l, respectively. For $n = kb^m + l$ we have

$$\vec{n} = (\overline{\lambda}_0, \ldots, \overline{\lambda}_{m-1}, \overline{\kappa}_1, \ldots, \overline{\kappa}_{r+1}, \ldots)^\top$$

and with the following representation of the matrices C_j:

$$C_j = \left(\begin{array}{c|c} C_j^{(m)} & F_j^{(m)} \\ \hline & H_j^{(m)} \end{array} \right) \in \mathbb{F}_b^{\mathbb{N} \times \mathbb{N}},$$

where $F_j^{(m)} \in \mathbb{F}_b^{m \times \mathbb{N}}$ and $H_j^{(m)} \in \mathbb{F}_b^{\mathbb{N} \times \mathbb{N}}$, we have

$$C_j \vec{n} = \begin{pmatrix} C_j^{(m)}\vec{l} \\ \overline{0} \\ \overline{0} \\ \vdots \end{pmatrix} + \begin{pmatrix} F_j^{(m)}\vec{k} \\ \overline{0} \\ \overline{0} \\ \vdots \end{pmatrix} + \begin{pmatrix} \overline{0} \\ \vdots \\ \overline{0} \\ H_j^{(m)}\vec{n} \end{pmatrix},$$

where $\vec{l} = (\overline{\lambda}_0, \ldots, \overline{\lambda}_{m-1})^\top$ and $\vec{k} = (\overline{\kappa}_1, \ldots, \overline{\kappa}_{r+1}, \overline{0}, \overline{0}, \ldots)^\top$.

In order to show that $\mathcal{P}_{k,m}$ is a (t, m, s)-net in base b we need to show that every elementary interval of the form

$$J = \prod_{j=1}^{s} \left[\frac{a_j}{b^{d_j}}, \frac{a_j + 1}{b^{d_j}} \right)$$

with $d_1, \ldots, d_s \in \mathbb{N}_0$ such that $d_1 + \cdots + d_s = m - t$, and $a_j \in \{0, \ldots, b^{d_j} - 1\}$ for $j = 1, 2, \ldots, s$ contains exactly b^t elements from $\mathcal{P}_{k,m}$. So we count the number of indices $n \in \{kb^m, kb^m + 1, \ldots, kb^m + b^m - 1\}$ for which $\boldsymbol{x}_n$ belongs to J, that is, the number of $l \in \{0, 1, \ldots, b^m - 1\}$ for which $\boldsymbol{x}_{kb^m + l}$ belongs to J. Let

$$\frac{a_j}{b^{d_j}} = \frac{e_1^{(j)}}{b} + \cdots + \frac{e_{d_j}^{(j)}}{b^{d_j}}.$$

Then we have, as in the proof of Theorem 5.28, that $\boldsymbol{x}_n \in J$ if and only if

$$x_{n,j} = \frac{e_1^{(j)}}{b} + \cdots + \frac{e_{d_j}^{(j)}}{b^{d_j}} + \cdots.$$

With the above considerations, this is equivalent to

$$\vec{c}_1^{\,(j)} \vec{l} = \varphi(e_1^{(j)}) - \vec{f}_1^{\,(j)} \vec{k}$$

$$\vdots$$

$$\vec{c}_{d_j}^{\,(j)} \vec{l} = \varphi(e_{d_j}^{(j)}) - \vec{f}_{d_j}^{\,(j)} \vec{k}$$

where $\vec{c}_i^{\,(j)}$ denotes the ith row vector of the matrix $C_j^{(m)}$ and where $\vec{f}_i^{\,(j)}$ denotes the ith row vector of the matrix $F_j^{(m)}$. For $j = 1, 2, \ldots, s$ this gives a linear system of $d_1 + \cdots + d_s = m - t$ linear equations in the m variables $\overline{\lambda}_0, \ldots, \overline{\lambda}_{m-1} \in \mathbb{F}_b$.

Since the system of row vectors $\vec{c}_i^{\,(j)}$ for $i = 1, \ldots, d_j$ and $j = 1, 2, \ldots, s$ is, by the definition of ρ_m, linearly independent over $\mathbb{F}_b$, the linear system has exactly $b^{m-(m-t)} = b^t$ solution and hence J contains b^t elements from $\mathcal{P}_{k,m}$. $\square$

Example 5.33 The van der Corput sequence can be generated by the digital construction scheme using the matrix

$$C_1 = \begin{pmatrix} \overline{1}\,\overline{0}\,\overline{0}\,\overline{0}\ldots \\ \overline{0}\,\overline{1}\,\overline{0}\,\overline{0}\ldots \\ \overline{0}\,\overline{0}\,\overline{1}\,\overline{0}\ldots \\ \overline{0}\,\overline{0}\,\overline{0}\,\overline{1}\ldots \\ \vdots\ \vdots\ \vdots\ \vdots\ \ddots \end{pmatrix}$$

over $\mathbb{F}_b$. Since $\rho_m(C_1) = m$ for every $m \in \mathbb{N}$, it follows from Theorem 5.32 that the van der Corput sequence in prime power base b is a digital $(0, 1)$-sequence over $\mathbb{F}_b$.

Interestingly, the construction of a second generating matrix to obtain a, say, digital $(0, 2)$-sequence over $\mathbb{F}_b$ is already non-trivial. This question is addressed in the following section.

5.4   Special Constructions of Digital Nets and Sequences

Although I.M. Sobol' was the first to construct digital (t, s)-sequences in base 2 in 1967 and H. Faure introduced constructions of digital $(0, s)$-sequences in prime base b with $b \geq s$ in 1982, it now seems most convenient to introduce digital (t, s)-sequences using H. Niederreiter's unifying approach based on polynomial arithmetic over finite fields. By considering generalized Niederreiter sequences, the constructions by Sobol' and Faure appear as special cases.

For a prime power b let $\mathbb{F}_b((x^{-1}))$ be the field of *formal Laurent series* over $\mathbb{F}_b$, that is, series of the form:

$$L(x) = \sum_{\ell=w}^{\infty} t_\ell x^{-\ell},$$

where w is an arbitrary integer and all $t_\ell \in \mathbb{F}_b$. Note that $\mathbb{F}_b((x^{-1}))$ contains the field of rational functions over $\mathbb{F}_b$ as a subfield. Furthermore, let $\mathbb{F}_b[x]$ be the set of all polynomials over $\mathbb{F}_b$. The *discrete exponential valuation* v on $\mathbb{F}_b((x^{-1}))$ is defined by $v(L) = -w$ if $L \neq 0$, where w is the smallest index with $t_w \neq 0$. For $L = 0$ we set $v(0) = -\infty$. Observe that we have $v(p/q) = \deg(p) - \deg(q)$ for all nonzero polynomials $p, q \in \mathbb{F}_b[x]$ (cf. Exercise 5.5).

Classical Niederreiter Sequence

Let $s \in \mathbb{N}$, b a prime power and let $p_1, \ldots, p_s \in \mathbb{F}_b[x]$ be distinct monic irreducible polynomials over $\mathbb{F}_b$. Recall that a polynomial $p \in \mathbb{F}_b[x]$ is called irreducible if it cannot be factored into the product of two or more non-trivial polynomials and it is called monic if its leading coefficient is equal to the identity $\overline{1}$ of $\mathbb{F}_b$. Let $e_j = \deg(p_j)$ for $j = 1, 2, \ldots, s$. For $j \in \{1, \ldots, s\}$, $i \in \mathbb{N}$ and $k \in \{0, \ldots, e_j - 1\}$, consider the expansions

$$\frac{x^k}{p_j(x)^i} = \sum_{r=0}^{\infty} a^{(j)}(i, k, r)x^{-r} \tag{5.9}$$

over the field of formal Laurent series $\mathbb{F}_b((x^{-1}))$. Then we define the matrix $C_j = (c_{i,r}^{(j)})_{i,r \in \mathbb{N}}$ by

$$c_{i,r}^{(j)} := a^{(j)}(Q + 1, k, r) \in \mathbb{F}_b \quad \text{for } j \in \{1, \ldots, s\}, \; i, r \in \mathbb{N}, \tag{5.10}$$

where the parameters Q and k are uniquely determined by i and j via the Euclidean division $i - 1 = Qe_j + k$ with $Q = Q(j, i) \in \mathbb{N}_0$ and remainder $k = k(j, i) \in \{0, \ldots, e_j - 1\}$. Hence $Q(j, i) = \lfloor (i - 1)/e_j \rfloor$.

Since the exponential evaluation v applied to the left-hand side of (5.9) gives $v(x^k/p_j(x)^i) = k - ie_j$ and this tends to $-\infty$ for growing i, it follows that $c_{i,r}^{(j)} = 0$ for i large enough. Thus condition **F** is satisfied.

Definition 5.34 A digital sequence over $\mathbb{F}_b$ generated by the $\mathbb{N} \times \mathbb{N}$ matrices $C_j = (c_{i,r}^{(j)})_{i,r \in \mathbb{N}}$ for $j = 1, 2, \ldots, s$ where the $c_{i,r}^{(j)}$ are given by (5.10) is called a *Niederreiter sequence*.

Theorem 5.35 (Niederreiter) *The Niederreiter sequence with generating matrices defined as above is a digital (t, s)-sequence over $\mathbb{F}_b$ with*

$$t = \sum_{j=1}^{s} (e_j - 1).$$

Proof According to Theorem 5.32, we need to show that for all integers $m > \sum_{j=1}^{s}(e_j - 1)$ and all $d_1, \ldots, d_s \in \mathbb{N}_0$ with $1 \le \sum_{j=1}^{s} d_j \le m - \sum_{j=1}^{s}(e_j - 1)$, the vectors

$$\vec{c}_i^{\,(j)} = (c_{i,1}^{(j)}, \ldots, c_{i,m}^{(j)}) \in \mathbb{F}_b^m \quad \text{for } i = 1, \ldots, d_j, \;\; j = 1, 2, \ldots, s \qquad (5.11)$$

are linearly independent over $\mathbb{F}_b$. Suppose that

$$\sum_{j=1}^{s} \sum_{i=1}^{d_j} f_i^{(j)} \vec{c}_i^{\,(j)} = \vec{0} \in \mathbb{F}_b^m$$

for some $f_i^{(j)} \in \mathbb{F}_b$, where, without loss of generality, we may assume that $d_j \ge 1$ for all $j = 1, 2, \ldots, s$. By comparing components, we obtain

$$\sum_{j=1}^{s} \sum_{i=1}^{d_j} f_i^{(j)} c_{i,r}^{(j)} = 0 \quad \text{for } r = 1, 2, \ldots, m. \qquad (5.12)$$

Consider the rational function

$$L := \sum_{j=1}^{s} \sum_{i=1}^{d_j} f_i^{(j)} \frac{x^{k(j,i)}}{p_j(x)^{Q(j,i)+1}} = \sum_{r=1}^{\infty} \left(\sum_{j=1}^{s} \sum_{i=1}^{d_j} f_i^{(j)} c_{i,r}^{(j)} \right) x^{-r},$$

where we used (5.9) and (5.10). From (5.12) we obtain that $v(L) < -m$, where $v(L)$ denotes the discrete exponential evaluation.

Recall that $Q(j, i) = \lfloor (i - 1)/e_j \rfloor$. Hence, if we put $Q_j = \lfloor (d_j - 1)/e_j \rfloor$ for $j = 1, 2, \ldots, s$, then a common denominator of L is $g(x) = \prod_{j=1}^{s} p_j(x)^{Q_j+1}$, which implies that Lg is a polynomial. On the other hand,

$$v(Lg) < -m + \deg(g) = -m + \sum_{j=1}^{s} (Q_j + 1)e_j \le -m + \sum_{j=1}^{s} (d_j - 1 + e_j) \le 0.$$

Thus $Lg = 0$, hence $L = 0$, and therefore

$$\sum_{j=1}^{s}\sum_{i=1}^{d_j} f_i^{(j)} \frac{x^{k(j,i)}}{p_j(x)^{Q(j,i)+1}} = 0.$$

The left-hand side is a partial fraction decomposition of a rational function, and because of its uniqueness all $f_i^{(j)} = 0$. This means that the vectors given in (5.11) are linearly independent over $\mathbb{F}_b$ and the proof is finished. $\square$

Remark 5.36 It was shown by J. Dick and H. Niederreiter in 2008 that the quality parameter $t = \sum_{j=1}^{s}(e_j - 1)$ is even exact, i.e., the Niederreiter sequence is a strict digital $(\sum_{j=1}^{s}(e_j - 1), s)$-sequence over $\mathbb{F}_b$.

Now we consider a special choice of polynomials $p_1, \ldots, p_s$. For fixed s and b, list all monic irreducible polynomials over $\mathbb{F}_b$ in a sequence according to nondecreasing degrees, and let $p_1, \ldots, p_s$ be the first s terms of this sequence. Then it has been shown by H. Niederreiter that for the strict quality parameter t of the corresponding Niederreiter sequence we have

$$t \le \begin{cases} 0 & \text{for } s \le b, \\ s(\log_b s + \log_b \log_b s + 1) & \text{for } s > b. \end{cases} \tag{5.13}$$

This shows that for prime power bases b the necessary condition from Theorem 5.16 for the existence of $(0, s)$-sequences in base b is also sufficient.

Corollary 5.37 *Let $s \in \mathbb{N}$ and let b be a prime power. Then a $(0, s)$-sequence in base b exists if and only if $s \le b$.*

Furthermore, also the necessary condition from Theorem 5.7 is sufficient for prime power bases.

Corollary 5.38 *Let $s \in \mathbb{N}$ and let b be a prime power. Then for $m \ge 2$ a $(0, m, s)$-net in base b exists if and only if $s \le b + 1$.*

Proof Assume that $s \le b + 1$. Then Corollary 5.37 implies the existence of a $(0, s - 1)$-sequence in base b. From this we obtain with Lemma 5.17 a $(0, m, s)$-net in base b. The other direction is just Theorem 5.7. $\square$

Generalized Niederreiter Sequence

In 1993 S. Tezuka proposed a generalization of Niederreiter's sequence. This sequence differs from the Niederreiter sequence introduced above, in that x^k in (5.9) are replaced by polynomials $y_{j,i,k}(x)$, where $j \in \{1, \ldots, s\}$, $i \in \mathbb{N}$, and

$k \in \{0, \ldots, e_j - 1\}$. In order for Theorem 5.35 to apply to these sequences, for each $i \in \mathbb{N}$ and $j = 1, 2, \ldots, s$ the set of polynomials $\{y_{j,i,k}(x) : 0 \leq k < e_j\}$ needs to be linearly independent modulo $p_j(x)$ over $\mathbb{F}_b$. The generalized Niederreiter sequence is then defined by the expansion

$$\frac{y_{j,i,k}(x)}{p_j(x)^i} = \sum_{r=1}^{\infty} a^{(j)}(i, k, r)x^{-r}$$

over the field $\mathbb{F}_b((x^{-1}))$. We then define the matrix $C_j = (c_{i,r}^{(j)})_{i,r \in \mathbb{N}}$ by

$$c_{i,r}^{(j)} := a^{(j)}(Q + 1, k, r) \in \mathbb{F}_b \quad \text{for} \quad j \in \{1, \ldots, s\}, \; i, r \in \mathbb{N}, \tag{5.14}$$

where $i - 1 = Qe_j + k$, with integers $Q = Q(j, i)$ and $k = k(j, i)$ satisfying $0 \leq k < e_j$.

Definition 5.39 A digital sequence over $\mathbb{F}_b$ generated by the matrices $C_j = (c_{i,r}^{(j)})_{i,r \in \mathbb{N}}$ for $j = 1, 2, \ldots, s$ where the $c_{i,r}^{(j)}$ are given by (5.14) is called a *generalized Niederreiter sequence*.

The proof of Theorem 5.35 still applies, hence a generalized Niederreiter sequence is a digital (t, s)-sequence over $\mathbb{F}_b$ with $t = \sum_{j=1}^{s}(e_j - 1)$. Under suitable conditions this quality parameter can be shown to be strict.

Sobol' Sequence

In 1967 I.M. Sobol' was the first to introduce a construction of (t, s)-sequences. This sequence, nowadays referred to as *Sobol' sequence*, is the generalized Niederreiter sequence where $b = 2$, $p_1(x) = x$, and, for $j = 2, \ldots, s$, $p_j(x)$ is the $(j - 1)$th primitive polynomial in a sequence of all primitive polynomials over $\mathbb{F}_2$ arranged according to nondecreasing degrees. Further, there are polynomials $g_{j,0}, \ldots, g_{j,e_j-1}$ with $\deg(g_{j,h}) = e_j - h + 1$ such that $y_{j,i,k} = g_{j,k}$ for all $i \in \mathbb{N}, k \in \{0, \ldots, e_j - 1\}$, and $j \in \{1, \ldots, s\}$.

Remark 5.40 Recall that a polynomial $p \in \mathbb{F}_b[x]$ of degree $m \in \mathbb{N}$ is called a *primitive polynomial* over $\mathbb{F}_b$ if it is the minimal polynomial over $\mathbb{F}_b$ of a primitive element of $\mathbb{F}_{b^m}$. In other words $p \in \mathbb{F}_b[x]$ of degree $m \in \mathbb{N}$ is primitive if it is monic and irreducible over $\mathbb{F}_b$ and it has a root $\alpha \in \mathbb{F}_{b^m}$ that generates the multiplicative group $\mathbb{F}_{b^m}^*$ of $\mathbb{F}_{b^m}$.

A Sobol' sequence can also be generated in the following way: Let $p_1, \ldots, p_s \in \mathbb{F}_2[x]$ be primitive polynomials ordered according to their degree, and let

$$p_j(x) = x^{e_j} + a_{1,j}x^{e_j-1} + a_{2,j}x^{e_j-2} + \cdots + a_{e_j-1}x + 1 \quad \text{for } j = 1, 2, \ldots, s.$$

Choose odd natural numbers $m_{1,j}, \ldots, m_{e_j,j}$ such that $m_{k,j} < 2^k$ for $1 \le k \le e_j$, and for all $k > e_j$ define $m_{k,j}$ recursively by

$$m_{k,j} = 2a_{1,j}m_{k-1,j} \oplus \cdots \oplus 2^{e_j-1}a_{e_j-1}m_{k-e_j+1,j} \oplus 2^{e_j}m_{k-e_j,j} \oplus m_{k-e_j,j},$$

where $\oplus$ is the bit-by-bit exclusive-or operator. The numbers

$$v_{k,j} := \frac{m_{k,j}}{2^k}$$

are called *direction numbers*. Then for $n \in \mathbb{N}_0$ with base 2 expansion $n = n_0 + 2n_1 + \cdots + 2^{r-1}n_{r-1}$ we define

$$x_{n,j} := n_0 v_{1,j} \oplus n_1 v_{2,j} \oplus \cdots \oplus n_{r-1} v_{r,j} \quad \text{and} \quad \boldsymbol{x}_n := (x_{n,1}, \ldots, x_{n,s}).$$

The Sobol' sequence is then the sequence of points $(\boldsymbol{x}_n)_{n \in \mathbb{N}_0}$.

Faure Sequence

In 1982 H. Faure introduced a construction of $(0, s)$-sequences over prime fields $\mathbb{F}_b$ with $s \le b$. This shows once more that for prime bases b the necessary condition from Theorem 5.16 for the existence of $(0, s)$-sequences in base b is also sufficient. These sequences, nowadays referred to as *Faure sequences*, correspond to the case where the base b is a prime number such that $b \ge s$, $p_j(x) = x - j + 1$ for $j = 1, 2, \ldots, s$ and all $y_{i,j,k}(x) = 1$.

The generating matrices of Faure sequences can also be written down explicitly in terms of the *Pascal matrix*, which is given by

$$P = \begin{pmatrix} \binom{0}{0} & \binom{0}{1} & \binom{0}{2} & \cdots \\ \binom{1}{0} & \binom{1}{1} & \binom{1}{2} & \cdots \\ \binom{2}{0} & \binom{2}{1} & \binom{2}{2} & \cdots \\ \vdots & \vdots & \vdots & \ddots \end{pmatrix},$$

where we set $\binom{k}{l} = 0$ for $l > k$. The generating matrices $C_1, \ldots, C_s$ of the Faure sequence are now given by

$$C_j = (P^\top)^{j-1} \pmod{b} \quad \text{for } j = 1, 2, \ldots, s. \tag{5.15}$$

This yields $C_j = (c_{i,r}^{(j)})_{i,r \in \mathbb{N}}$, where

$$c_{i,r}^{(j)} = \begin{cases} 0 & \text{if } 1 \le r < i, \\ \binom{r-1}{i-1}(j-1)^{r-i} & \text{if } l \ge k, \end{cases}$$

where $\binom{r-1}{i-1}(j-1)^{r-i}$ is evaluated modulo b and where $0^0 := 1$ by convention.

Fig. 5.3 The first 1024 elements of the Faure sequence for $s = b = 2$

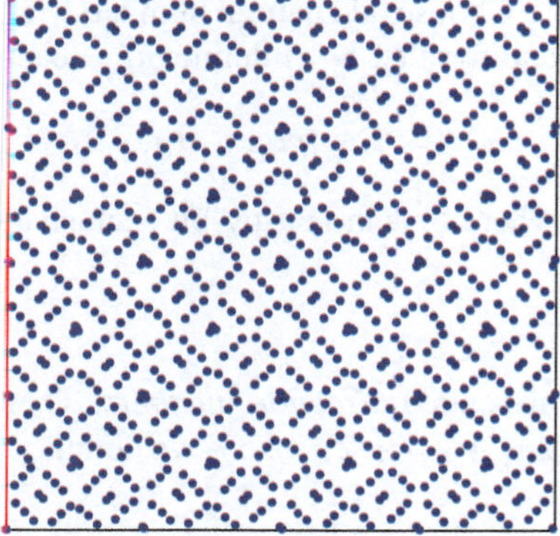

Example 5.41 In the case $s = b = 2$ we obtain the matrices

$$C_1 = \begin{pmatrix} 1 & 0 & 0 & 0 & \cdots \\ 0 & 1 & 0 & 0 & \cdots \\ 0 & 0 & 1 & 0 & \cdots \\ 0 & 0 & 0 & 1 & \cdots \\ \vdots & \vdots & \vdots & \vdots & \ddots \end{pmatrix} \in \mathbb{Z}_2^{\mathbb{N} \times \mathbb{N}}$$

and

$$C_2 = \begin{pmatrix} \binom{0}{0} & \binom{1}{0} & \binom{2}{0} & \binom{3}{0} & \cdots \\ 0 & \binom{1}{1} & \binom{2}{1} & \binom{3}{1} & \cdots \\ 0 & 0 & \binom{2}{2} & \binom{3}{2} & \cdots \\ 0 & 0 & 0 & \binom{3}{3} & \cdots \\ \vdots & \vdots & \ddots & \ddots & \ddots \end{pmatrix} = \begin{pmatrix} 1 & 1 & 1 & 1 & 1 & \cdots \\ 0 & 1 & 0 & 1 & 0 & \cdots \\ 0 & 0 & 1 & 1 & 0 & \cdots \\ 0 & 0 & 0 & 1 & 0 & \cdots \\ 0 & 0 & 0 & 0 & 1 & \cdots \\ \vdots & \vdots & \vdots & \vdots & \vdots & \ddots \end{pmatrix} \in \mathbb{Z}_2^{\mathbb{N} \times \mathbb{N}}.$$

The first 1024 points of the resulting Faure sequence are shown in Fig. 5.3.

5.5 Polynomial Lattice Point Sets

Polynomial lattice point sets are special types of (t, m, s)-nets, which are in their overall structure very similar to classical lattice point sets as discussed in Chap. 4. The main difference is that here one uses polynomial arithmetic over a finite field instead of the usual integer arithmetic.

Throughout this section, we assume that b is a prime number. Let $\mathbb{F}_b$ be the finite field with b elements, which we identify with the set $\{0, 1, \ldots, b-1\}$ equipped with arithmetic operations modulo b. By $\mathbb{F}_b[x]$ we denote the set of polynomials over $\mathbb{F}_b$. Define $\mathbb{F}_{b,m}[x] := \{h \in \mathbb{F}_b[x] : \deg(h) < m\}$. Obviously, $|\mathbb{F}_{b,m}[x]| = b^m$. As before, $\mathbb{F}_b((x^{-1}))$ denotes the field of formal Laurent series over $\mathbb{F}_b$, whose elements are of the form $L(x) = \sum_{\ell=w}^{\infty} t_\ell x^{-\ell}$, where w is an arbitrary integer and all $t_\ell \in \mathbb{F}_b$. Note that, although L is only a "formal" series, $L(b)$ does converge, i.e., is a real number, and the fractional part of $L(b)$ is given by $\sum_{\ell=\max(1,w)}^{\infty} t_\ell b^{-\ell}$ (save for the case where $t_\ell = b$ for all $\ell > 1$). For $m \in \mathbb{N}$ let $v_m : \mathbb{F}_b((x^{-1})) \to [0, 1)$ be defined by

$$v_m \left(\sum_{\ell=w}^{\infty} t_\ell x^{-\ell} \right) := \sum_{\ell=\max(1,w)}^{m} t_\ell b^{-\ell}.$$

In the following, we sometimes identify a non-negative integer h with b-adic expansion $h = h_0 + h_1 b + \cdots + h_a b^a$, with the polynomial $h(x) = h_0 + h_1 x + \cdots + h_a x^a \in \mathbb{F}_b[x]$ and vice versa. Further, for arbitrary $\boldsymbol{k} = (k_1, \ldots, k_s) \in \mathbb{F}_b[x]^s$ and $\boldsymbol{q} = (q_1, \ldots, q_s) \in \mathbb{F}_b[x]^s$, we define the "inner product"

$$\boldsymbol{k} \cdot \boldsymbol{q} := \sum_{j=1}^{s} k_j q_j \in \mathbb{F}_b[x]$$

and we write $q \equiv 0 \pmod{p}$ if p divides q in $\mathbb{F}_b[x]$.

Now we state the formal definition of polynomial lattice point sets and polynomial lattice rules.

Definition 5.42 For $s, m \in \mathbb{N}$, $p \in \mathbb{F}_b[x]$, with $\deg(p) = m$, and $\boldsymbol{q} \in \mathbb{F}_b[x]^s$ the point set $\mathcal{P}(\boldsymbol{q}, p)$ consisting of the b^m elements

$$\boldsymbol{x}_h := v_m \left(\frac{h(x)}{p(x)} \boldsymbol{q}(x) \right) \quad \text{for all } h \in \mathbb{F}_{b,m}[x]$$

is called a *polynomial lattice point set*. A QMC rule using $\mathcal{P}(\boldsymbol{q}, p)$ as underlying node set is called a *polynomial lattice rule*.

When one compares Definitions 4.1 and 5.42, the analogy between the construction of lattice point sets and polynomial lattice point sets becomes evident. But also the differences become evident. In any case, if one chooses $b = N$, $m = 1$, and $p(x) = x$, then one obtains a lattice point set in the classical sense.

Example 5.43 The polynomial lattice point set $\mathcal{P}(\boldsymbol{q}, p)$ with $p(x) = x^4 + x^2 + 1$ and $\boldsymbol{q} = (1, x^3)$ over $\mathbb{F}_2$ is shown in Fig. 5.4.

Polynomial Lattice Point Sets Considered as Digital Nets

While classical lattice point sets show a very regular lattice structure, such geometry cannot be observed for polynomial lattice point sets. Nevertheless, also polynomial

Fig. 5.4 The polynomial
lattice point set $\mathcal{P}(\mathbf{q}, p)$ with
$p(x) = x^4 + x^2 + 1$ and
$\mathbf{q} = (1, x^3)$ over $\mathbb{F}_2$

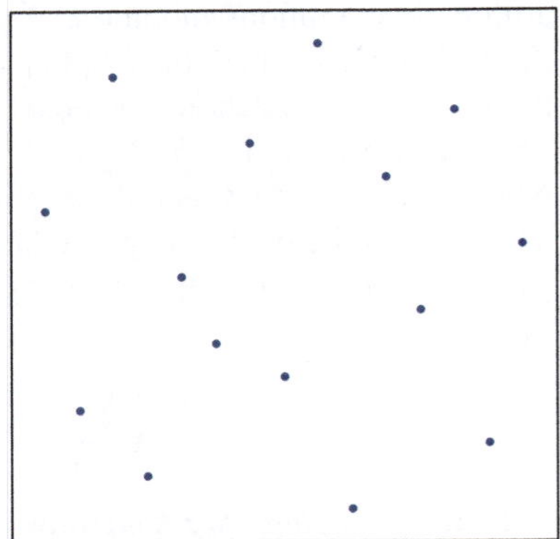

lattice point sets have some inherent structure, namely, they form digital (t, m, s)-nets over $\mathbb{F}_b$, which also constitutes a nice geometric property.

Let us start with a very helpful lemma.

Lemma 5.44 *For $s, m \in \mathbb{N}$, $p \in \mathbb{F}_b[x]$, with $\deg(p) = m$, and $q \in \mathbb{F}_b[x]$ consider the Laurent series expansions of the rational function*

$$\frac{q(x)}{p(x)} = \sum_{\ell=w}^{\infty} u_\ell x^{-\ell} \in \mathbb{F}_b((x^{-1}))$$

and let $h \in \mathbb{F}_{b,m}[x]$ with $h(x) = h_0 + h_1 x + \cdots + h_{m-1} x^{m-1}$. Then for $i \in \mathbb{N}$ the coefficient of x^{-i} in $\frac{h(x)q(x)}{p(x)}$ equals $\sum_{r=0}^{m-1} u_{r+i} h_r$ (using arithmetic in $\mathbb{F}_b$).

Proof We calculate

$$\frac{h(x)q(x)}{p(x)} = \left(\sum_{r=0}^{m-1} h_r x^r\right)\left(\sum_{\ell=w}^{\infty} u_\ell x^{-\ell}\right) = \sum_{r=0}^{m-1} h_r \sum_{\ell=w}^{\infty} u_\ell x^{r-\ell}$$

$$= \sum_{r=0}^{m-1} h_r \sum_{i=w-r}^{\infty} u_{r+i} x^{-i}$$

$$= \sum_{r=0}^{m-1} h_r \sum_{i=w-r}^{0} u_{r+i} x^{-i} + \sum_{r=0}^{m-1} h_r \sum_{i=1}^{\infty} u_{r+i} x^{-i}$$

$$= \sum_{r=0}^{m-1} h_r \sum_{i=w-r}^{0} u_{r+i} x^{-i} + \sum_{i=1}^{\infty} \left(\sum_{r=0}^{m-1} h_r u_{r+i}\right) x^{-i},$$

where the sum $\sum_{i=w-r}^{0}$ is considered as an empty sum, and therefore equals 0, whenever $w - r > 0$. This gives the result. $\qquad\square$

Proposition 5.45 *For $s, m \in \mathbb{N}$, $p \in \mathbb{F}_b[x]$, with $\deg(p) = m$, and $\mathbf{q} = (q_1, \ldots, q_s) \in \mathbb{F}_b[x]^s$ consider the Laurent series expansions of the rational functions*

$$\frac{q_j(x)}{p(x)} = \sum_{\ell=w_j}^{\infty} u_\ell^{(j)} x^{-\ell} \in \mathbb{F}_b((x^{-1})).$$

Then the polynomial lattice point set $\mathcal{P}(\mathbf{q}, p)$ is a digital net over $\mathbb{F}_b$ with generating $m \times m$ matrices $C_1, \ldots, C_s$, where for $j = 1, \ldots, s$ the elements $c_{i,r+1}^{(j)}$ of the matrix C_j are given by

$$c_{i,r+1}^{(j)} = u_{r+i}^{(j)} \in \mathbb{F}_b \quad \text{for } i = 1, \ldots, m \text{ and } r = 0, \ldots, m-1.$$

That is, the generating matrices are of the form:

$$C_j = \begin{pmatrix} u_1^{(j)} & u_2^{(j)} & \cdots & u_m^{(j)} \\ u_2^{(j)} & u_3^{(j)} & \cdots & u_{m+1}^{(j)} \\ \vdots & \vdots & & \vdots \\ u_m^{(j)} & u_{m+1}^{(j)} & \cdots & u_{2m-1}^{(j)} \end{pmatrix}. \tag{5.16}$$

Remark 5.46 A matrix of the form (5.16) is a so-called *Hankel matrix* associated with the sequence $(u_1^{(j)}, u_2^{(j)}, \ldots)$. It can be shown that if $\gcd(q_j, p) = 1$, then the matrix C_j is non-singular.

Proof of Proposition 5.45. Let $q, p \in \mathbb{F}_b[x]$ with $\deg(p) = m$ and write $\frac{q(x)}{p(x)} = \sum_{\ell=w}^{\infty} u_\ell x^{-\ell} \in \mathbb{F}_b((x^{-1}))$. Then for $h \in \mathbb{F}_b[x]$ with $h(x) = h_0 + h_1 x + \cdots + h_{m-1} x^{m-1}$ we have

$$v_m\left(\frac{h(x)q(x)}{p(x)}\right) = \sum_{i=1}^{m} b^{-i} \sum_{r=0}^{m-1} h_r u_{r+i},$$

by Lemma 5.44. But

$$\sum_{r=0}^{m-1} h_r u_{r+i} = (u_i, \ldots, u_{i+m-1}) \cdot \vec{h},$$

where $\vec{h} = (h_0, \ldots, h_{m-1})^\top$ is the column vector formed from the coefficients of h. From this the result follows. $\qquad\square$

Knowing that polynomial lattice point sets are digital nets, the question about their quality parameter arises. Before we answer this question we need an analog to the dual lattice of classical lattice point sets.

Definition 5.47 The *dual net* of the polynomial lattice point set $\mathcal{P}(q, p)$ from Definition 5.42 is defined as

$$\mathcal{D}_{q,p} := \{h \in \mathbb{F}_{b,m}[x]^s \ : \ h \cdot q \equiv 0 \pmod{p}\}.$$

The quantity introduced in the next definition is needed for the determination of the quality parameter of a polynomial lattice point set.

Definition 5.48 For $s, m \in \mathbb{N}$, $p \in \mathbb{F}_b[x]$ with $\deg(p) = m$, and $q \in \mathbb{F}_b[x]^s$ the *figure of merit* $\rho(q, p)$ is defined as

$$\rho(q, p) := s - 1 + \min_{h \in \mathcal{D}_{q,p} \setminus \{0\}} \sum_{i=1}^{s} \deg(h_i),$$

where $h = (h_1, \ldots, h_s)$.

The figure of merit $\rho(q, p)$ is closely related to the quality parameter t of $\mathcal{P}(q, p)$ considered as a (t, m, s)-net over $\mathbb{F}_b$.

Theorem 5.49 *For $s, m \in \mathbb{N}$, $p \in \mathbb{F}_b[x]$ with $\deg(p) = m$ and $q \in \mathbb{F}_b[x]^s$ the polynomial lattice point set $\mathcal{P}(q, p)$ is a strict digital (t, m, s)-net over $\mathbb{F}_b$ with $t = m - \rho(q, p)$.*

For the proof of this result we need a further lemma, which relates the elements from the dual net of a polynomial lattice point set to the solution of a linear system involving the generating matrices of the polynomial lattice point set represented as a digital net.

Lemma 5.50 *For $s, m \in \mathbb{N}$ let $C_1, \ldots, C_s$ be the generating $m \times m$ matrices of a polynomial lattice point set $\mathcal{P}(q, p)$ as given in Definition 5.42. Then for $k = (k_1, \ldots, k_s) \in \{0, \ldots, b^m - 1\}^s$ with b-adic digit vectors $\vec{k}_1, \ldots, \vec{k}_s$ we have*

$$C_1^\top \vec{k}_1 + \cdots + C_s^\top \vec{k}_s = \vec{0}, \tag{5.17}$$

where $\vec{0} = (0, \ldots, 0)^\top$, if and only if

$$k \cdot q \equiv 0 \pmod{p},$$

where in the last expression k is the associated vector of polynomials in $\mathbb{F}_b[x]$.

Proof For $j = 1, \ldots, s$ let $\frac{q_j(x)}{p(x)} = \sum_{\ell=w_j}^{\infty} u_\ell^{(j)} x^{-\ell} \in \mathbb{F}_b((x^{-1}))$ and $k_j(x) = \kappa_{j,0} + \kappa_{j,1} x^1 + \ldots + \kappa_{j,m-1} x^{m-1} \in \mathbb{F}_{b,m}[x]$. For $i \in \mathbb{N}$ the coefficient of x^{-i} in $\frac{k_j(x) q_j(x)}{p(x)}$ is $\sum_{r=0}^{m-1} u_{r+i}^{(j)} \kappa_{j,r}$ by Lemma 5.44. Summing up, we obtain that for $i \in \mathbb{N}$

the coefficient of x^{-i} in $\frac{1}{p}\mathbf{k}\cdot\mathbf{q}$ is given by $\sum_{j=1}^{s}\sum_{r=0}^{m-1}u_{r+i}^{(j)}\kappa_{j,r}$. However, condition (5.17) is equivalent to

$$\sum_{j=1}^{s}\sum_{r=0}^{m-1}u_{r+i}^{(j)}\kappa_{j,r} = 0 \in \mathbb{F}_b$$

for all $i \in \{1,\ldots,m\}$. Therefore we obtain

$$\frac{1}{p}\mathbf{k}\cdot\mathbf{q} = g + L$$

for some $g \in \mathbb{F}_b[x]$ and for some $L \in \mathbb{F}_b((x^{-1}))$ of the form $\sum_{\ell=m+1}^{\infty}f_\ell x^{-\ell}$. Equivalently, we have $\mathbf{k}\cdot\mathbf{q} - gp = Lp$. Here, on the left-hand side, we have a polynomial over $\mathbb{F}_b$, whereas on the right-hand side we have a Laurent series Lp which is of the form

$$(Lp)(x) = \sum_{\ell=w}^{\infty}v_\ell x^{-\ell}$$

with $w > 0$, since $\deg(p) = m$. However, this is only possible if $Lp = 0$, which means that $\mathbf{k}\cdot\mathbf{q} - gp = 0$ or equivalently $\mathbf{k}\cdot\mathbf{q} \equiv 0 \pmod{p}$. $\qquad\square$

We are now in a position to give the proof of Theorem 5.49.

Proof of Theorem 5.49 It suffices to show that we have $\rho(C_1,\ldots,C_s) = \rho(\mathbf{q},p)$, where $\rho(C_1,\ldots,C_s)$ is the linear independence parameter from Definition 5.26. The result then follows from Theorem 5.28.

According to the definition of $\rho(C_1,\ldots,C_s)$ there exist $d_1,\ldots,d_s \in \mathbb{N}_0$ with $d_1 + \cdots + d_s = \rho(C_1,\ldots,C_s) + 1$ such that the system consisting of the union of the first d_j row vectors $\vec{c}_1^{(j)},\ldots,\vec{c}_{d_j}^{(j)}$ of the matrix C_j, where $j = 1,\ldots,s$, is linearly dependent over $\mathbb{F}_b$. That is, there exist $\kappa_{j,r} \in \mathbb{F}_b$ for $r = 0,\ldots,d_j - 1$, $j = 1,\ldots,s$, not all of them zero, such that

$$\sum_{j=1}^{s}\sum_{r=0}^{d_j-1}\kappa_{j,r}\vec{c}_{r+1}^{(j)} = \vec{0} \in \mathbb{F}_b^m. \tag{5.18}$$

Putting $\kappa_{j,r} = 0$ for $r = d_j,\ldots,m-1$ and $j = 1,\ldots,s$ and $k_j = \kappa_{j,0} + \kappa_{j,1}b + \cdots + \kappa_{j,m-1}b^{m-1}$ and correspondingly $\vec{k}_j = (\kappa_{j,0},\ldots,\kappa_{j,m-1})^\top$ for $j = 1,\ldots,s$, we may write (5.18) equivalently as

$$C_1^\top\vec{k}_1 + \cdots + C_s^\top\vec{k}_s = \vec{0}.$$

By Lemma 5.50 this is equivalent to $\mathbf{k}\cdot\mathbf{q} \equiv 0 \pmod{p}$ for $\mathbf{k} = (k_1,\ldots,k_s) \in \mathbb{F}_b[x]^s \setminus \{\mathbf{0}\}$ where $k_j(x) = \kappa_{j,0} + \kappa_{j,1} + \cdots + \kappa_{j,m-1}x^{m-1} \in \mathbb{F}_b[x]$. Hence, from the definition of $\rho(\boldsymbol{q}, p)$ we obtain

$$\rho(\boldsymbol{q}, p) \leq s - 1 + \sum_{j=1}^{s} \deg(k_j) \leq s - 1 + \sum_{j=1}^{s} (d_j - 1) = \rho(C_1, \ldots, C_s).$$

On the other hand, from the definition of $\rho(\boldsymbol{q}, p)$ we find that there exists a nonzero $\boldsymbol{k} = (k_1, \ldots, k_s) \in \mathbb{F}_b[x]^s$ with $\deg(k_j) < m$ for $j = 1, \ldots, s$ and $\boldsymbol{k} \cdot \boldsymbol{q} \equiv 0 \pmod{p}$ such that $\rho(\boldsymbol{q}, p) = s - 1 + \sum_{j=1}^{s} \deg(k_j)$. From Lemma 5.50 we obtain $C_1^{\top} \vec{k}_1 + \cdots + C_s^{\top} \vec{k}_s = \vec{0}$, where each column vector $\vec{k}_j$ is determined by the coefficients of the polynomials k_j, $j = 1, \ldots, s$. For $j = 1, \ldots, s$ let $d_j = \deg(k_j) + 1$. Then the system consisting of the union of the first d_j row vectors of C_j, $j = 1, \ldots, s$, is linearly dependent over $\mathbb{F}_b$. Hence

$$\rho(C_1, \ldots, C_s) \leq -1 + \sum_{j=1}^{s} d_j = -1 + \sum_{j=1}^{s} (\deg(k_j) + 1) = \rho(\boldsymbol{q}, p).$$

Both inequalities together imply $\rho(C_1, \ldots, C_s) = \rho(\boldsymbol{q}, p)$ and this finishes the proof. $\qquad\square$

Example 5.51 For the polynomial lattice point set $\mathcal{P}(\boldsymbol{q}, p)$ from Example 5.43, i.e., $p(x) = x^4 + x^2 + 1$ and $\boldsymbol{q}(x) = (1, x^3)$, it can easily be shown with a computer algebra system that the elements in $\mathcal{D}_{\boldsymbol{q},p} \setminus \{\boldsymbol{0}\}$ are $(1, x^3)$, $(x, 1 + x^2)$, $(1 + x, 1 + x^2 + x^3)$, $(x^2, x + x^3)$, $(1 + x^2, x)$, $(x + x^2, 1 + x + x^2 + x^3)$, $(1 + x + x^2, 1 + x + x^2)$, $(x^3, 1)$, $(1 + x^3, 1 + x^3)$, $(x + x^3, x^2)$, $(1 + x + x^3, x^2 + x^3)$, $(x^2 + x^3, 1 + x + x^3)$, $(1 + x^2 + x^3, 1 + x)$, $(x + x^2 + x^3, x + x^2 + x^3)$, $(1 + x + x^2 + x^3, x + x^2)$. This yields $\rho(\boldsymbol{q}, p) = 4$. According to Theorem 5.49 $\mathcal{P}(\boldsymbol{q}, p)$ is therefore a digital $(0, 4, 2)$-net over $\mathbb{F}_2$. Indeed, we observe from the pictures in Fig. 5.5 that every 2-adic elementary interval of area 2^{-4} contains exactly one element of the point set $\mathcal{P}(\boldsymbol{q}, p)$.

Notes on the Discrepancy of $\mathcal{P}(\boldsymbol{q}, p)$

The analysis of the discrepancy of polynomial lattice point sets follows a parallel track to the study of discrepancy of classical lattice point sets as discussed in Sects. 4.1 and 4.3. This is briefly illustrated in the following paragraphs, but without going into technical details.

For $p \in \mathbb{F}_b[x]$ with $\deg(p) = m$ and for $\boldsymbol{q} \in \mathbb{F}_{b,m}[x]^s$ define the quality measure

$$T_b(\boldsymbol{q}, p) := \sum_{\boldsymbol{h} \in \mathcal{D}_{\boldsymbol{q},p} \setminus \{\boldsymbol{0}\}} \frac{1}{\varrho_b(\boldsymbol{h})},$$

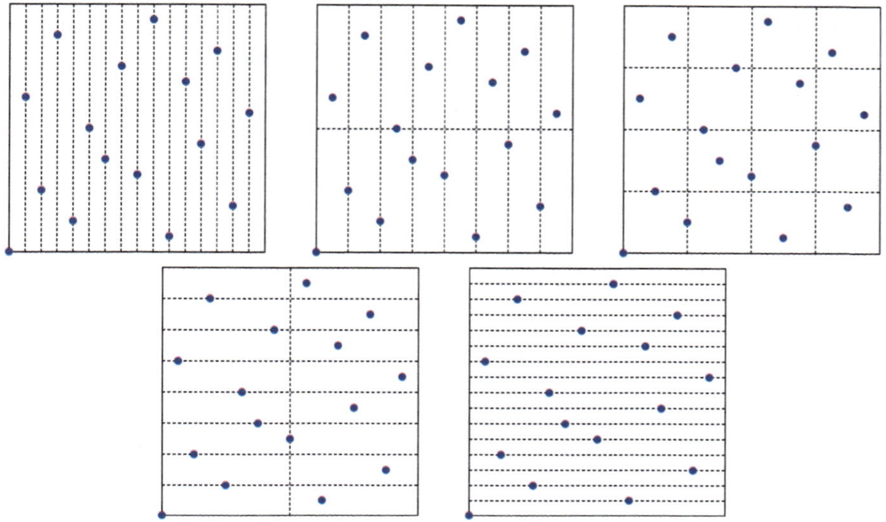

Fig. 5.5 The polynomial lattice point set $\mathcal{P}(q, p)$ from Fig. 5.4 as digital $(0, 4, 2)$-net over $\mathbb{F}_2$; every 2-adic elementary interval of area 2^{-4} contains exactly one point (recall that elementary intervals are left-closed and right-open)

where for $h \in \mathbb{F}_b[x]$,

$$\varrho_b(h) := \begin{cases} 1 & \text{if } h = 0, \\ b^{r+1} \sin^2(\frac{\pi \kappa_r}{b}) & \text{if } h(x) = \kappa_0 + \kappa_1 x + \cdots + \kappa_r x^r \\ & \quad \text{with } r \in \mathbb{N}_0 \text{ and } \kappa_r \neq 0, \end{cases}$$

and $\varrho_b(\boldsymbol{h}) := \prod_{j=1}^{s} \varrho_b(h_j)$ for $\boldsymbol{h} = (h_1, \ldots, h_s) \in \mathbb{F}_b[x]^s$. Likewise, define the quality measure

$$\widetilde{T}_{b,\boldsymbol{\gamma}}(q, p) := \sum_{\boldsymbol{h} \in \mathcal{D}_{q,p} \setminus \{0\}} \frac{1}{\widetilde{\varrho}_{b,\boldsymbol{\gamma}}(\boldsymbol{h})},$$

where, for a generic $\gamma > 0$ and $h \in \mathbb{F}_b[x]$,

$$\widetilde{\varrho}_{b,\gamma}(h) := \begin{cases} (1 + \gamma)^{-1} & \text{if } h = 0, \\ \varrho_b(h)\gamma^{-1} & \text{if } h \neq 0, \end{cases}$$

and $\widetilde{\varrho}_{b,\boldsymbol{\gamma}}(\boldsymbol{h}) := \prod_{j=1}^{s} \widetilde{\varrho}_{b,\gamma_j}(h_j)$ for $\boldsymbol{h} = (h_1, \ldots, h_s) \in \mathbb{F}_b[x]^s$ and a sequence $(\gamma_j)_{j \geq 1}$ of coordinate weights.

These quantities can be used to bound the (weighted) star discrepancy of polynomial lattice point sets in an analog fashion like $R_N(\boldsymbol{g})$ and $R_{N,\boldsymbol{\gamma}}(\boldsymbol{g})$ were used to bound the (weighted) star discrepancy of classical lattice point sets in Chap. 4.

Proposition 5.52 *For $s, m \in \mathbb{N}$, $p \in \mathbb{F}_b[x]$ with $\deg(p) = m$ and $q \in \mathbb{F}_b[x]^s$ the star discrepancy of the polynomial lattice point set $\mathcal{P}(q, p)$ satisfies*

$$D^*_{b^m}(\mathcal{P}(q, p)) \le \frac{s}{b^m} + T_b(q, p),$$

and the weighted star discrepancy, for product weights with coordinate weights $(\gamma_j)_{j \ge 1}$, satisfies

$$D^*_{b^m, \boldsymbol{\gamma}}(\mathcal{P}(q, p)) \le \frac{1}{b^m} \prod_{j=1}^{s} (1 + 2\gamma_j) + \widetilde{T}_{b, \boldsymbol{\gamma}}(q, p).$$

A proof of the bound on the star discrepancy is very similar to the proof of Proposition 4.3 for classical lattice point sets, which relies on the general discrepancy estimate in terms of exponential sums from Theorem 2.30. Quite analogously, one can show a general discrepancy bound also in terms of Walsh functions (see Exercise 3.5 for the definition of Walsh functions). This was first done by Hellekalek [55, Theorem 1]. Using this result and similar reasoning as in the proof of Propsition 4.3, the present bound on the star discrepancy of polynomial lattice point sets can be deduced. The proof of the bound on the weighted star discrepancy can be deduced from the one for the star discrepancy like in the proof of Proposition 4.12. We omit the details here.

The quality measures $T_b(q, p)$ and $\widetilde{T}_{b, \boldsymbol{\gamma}}(q, p)$ can be computed very efficiently. Indeed, if $\mathcal{P}(q, p) = \{x_0, \dots, x_{b^m - 1}\}$ where $x_n = (x_{n,1}, \dots, x_{n,s})$, then we have

$$T_b(q, p) = -1 + \frac{1}{b^m} \sum_{n=0}^{N-1} \prod_{j=1}^{s} \theta_{b,m}(x_{n,j})$$

and

$$\widetilde{T}_{b, \boldsymbol{\gamma}}(q, p) = - \prod_{j=1}^{s} (1 + \gamma_j) + \frac{1}{b^m} \sum_{n=0}^{N-1} \prod_{j=1}^{s} (1 + \gamma_j \theta_{b,m}(x_{n,j})),$$

where for $x = \sum_{i=1}^{\infty} \xi_i b^{-i}$ with b-adic digits $\xi_i \in \{0, \dots, b-1\}$

$$\theta_{b,m}(x) := \begin{cases} 1 + i_0 \frac{b^2 - 1}{3b} + \frac{2}{b} \xi_{i_0}(\xi_{i_0} - b) & \text{if } \xi_1 = \dots = \xi_{i_0 - 1} = 0 \text{ and} \\ & \qquad \xi_{i_0} \ne 0 \text{ with } 1 \le i_0 \le m, \\ 1 + m \frac{b^2 - 1}{3b} & \text{otherwise.} \end{cases}$$

Using a CBC algorithm with T_b or $\widetilde{T}_{b, \boldsymbol{\gamma}}$ as underlying quality measure one can construct polynomial lattice point sets with low discrepancy.

Algorithm 5.53 *(CBC algorithm)* Let $s, m \in \mathbb{N}$ and let $p \in \mathbb{F}_b[x]$ with $\deg(p) = m$.

1. Choose $q_1 = 1$.

2. For $d = 1, 2, \ldots, s - 1$, choose $q_{d+1} \in \mathbb{F}_{b,m}[x]$ to minimize $T_b((q_1, \ldots, q_d, q), p)$ as a function of $q \in \mathbb{F}_{b,m}[x]$.

Likewise, one can employ $\widetilde{T}_{b,\boldsymbol{\gamma}}$ as criterion in Algorithm 5.53.

Theorem 5.54 *Let $s, m \in \mathbb{N}$ and let $p \in \mathbb{F}_b[x]$ be irreducible with $\deg(p) = m$. If $\boldsymbol{q} = (q_1, \ldots, q_s)$ is constructed using Algorithm 5.53 using T_b (or $\widetilde{T}_{b,\boldsymbol{\gamma}}$ in the weighted case), then we have*

$$T_b(\boldsymbol{q}, p) \le \frac{1}{b^m - 1} \left(1 + m \frac{b^2 - 1}{3b} \right)^s,$$

and

$$\widetilde{T}_{b,\boldsymbol{\gamma}}(\boldsymbol{q}_d, p) \le \frac{1}{b^m - 1} \prod_{j=1}^{s} \left(1 + \gamma_j \left(1 + m \frac{b^2 - 1}{3b} \right) \right),$$

respectively.

From Proposition 5.52 and Theorem 5.54 we obtain the following result.

Corollary 5.55 *Let $s, m \in \mathbb{N}$ and let $p \in \mathbb{F}_b[x]$ be irreducible with $\deg(p) = m$. If $\boldsymbol{q} = (q_1, \ldots, q_s)$ is constructed using Algorithm 5.53 using T_b (or $\widetilde{T}_{b,\boldsymbol{\gamma}}$ in the weighted case), then we have*

$$D_{b^m}^*(\mathcal{P}(\boldsymbol{q}, p)) \le \frac{s}{b^m} + \frac{1}{b^m - 1} \left(1 + m \frac{b^2 - 1}{3b} \right)^s,$$

and

$$D_{b^m, \boldsymbol{\gamma}}^*(\mathcal{P}(\boldsymbol{q}, p)) \le \frac{1}{b^m} \prod_{j=1}^{s} (1 + 2\gamma_j) + \frac{1}{b^m - 1} \prod_{j=1}^{d} \left(1 + \gamma_j \left(1 + m \frac{b^2 - 1}{3b} \right) \right),$$

respectively.

Like in Corollary 4.16 for classical lattice point sets one can deduce the following result.

Corollary 5.56 *Let $s, m \in \mathbb{N}$ and $p \in \mathbb{F}_b[x]$ be irreducible with $\deg(p) = m$. Assume that the coordinate weights $(\gamma_j)_{j \ge 1}$ are summable, i.e., $\sum_{j=1}^{\infty} \gamma_j < \infty$. Then for every $\delta > 0$ there exists a quantity $C_{\delta, \boldsymbol{\gamma}} > 0$, independent of s and m, such that the weighted star discrepancy of the polynomial lattice point set $\mathcal{P}(\boldsymbol{q}, p)$ whose generating vector $\boldsymbol{q}$ is constructed using Algorithm 5.53 satisfies*

$$D_{b^m, \boldsymbol{\gamma}}^*(\mathcal{P}(\boldsymbol{q}, p)) \le \frac{C_{\delta, \boldsymbol{\gamma}}}{b^{m(1-\delta)}}.$$

Again, $\delta > 0$ can be chosen arbitrarily close to 0 and hence the convergence rate in $N = b^m$ is essentially best possible. Furthermore, the bound holds uniformly in s.

Further Reading

The theory of (digital) (t, m, s)-nets and (t, s)-sequences, respectively, was developed by Niederreiter in 1980s. The original paper is [108]. The early development is very well presented in Niederreiter's book [110]. A more current reference is the book by Dick and Pillichshammer [33] which, besides a detailed introduction, also describes the development since the early 1990s. An overview over digital nets and sequences worth reading is the article of Larcher [91]. An introduction into the theory of finite fields can be found in the book by Lidl and Niederreiter [104].

For the proof of the result in Remark 5.9, item 6, we refer to [111] for the upper bound on $t_b(s)$ and to [112] for the lower bound. The currently best asymptotic discrepancy bound for (t, m, s)-nets and (t, s)-sequences was shown by Faure and Kritzer [43]. Lower bounds on the discrepancy of (t, s)-sequences are discussed in [103]. For the definition of digital sequences in the broad sense we refer to the book [113] by Niederreiter and Xing. First examples of digital sequences were already introduced, some years before Niederreiter's unifying approach, by Sobol' in [139], called binary LP_τ-sequences, and by Faure [42]. The currently best constructions for digital nets and sequences with respect to the quality parameter are based on methods from Algebraic Geometry. These constructions were developed in a series of papers by Niederreiter and Xing. For an introduction into this subject and an overview we refer to the book by Niederreiter and Xing [113, Chap. 8]. Section 5.4 is based on parts of [33, Chap. 8] to which we refer for further constructions of digital sequences such as the ones by Niederreiter and Xing. A discussion of the quality parameter t of the digital sequences in Sect. 5.4 can be found in the paper [31] by Dick and Niederreiter and a proof for the bound (5.13) on the quality parameter of Niederreiter sequences can be found in [110, Theorem 4.54]. Polynomial lattice point sets have been introduced by Niederreiter [109]. In our exposition we only discussed polynomial lattice point sets in prime base b for simplicity. For the more general case of prime power bases we refer to [33, 110]. For a survey about polynomial lattice point sets and a comparison to classical lattice point sets we refer to [127]. The results about star discrepancy of polynomial lattice point sets were first shown in [30]. An overview and full proofs can also be found in [33, Chap. 10]. Here we formulated the results on discrepancy only for irreducible polynomials p, which is convenient for technical reasons, but in principle not necessary. For the general case see [25]. For the star discrepancy of polynomial lattice point sets an improved existence result is known. For prime b and for every monic $p \in \mathbb{F}_b[x]$ of degree $\deg(p) = m$ satisfying $p(x) = x^m$ (see [90]) or $\gcd(p, x) = 1$ (see [84]) there exists a $\boldsymbol{q} \in \mathbb{F}_{b,m}[x]^s$ such that $D^*_{b^m}(\mathcal{P}(\boldsymbol{q}, p)) = O(m^{s-1}(\log m)/b^m)$. This improves the order $O(m^s/b^m)$ in Corollary 5.55. Polynomial lattice rules can be also successfully employed for the integration in Walsh and Sobolev spaces, see [27,33]. For the fast CBC construction of polynomial lattice rules see [122,124].

An online database for optimal parameters of (t, m, s)-nets and (t, s)-sequences called "MinT" (see [132,133]) is available under: http://mint.sbg.ac.at/index.php.

Exercises

5.1 Construct "by hand" a $(0, 2, 2)$-net in base 3.

5.2 Let $b \in \mathbb{N}$, $b \geq 2$. Show that for any $m, s \in \mathbb{N}$, $s \geq 2$ and $m \geq 2$ there exists an $(m - 1, m, s)$-net in base b.

5.3 For $j = 1, \ldots, r$ let $\mathcal{P}_j$ be a (t_j, m_j, s)-net in base b and let $b^{m_1} + \cdots + b^{m_r} = b^m$. Show that the point set $\mathcal{P}$ consisting of the elements of $\mathcal{P}_1, \ldots, \mathcal{P}_r$ is a (t, m, s)-net in base b with quality parameter

$$t = m - \min_{j=1,2,\ldots,r} (m_j - t_j).$$

5.4 Let $b, k \geq 2$ be integers. Show that every s-dimensional elementary interval in base b^k is also an s-dimensional elementary interval in base b.

5.5 Show that every $(t, \mu k, s)$-net in base b is also a $(\lceil t/k \rceil, \mu, s)$-net in base b^k. *Hint*: Use Exercise 5.5.

5.6 Deduce the assertion of Theorem 5.18 for the special case $s = 1$ and $t = 0$ from Proposition 2.31.

5.7 Construct the finite field $\mathbb{F}_{25}$ with 25 elements and determine the operation tables for addition and multiplication.

5.8 Show that the 4×4 matrices

$$C_1 = \begin{pmatrix} 1 & 0 & 0 & 0 \\ 0 & 1 & 0 & 0 \\ 0 & 0 & 1 & 0 \\ 0 & 0 & 0 & 1 \end{pmatrix}, \quad C_2 = \begin{pmatrix} 0 & 0 & 0 & 1 \\ 0 & 0 & 1 & 0 \\ 0 & 1 & 0 & 0 \\ 1 & 0 & 0 & 0 \end{pmatrix},$$

$$C_3 = \begin{pmatrix} 1 & 1 & 1 & 1 \\ 0 & 1 & 0 & 1 \\ 0 & 0 & 1 & 1 \\ 0 & 0 & 0 & 1 \end{pmatrix}, \quad C_4 = \begin{pmatrix} 0 & 1 & 1 & 0 \\ 1 & 1 & 0 & 1 \\ 0 & 0 & 0 & 1 \\ 0 & 0 & 1 & 0 \end{pmatrix},$$

over $\mathbb{F}_2$ generate a digital $(1, 4, 4)$-net over $\mathbb{F}_2$.

5.9 Let b be a prime power and let the $m \times m$ matrices $C_1, \ldots, C_s$ over $\mathbb{F}_b$ be the generator matrices of a (strict) digital (t, m, s)-net over $\mathbb{F}_b$. Let Z be a non-singular $m \times m$ matrix over $\mathbb{F}_b$. Show that the matrices $C_1', \ldots, C_s'$ where $C_j' := C_j Z$ for $j = 1, 2, \ldots, s$ generate a (strict) digital (t, m, s)-net over $\mathbb{F}_b$, namely, the same net but with a different ordering of points.

5.10 Let b be a prime power and $\varphi : \{0, \ldots, b - 1\} \to \mathbb{F}_b$ be a bijection with $\varphi(0) = \bar{0}$. For $x = \sum_{i=1}^{\infty} \frac{\xi_i}{b^i} \in [0, 1)$ and $\sigma = \sum_{i=1}^{\infty} \frac{\varsigma_i}{b^i} \in [0, 1)$, where $\xi_i, \varsigma_i \in \{0, \ldots, b - 1\}$, we define the ($b$-adic) *digital shifted point* y by $y =$

$x \oplus_{b,\varphi} \sigma := \sum_{i=1}^{\infty} \frac{\eta_i}{b^i}$ where $\eta_i = \varphi^{-1}(\varphi(\xi_i) + \varphi(\varsigma_i))$, and where the "$+$" is addition in $\mathbb{F}_b$.

For dimensions $s > 1$ let $\sigma = (\sigma_1, \ldots, \sigma_s) \in [0,1)^s$. For $x = (x_1, \ldots, x_s) \in [0,1)^s$ we define the (b-adic) *digital shifted point* y by $y = x \oplus_{b,\varphi} \sigma = (x_1 \oplus_{b,\varphi} \sigma_1, \ldots, x_s \oplus_{b,\varphi} \sigma_s)$.

Let $\{x_0, \ldots, x_{b^m-1}\}$ be a (strict) (t, m, s)-net in base b and let $\sigma = (\sigma_1, \ldots, \sigma_s) \in [0,1)^s$. Show that the digitally shifted point set formed by the points $y_n = x_n \oplus_{b,\varphi} \sigma$ for $n = 0, \ldots, b^m - 1$, is again a (strict) (t, m, s)-net in base b with probability one with respect to the Lebesgue measure of σ's. (If the σ_i's have only finitely many b-adic digits different from zero, then the assertion is always true.) *Hint:* This is [33, Lemma 4.67].

5.11 Assume that $\mathcal{P} = \{x_0, x_1, \ldots, x_{b^m-1}\}$ is a digital net over $\mathbb{F}_b$ and let $\oplus_{b,\varphi}$ be the addition from Exercise 5.5. Show that $(\mathcal{P}, \oplus_{b,\varphi})$ is a group. If the points of $\mathcal{P}$ are pairwise different, then this group is isomorphic to $\mathbb{F}_b^m$.

5.12 Remember the definition of Walsh functions $_b\mathrm{wal}_k(x)$ from Exercise 3.5. Let b be a prime number and let $\{x_0, x_1, \ldots, x_{b^m-1}\}$ be a digital net over $\mathbb{F}_b$ (which we identify with $\mathbb{Z}_b$) with generating matrices $C_1, \ldots, C_s \in \mathbb{F}_b^{m \times m}$. Show that for all $k_1, \ldots, k_s \in \{0, 1, \ldots, b^m - 1\}$ we have

$$\sum_{n=0}^{b^m-1} {}_b\mathrm{wal}_{k_1}(x_{n,1}) \cdots {}_b\mathrm{wal}_{k_s}(x_{n,s}) = \begin{cases} b^m & \text{if } C_1^\top \vec{k}_1 + \cdots + C_s^\top \vec{k}_s = \vec{0}, \\ 0 & \text{otherwise,} \end{cases}$$

where $x_{n,j}$ denotes the jth component of x_n and where for $k = \kappa_0 + \kappa_1 b + \cdots + \kappa_{m-1} b^{m-1}$ we denote $\vec{k} = (\kappa_0, \kappa_1, \ldots, \kappa_{m-1})^\top$. *Hint:* This is [33, Lemma 4.75] where one can find a proof. *Remark*: This relation is a very crucial tool in the analysis of digital nets and their application for numerical integration. Further information in this direction can be found in [33].

5.13 Let b be a prime power. Show that every upper triangular $\mathbb{N} \times \mathbb{N}$ matrix over $\mathbb{F}_b$ with nonzero diagonal entries generates a digital $(0, 1)$-sequence over $\mathbb{F}_b$.

5.14 Let b be a prime power and let the $\mathbb{N} \times \mathbb{N}$ matrices $C_1, \ldots, C_s$ generate a digital (t, s)-sequence over the finite field $\mathbb{F}_b$. For any $m \in \mathbb{N}$ consider the left upper $m \times m$ sub-matrices $C_1^{(m)}, \ldots, C_s^{(m)}$. Take

$$C_{s+1}^{(m)} := E_m' = \begin{pmatrix} \overline{0} & \overline{0} & \ldots & \overline{0} & \overline{1} \\ \overline{0} & & \cdot\cdot\cdot & \overline{1} & \overline{0} \\ \vdots & \cdot\cdot & \cdot\cdot & \cdot\cdot & \vdots \\ \overline{0} & \overline{1} & \cdot\cdot\cdot & & \overline{0} \\ \overline{1} & \overline{0} & \ldots & \overline{0} & \overline{0} \end{pmatrix} \in \mathbb{F}_b^{m \times m}.$$

Show that the $m \times m$ matrices $C_1^{(m)}, \ldots, C_s^{(m)}, C_{s+1}^{(m)}$ generate a digital $(t, m, s + 1)$-net over $\mathbb{F}_b$. *Remark*: Note that this is a "digital version" of Lemma 5.17. Note also the increase of the dimension from s to $s + 1$.

5.15 Prove that for the exponential discrete evaluation we have $v(p/q) = \deg(p) - \deg(q)$ for all nonzero polynomials $p, q \in \mathbb{F}_b[x]$. In particular, $v(p) = \deg(p)$ for all nonzero $p \in \mathbb{F}_b[x]$

5.16 Consider the first 2^4 points of a Niederreiter sequence in base 2 and dimension 3. Find the smallest value of t such that these points form a digital (t, m, s)-net.

5.17 Show that the Sobol' sequence defined via the recurrence relation is a special case of a generalized Niederreiter sequence as stated in the section on Sobol' sequences.

5.18 Show that the generating matrices of a Faure sequence defined via the polynomials can be written in terms of Pascal matrices as stated in (5.15).

5.19 Prove Theorem 5.54 for the unweighted case. *Hint 1*: Use $\sum_{h \in \mathbb{F}_{b,m}[x]^s} \frac{1}{\varrho_b(h)} = (1 + m\frac{b^2-1}{3b})^s$. *Hint 2*: This is [33, Theorem 10.28] where one can find a proof.

5.20 Prove Corollary 5.56.

A Brief Discussion of the Discrepancy Bounds

6.1 The Curse of Dimensionality

In many applications the dimension s can be rather large. But, in this case, the asymptotically almost optimal bounds on the discrepancy which we obtained, e.g., for the Hammersley point set or for (t, m, s)-nets soon become useless for a modest number N of points. For example, assume that for every $s, N \in \mathbb{N}$ we have a point set $\mathcal{P}_{s,N}$ in the s-dimensional unit cube of cardinality N with star discrepancy of at most

$$D_N^*(\mathcal{P}_{s,N}) \le c_s \frac{(\log N)^{s-1}}{N} \tag{6.1}$$

with some quantity $c_s > 0$ that is independent of N. Hence for any $\delta > 0$ the star discrepancy behaves asymptotically like $N^{-1+\delta}$, which is the optimal rate of convergence since for dimension $s = 1$ we already have $D_N^*(\mathcal{P}_{1,N}) \ge 1/(2N)$. However, the function $N \mapsto (\log N)^{s-1}/N$ does not start to decrease to zero until $N \ge \exp(s - 1)$. For $N \le \exp(s - 1)$ this function is increasing, which means that for cardinality N in this range our discrepancy bounds are useless. But even for moderately large dimension s, the value of $\exp(s - 1)$ is huge, such that point sets with cardinality $N \ge \exp(s - 1)$ cannot be used for practical applications. Therefore, the bound (6.1) is useful only if N is large compared to the dimension s.

For practical applications one is interested in the discrepancy of point sets with cardinality N not too large (compared to s). To analyze this problem systematically one considers the following quantity.

Definition 6.1 For $s, N \in \mathbb{N}$ define the Nth *minimal star discrepancy* by

$$\operatorname{disc}_\infty(N, s) := \inf_{\substack{\mathcal{P} \subseteq [0,1)^s \\ \#\mathcal{P} = N}} D_N^*(\mathcal{P}).$$

© The Author(s), under exclusive license to Springer Nature Switzerland AG 2026
G. Leobacher and F. Pillichshammer, *Introduction to Quasi-Monte Carlo Integration and Applications*, Compact Textbooks in Mathematics,
https://doi.org/10.1007/978-3-032-05446-3_6

For example, one may ask if

$$\lim_{s \to \infty} \mathrm{disc}_\infty(2^s, s) = 0? \tag{6.2}$$

In other words, do there exist point sets in dimension s of cardinality 2^s (which is already dramatically large for dimensions $s \geq 30$) for which the star discrepancy approaches zero when s tends to infinity? For example, the discrepancy bound from Corollary 5.20 for (t, m, s)-nets $\mathcal{P}_s$ in base 2 gives for $m = s$ and $t = t(s) \asymp s$ (which is optimal)

$$D^*_{2^s}(\mathcal{P}_s) \leq \frac{2^{t(s)}}{2^s} \left(1 + \frac{s^{s-1}}{(s-2)!} \right) \sim 2^{t(s)} \left(\frac{e}{2} \right)^s \sqrt{\frac{2s}{\pi}} \to \infty \quad \text{for} \quad s \to \infty.$$

So the asymptotically excellent discrepancy bounds for $(0, m, s)$-nets cannot help to answer the question (6.2).

As another attempt, let $s, m \in \mathbb{N}$ where $m \geq 2$ and consider the regular lattice $\Gamma_{m,s}$. According to Theorem 2.19, its star discrepancy is given by

$$D^*_N(\Gamma_{m,s}) = 1 - \left(1 - \frac{1}{2m} \right)^s,$$

where $N = m^s$. Let $\varepsilon \in (0, 1)$. Then to achieve a star discrepancy of at most ε we would require a regular lattice consisting of

$$N \approx \left(\frac{s}{2|\log(1 - \varepsilon)|} \right)^s$$

points (i.e., choose $m \approx \frac{s}{2|\log(1-\varepsilon)|}$). This number grows superexponentially in s. For example, for $\varepsilon = 1/3$ one requires $N \approx (1.23s)^s$. Hence, for a star discrepancy of at most ε we require a cardinality which is exponential in s. Such an exponential dependence is called the *curse of dimensionality*, a notion coined by R. Bellman in 1957.

The above discussion shows that although we have excellent asymptotic results for the star discrepancy of many sequences, these results do not help to answer question (6.2). This means that, according to the classical theory, QMC methods could not be expected to work for very high dimensions. Nevertheless and surprisingly we know from practical applications that QMC rules often do very well and even work much better than we have any right to expect. As already mentioned in the preface, I.H. Sloan[1] spoke in this context of "The unreasonable effectiveness of QMC." A spectacular and surprising example in this direction was reported by S.H. Paskov and J.F. Traub in 1995. They valuated a financial derivative, a 30-year Collateralized Mortgage Obligation, which required the computation of ten 360-dimensional integrals. They tested two QMC rules based on Halton and Sobol' sequences and compared them with the MC rule. Their experiments showed that both QMC rules

---

[1] Talk at the MCQMC conference in Warsaw, August 15, 2010.

outperform the MC rule and that the convergence of the QMC rules is much smoother than that of the MC rule. Following the success of those experiments, it is nowadays an important stream of research to explain *why* QMC rules do so well for many high-dimensional applications. Motivated by the idea that perhaps the reason for the success of QMC is that some coordinate directions are more important than others I.H. Sloan and H. Woźniakowski began to study weighted function spaces. The weights model the behavior of different coordinate directions and under suitable conditions one can break the curse of dimensionality. As examples we already discussed QMC integration in weighted Korobov spaces in Sect. 4.6 and weighted star discrepancy in Sects. 3.5, 4.3, and 5.5.

Furthermore, we nowadays also know that the situation for the star discrepancy is not so bad as it seems from the discussion above. Actually the question (6.2) can be answered in the affirmative, and even more is possible. This will be explained in the following section.

6.2 Tractability of the Star Discrepancy

In 2001 S. Heinrich, E. Novak, G.W. Wasilkowski, and H. Woźniakowski established the following very surprising result.

Theorem 6.2 (Heinrich, Novak, Wasilkowski, and Woźniakowski) *There exists a constant $c > 0$ with the property that*

$$\mathrm{disc}_\infty(N, s) \le c \sqrt{\frac{s}{N}} \quad \textit{for all} \quad N, s \in \mathbb{N}. \tag{6.3}$$

The proof of (6.3) uses deep results from probability theory and is beyond the scope of this book. Later, Ch. Aistleitner showed by a simplified argument that in (6.3) one can choose $c = 10$. Both proofs are unfortunately non-constructive, and until now we do not know point sets for which the bound (6.3) holds. However, using a probabilistic argument, it can be shown that many N-element point sets in $[0, 1)^s$ satisfy the bound modulo a multiplicative factor greater than one.

Definition 6.3 For $s \in \mathbb{N}$ and $\varepsilon > 0$ the *inverse of the star discrepancy* is defined as

$$N_\infty(s, \varepsilon) := \min\{N \in \mathbb{N} : \mathrm{disc}_\infty(N, s) \le \varepsilon\}.$$

The inverse of the star discrepancy is the minimal cardinality N that a point set in $[0, 1)^s$ must have so that we can achieve a star discrepancy not larger than ε. It follows from (6.3) that

$$N_\infty(s, \varepsilon) \le C s \varepsilon^{-2} \tag{6.4}$$

for some positive constant C. Hence, the inverse of the star discrepancy depends only polynomially on s and ε^{-1}. In Information-Based Complexity (IBC) theory such a behavior is called *polynomial tractability* (cf. Definition 4.28).

Furthermore, it is known that the dependence on the dimension s of the inverse of the star discrepancy in (6.4) cannot be improved. For example, it was shown by A. Hinrichs in 2004 that there exist constants $c, \varepsilon_0 > 0$ such that

$$N_\infty(s, \varepsilon) \geq cs\varepsilon^{-1} \quad \text{for all } \varepsilon \in (0, \varepsilon_0), \ s \in \mathbb{N}$$

and $\mathrm{disc}_\infty(N, s) \geq \min(\varepsilon_0, cs/N)$. The exact dependence of $N_\infty(s, \varepsilon)$ on ε^{-1} is still an open question which seems to be very difficult.

We will prove a result which also implies polynomial tractability of the star discrepancy, although it is slightly weaker than the one presented in Theorem 6.2. Also this result was first shown by Heinrich, Novak, Wasilkowski, and Woźniakowski in 2001.

Theorem 6.4 (Heinrich, Novak, Wasilkowski, and Woźniakowski) *We have*

$$\mathrm{disc}_\infty(N, s) \leq \frac{2\sqrt{2}}{\sqrt{N}} \left(s \log \left\lceil \frac{s\sqrt{N}}{2(\log 2)^{1/2}} \right\rceil + \log 2 \right)^{1/2} \quad \textit{for all } N, s \in \mathbb{N}, \quad (6.5)$$

and

$$N_\infty(s, \varepsilon) \leq \lceil 8\varepsilon^{-2}(s \log(\lceil 2s/\varepsilon \rceil + 1) + \log 2) \rceil \quad \textit{for all } s \in \mathbb{N} \textit{ and } \varepsilon > 0. \quad (6.6)$$

Again this result is a pure existence result. The idea of the proof of Theorem 6.4 is to show that the probability of randomly picking an N-element point set in $[0, 1)^s$ with star discrepancy at most the right-hand side of (6.5) is positive. To this end we need an auxiliary result from probability theory which we state here without proof.

Lemma 6.5 (Höffding's inequality) Let $Z_1, \ldots, Z_N$ be independent random variables satisfying $u_n \leq Z_n \leq v_n$ for some $u_n < v_n$ and with expectation $\mathbb{E}[Z_n] = 0$ for all $n = 1, \ldots, N$. Then for all $\eta > 0$ we have

$$\mathbb{P}[|Z_1 + \cdots + Z_N| \geq N\eta] \leq 2 \exp \left(-2\eta^2 N^2 / \sum_{n=1}^{N}(v_n - u_n)^2 \right),$$

where $\mathbb{P}[\cdot]$ is the probability on the probability space supporting $Z_1, \ldots, Z_N$.

Proof of Theorem 6.4. For given $\delta > 0$ we define $m := \lceil s/\delta \rceil$. Then Remark 2.17 implies that for every N-element point set $\mathcal{P} = \{x_0, \ldots, x_{N-1}\}$ in $[0, 1)^s$ we have

$$D_N^*(\mathcal{P}) \leq \max_{y \in \mathcal{G}_{m,s}} \left| y_1 \cdots y_s - \frac{1}{N} \sum_{i=0}^{N-1} \chi_{[0, y)}(x_i) \right| + \delta, \quad (6.7)$$

where $\mathcal{G}_{m,s} := \left\{0, \frac{1}{m}, \frac{2}{m}, \ldots, \frac{m}{m}\right\}^s$. Note that $\#\mathcal{G}_{m,s} = (m+1)^s$.

Let $X_0, \ldots, X_{N-1}$ be independent and uniformly distributed random variables in $[0,1)^s$. For $\boldsymbol{y} \in [0,1]^s$ let

$$Z_{\boldsymbol{y}}^{(n)} := y_1 \cdots y_s - \chi_{[\boldsymbol{0},\boldsymbol{y})}(X_n) \quad \text{for all} \quad n = 0, \ldots, N-1.$$

Then $\mathbb{E}[Z_{\boldsymbol{y}}^{(n)}] = 0$ and $|Z_{\boldsymbol{y}}^{(n)}| \le 1$ for all $n = 0, \ldots, N-1$. Thus we may apply Lemma 6.5 to the sum of the $Z_{\boldsymbol{y}}^{(n)}$'s from which we obtain

$$\mathbb{P}\left[\left|\frac{1}{N} \sum_{n=0}^{N-1} Z_{\boldsymbol{y}}^{(n)}\right| \ge \delta\right] \le 2\exp\left(-\frac{\delta^2 N}{2}\right) \quad \text{for} \quad \boldsymbol{y} \in [0,1]^s.$$

Using the estimate (6.7) it follows that

$$\mathbb{P}\left[D_N^*(\{X_0, \ldots, X_{N-1}\}) \le 2\delta\right] \ge \mathbb{P}\left[\max_{\boldsymbol{y} \in \mathcal{G}_{m,s}} \left|y_1 \cdots y_s - \frac{1}{N} \sum_{n=0}^{N-1} \chi_{[\boldsymbol{0},\boldsymbol{y})}(X_n)\right| \le \delta\right]$$

$$\ge 1 - 2(m+1)^s \exp(-\delta^2 N/2).$$

The last term is strictly positive if

$$\log 2 + s \log\left(\left\lceil \frac{s}{\delta} \right\rceil + 1\right) - \delta^2 \frac{N}{2} < 0.$$

This inequality is valid for all $\delta > \delta_0 = \delta_0(N, s)$, where

$$\delta_0^2 = \frac{2}{N}\left(s \log\left(\left\lceil \frac{s}{\delta_0} \right\rceil + 1\right) + \log 2\right). \tag{6.8}$$

This implies $\frac{1}{\delta_0} \le \left(\frac{N}{4\log 2}\right)^{1/2}$. Re-inserting into (6.8) leads to

$$\delta_0^2 \le \frac{2}{N}\left(s \log\left\lceil \frac{s\sqrt{N}}{2(\log 2)^{1/2}} \right\rceil + \log 2\right). \tag{6.9}$$

Thus, for all $\delta > \delta_0$ we have

$$\mathbb{P}\left[D_N^*(\{X_0, \ldots, X_{N-1}\}) \le 2\delta_0\right] > 0$$

and hence there exists a realization $\boldsymbol{x}_0, \ldots, \boldsymbol{x}_{N-1}$ of $X_0, \ldots, X_{N-1}$ in $[0,1)^s$ such that $D_N^*(\{\boldsymbol{x}_0, \ldots, \boldsymbol{x}_{N-1}\}) \le 2\delta_0$. This shows $\mathrm{disc}_\infty(N, s) \le 2\delta_0$, which implies, together with (6.9), the bound (6.5). The proof of (6.6) is left as exercise. $\square$

6.3 Strong Polynomial Tractability of the Weighted Star Discrepancy

The positive result Theorem 6.4 can be improved for the weighted star discrepancy with sufficiently small product weights. In accordance with Definition 6.1 we define the Nth minimal weighted star discrepancy.

Definition 6.6 For $s, N \in \mathbb{N}$ and weights γ define the *Nth minimal weighted star discrepancy* by

$$\mathrm{disc}_{\infty,\gamma}(N, s) := \inf_{\substack{\mathcal{P} \subseteq [0,1)^s \\ \#\mathcal{P}=N}} D^*_{N,\gamma}(\mathcal{P}).$$

Furthermore, in accordance with Definition 6.3 we define the inverse of the weighted star discrepancy.

Definition 6.7 For $s \in \mathbb{N}$, $\varepsilon \in (0, 1)$, and (product) weights γ the *inverse of the weighted star discrepancy* is defined as

$$N_{\infty,\gamma}(s, \varepsilon) := \min\{N \in \mathbb{N} \ : \ \mathrm{disc}_{\infty,\gamma}(N, s) \le \varepsilon\}.$$

Then we have the following exciting result.

Theorem 6.8 *Consider product weights* $\gamma = \{\gamma_{\mathfrak{u}} = \prod_{j\in\mathfrak{u}} \gamma_j \ : \ \mathfrak{u} \subseteq [s]\}$ *with a real summable sequence* $(\gamma_j)_{j\ge 1}$, *i.e.,* $\sum_{j=1}^{\infty} \gamma_j < \infty$. *For arbitrary small* $\delta \in (0, 1)$ *there exists a positive real* $\widetilde{C}_{\delta,\gamma}$ *with the following property: for every* $s \in \mathbb{N}$ *and* $\varepsilon \in (0, 1)$ *we have*
$$N_{\infty,\gamma}(s, \varepsilon) \le \widetilde{C}_{\delta,\gamma}\varepsilon^{-1/(1-\delta)}.$$

Proof The proof relies on Corollary 4.16. Fix $\delta > 0$ and let $C_{\delta,\gamma} > 0$ be the quantity whose existence is guaranteed by Corollary 4.16. Now let N be the smallest prime number greater than or equal to

$$M := \left\lceil (C_{\delta,\gamma}\varepsilon^{-1})^{1/(1-\delta)} \right\rceil.$$

According to Bertrand's postulate we have $N \le 2M$. Then for the generating vector $\boldsymbol{g}$ in Corollary 4.16 we have

$$\mathrm{disc}_{\infty,\gamma}(N, s) \le D^*_{N,\gamma}(\mathcal{P}(\boldsymbol{g}, N)) \le \varepsilon.$$

Hence it follows that

$$N_{\infty,\gamma}(s, \varepsilon) \le N \le 2M = 2\left\lceil (C_{\delta,\gamma}\varepsilon^{-1})^{1/(1-\delta)} \right\rceil \le \widetilde{C}_{\delta,\gamma}\varepsilon^{-1/(1-\delta)},$$

with a suitably chosen positive real $\widetilde{C}_{\delta,\gamma}$ that is independent of ε, s, and N. $\qquad\square$

In Information-Based Complexity (IBC) theory the behavior from Theorem 6.8 is called *strong polynomial tractability* (cf. Definition 4.28). In this sense, the weighted star discrepancy is strongly polynomially tractable. Since $\delta > 0$ can be chosen arbitrarily close to zero, the exponent of ε^{-1} can be arbitrarily close to 1. One says the that the so-called ε-exponent of strong polynomial tractability equals 1. Furthermore, we have seen that strong polynomial tractability for the weighted star discrepancy can be achieved by means of lattice point sets whose generating vector is obtained from a CBC construction (see Sect. 4.3). From this point of view, our result is even constructive and not just an existence result like for the unweighted star discrepancy.

Further Reading

According to [54, p. 280], the question about the dependence on the dimension for the star discrepancy was first raised by G. Larcher in 1998. Nice introductions into the topic offer the overview articles by Novak [114, 115] and the book by Traub and Werschulz [143]. Exhaustive information about tractability theory is summarized in the three volumes of Novak and Woźniakowski [118–120]. Concerning the results presented in this chapter we particularly refer to [119, Chap. 9]. A report on the 360-dimensional experiment of Paskov and Traub can be found in [126], see also [143, Sect. 4]. An introduction to the study of weighted spaces in the context of QMC and warmy recommended is the paper by Sloan and Woźniakowski [137]. The original proof of Theorem 6.4 can be found in [54]. A simplified version according to Aistleitner with the explicit constant $c = 10$ in (6.3) (which means $C = 100$ in (6.4)) can be found in [1], while currently, the best known possible value is $c = 2.4968$ (which means $C = 6.2341$ in (6.4)), as shown in [48]. The mentioned lower bound on the inverse of the star discrepancy was shown by Hinrichs [61, Theorem 1]. A proof of Höffding's inequality can be found in [69]. While the usual star discrepancy is polynomially tractable, it is known that the L_p discrepancy for $p \in (1, \infty)$ suffers from the course of dimensionality. A proof and more information in this regard can be found in [117]. Tractability of the weighted star discrepancy is studied in great detail in [2,37,65,66]. Information about tractability of the weighted L_p discrepancy can be found in [116].

Exercises

6.1 Show that $\lim_{s \to \infty} D_{2^s}^*(\Gamma_{2,s}) = 1$.

6.2 Revisit the proof of Theorem 6.4 and show that the probability of randomly picking an N-element point set $\mathcal{P}_{N,s}$ in $[0, 1)^s$ with star discrepancy

$$D_N^*(\mathcal{P}_{N,s}) \leq \frac{2\sqrt{2}}{\sqrt{N}} \left(s \log \left\lceil \frac{s\sqrt{N}}{2(\log 2)^{1/2}} \right\rceil + \log 2 + \eta \right)^{1/2}$$

for arbitrary $\eta > 0$ is larger than $1 - \exp(-\eta)$.

6.3 Prove (6.6). *Hint*: The solution can be found in [33, Theorem 3.54].

6.4 For $s, N \in \mathbb{N}$ define the *Nth minimal L_2 discrepancy* by

$$\mathrm{disc}_2(N, s) := \inf_{\substack{\mathcal{P} \subseteq [0,1)^s \\ \#\mathcal{P} = N}} L_{2,N}(\mathcal{P}).$$

Show that $\mathrm{disc}_2(0, s) = 3^{-s/2}$ and for $N \in \mathbb{N}$

$$\mathrm{disc}_2(N, s) \le \frac{1}{\sqrt{N}} \left(\frac{1}{2^s} - \frac{1}{3^s} \right)^{1/2}.$$

Hint: Use Exercise 2.10.

6.5 For $s \in \mathbb{N}$ and $\varepsilon > 0$ the *inverse of the L_2 discrepancy* is defined as

$$N_2(s, \varepsilon) := \min \left\{ N \in \mathbb{N} \ : \ \mathrm{disc}_2(N, s) \le \varepsilon \, \mathrm{disc}_2(0, s) \right\}.$$

Show that

$$N_2(s, \varepsilon) \le \left(\frac{3}{2} \right)^s \varepsilon^{-2}.$$

Remark: It is known, see for example [137] or [33, Proposition 3.58], that $N_2(s, \varepsilon) \ge (9/8)^s (1 - \varepsilon^2)$. Hence $N_2(s, \varepsilon)$ grows exponentially fast in s which means that the L_2 discrepancy suffers from the curse of dimensionality. The same is known for the L_p discrepancy for all $p \in (1, \infty)$ as shown in [117].

Basics of Financial Mathematics

7

In this chapter, we will give some background on mathematical finance, or, to be precise, on the mathematical theory that lies behind derivative pricing. Since the 1980s, financial mathematics has become a huge field of applied mathematics that uses methods from many other branches of mathematics, most notably from probability theory. The reliance on probability theory provides us with a wealth of applications for simulation techniques.

We want to give a quick entry into topics of Monte Carlo and quasi-Monte Carlo pricing of financial derivatives. Obviously, a rigorous treatment of the probabilistic theory is well beyond the scope of this introductory text on QMC methods.

7.1 Bonds, Stocks, and Derivatives

Since we want to discuss valuation of financial derivatives, we have to define at least the most basic ones as well as some of the primary instruments from which they "derive." The explanations given below are more general than would be necessary to achieve our primary goal of giving a quick introduction into derivative pricing. But we do not want to hide from the reader the fact that most mathematical models in finance are rather coarse, and this fact is illustrated best by showing what *does not* enter the models.

- A *bond* is a financial instrument that pays its owner a fixed amount of money at a pre-specified date in the future. The writer of the bond is usually a big company or a government. The owner effectively becomes a creditor to the writer. If the quality of the debtor is very high, as is the case for many government bonds, the

© The Author(s), under exclusive license to Springer Nature Switzerland AG 2026
G. Leobacher and F. Pillichshammer, *Introduction to Quasi-Monte Carlo Integration and Applications*, Compact Textbooks in Mathematics,
https://doi.org/10.1007/978-3-032-05446-3_7

bond can be modeled as a deterministic payment. The bond usually sells at a lower price than its payoff and thus pays *interest*.

Bond payments may be subject to credit risk, i.e., for some debtors the probability that the debtor will not be able to make the payment in full at the required date cannot be ignored. Fruitful theories have been put forward for valuation of credit risky bonds, and those theories provide fields of applications for simulation methods, too.

However, we will not consider that topic and for us a bond will always be non-defaultable. Nevertheless, in practice, even then the price of the bond may and will vary over time, due to a change of inflation and other parameters of the general economy.

- A *share* is a financial instrument that warrants its holder ownership of a fraction of a corporation. In particular, the shareholder participates in the business revenue by the means of dividend payments.

  However, dividend payments are not the only possible source of income through a share. At least equally important is the possible gain due to a price change: an investor may buy a share and sell it shortly afterward if the price has risen. The possible gain is usually one or two orders of magnitude higher than that through interest payments. On the downside, the price change may as well result in a loss. If the shares of a company are traded at a stock exchange then buying and selling them is particularly convenient and high-frequency traders may buy and sell large contingents of shares several times per second.

  The value of a share depends on many things, such as the preferences of the individual agent, the assets of the company, the state of the economy, the future dividend payments and future interest rates and inflation.

  The so-called *efficient market hypothesis* assumes that the value of the share at a given time is just the market price at that very time. Under this hypothesis it does not make sense to compute the objective value of a share in a mathematical model and compare it to the market price. The only way that a computed value of a share can differ from its market price is that our preferences and/or expectations differ from that of the majority of the market, thus giving a subjective price.

- A *contingent claim* is a financial instrument whose value at a future date can be completely described in terms of the prices of other financial instruments, its so-called underlyings. A typical example is an option on a share. A *European call option* on a share with *maturity* T is a contract which gives its holder the right (but not the obligation) to buy one share from the option writer at some fixed time T in the future at the previously agreed price K. In exchange, the option writer is paid a *premium* at the time of writing the option. One also says that the holder has the long position of the option, the writer has the short position.

  Denote the price of the share at the future date T by S_T. Since the option holder may sell the share instantly at the stock exchange, the value of the option at time T is $S_T - K$ if $S_T > K$, and 0 if $S_T \leq K$.

  The left-hand side of Fig. 7.1 shows the payoff of a European call option dependent on the price of the share at maturity. An important feature is the kink at the strike

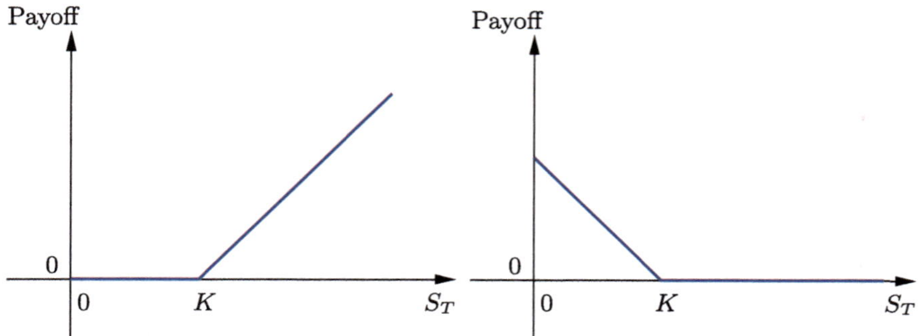

Fig. 7.1 Payoff of a European call (left) and put (right) option

price K. It introduces a non-differentiability that makes estimates of the QMC error substantially more difficult.

Note that the buyer of the option swaps a fixed amount of money—the price of the option at time 0, i.e., the premium—for an uncertain payment in the future. This is a feature that an option shares with an insurance contract, and indeed an option can be used to insure against "unfavorable" events. For example, an investor who intends to buy a share in the future can buy an option now and is therefore insured against any rise of the share price.

On the right-hand side of Fig. 7.1 we plot the payoff of a European put option. This is an option which gives its holder the right (but not the obligation) to sell one share to the option writer at some fixed time T in the future, at the previously agreed price K. If the share price satisfies $S_T \geq K$ at time T, then the option is worthless. But if $S_T < K$, then the option holder may buy the share at the stock exchange at price S_T and sell it immediately to the writer at price K, thus realizing a gain of $K - S_T$.

In Fig. 7.2, we show the payoff of another contingent claim, a so-called *digital cash-or-nothing call option*. This option pays a fixed amount of cash at expiry if at that time the price S_T of the underlying is above the strike K. We also plot the payoff of the corresponding put option. The digital option serves as an example of a contingent claim with discontinuous payoff.

Since the value of an option is strongly tied to that of the underlying, and in simple models is completely determined by the parameters of the model, an objective value of the option can be computed in these models using arbitrage arguments. Even in more general models a—not necessarily unique—value can be computed with the property, that trading the option at that price does not introduce the possibility of arbitrage into the market, that is, the possibility of making riskless profit above the interest rate.

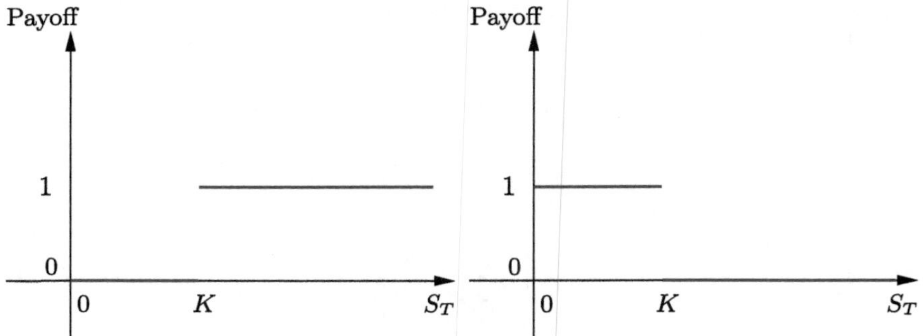

Fig. 7.2 Payoff of a digital cash-or-nothing call and put options

7.2 Arbitrage and the No-Arbitrage Principle

Suppose you are given an option on a stock. You know the specifications of the option and therefore you know the uncertain payoff at its maturity T given the uncertain value of the stock at that particular time. That is, the payoff is of the form $f(S_T)$, where S_T is the share price at time T. One is tempted to model S_T as a random variable, then use statistical methods to estimate its parameters from historical stock prices, and *a fortiori* estimate the value of the option as its discounted expected value. Indeed this is exactly how one usually proceeds in determining the fair stake in a game of chance. We will show in this section that this reasonable program will in general yield a price that is unreasonable from a more basic perspective, in that it allows for riskless profit.

While general arbitrage theory is well beyond the scope of this book, the underlying principle can be illustrated rather quickly. Assume the following simple market model where we have only two times, 0 and 1, and three instruments, a bond, a share, and a European call option with strike $K = 1$ and maturity $T = 1$. Let $B = (B_t)_{t \in \{0,1\}}$, $S = (S_t)_{t \in \{0,1\}}$, $C = (C_t)_{t \in \{0,1\}}$ denote the price processes of the bond, share, and option, respectively, and assume the following parameters: $B_0 > 0$, $B_1 = B_0(1 + r)$, $r \geq 0$, $S_0 > 0$, $S_1 = S_0 u$ with probability p and $S_1 = S_0 d$ with probability $1 - p$, where $0 < d < 1 + r < u$. The value of the option at time 1 is $\max(S_1 - K, 0)$, thus $C_1 = \max(S_0 u - K, 0)$ with probability p, and $C_1 = \max(S_0 d - K, 0)$ with probability $1 - p$. Suppose we know, for example from statistical studies, the value of p.

Following the program proposed above, we would get as the price of the option at time 0 the value

$$\widehat{C}_0 = \frac{B_0}{B_1} \mathbb{E}[\max(S_1 - K, 0)] = \frac{1}{1 + r}(p \max(S_0 u - K, 0) + (1 - p) \max(S_0 d - K, 0)).$$

However, this formula can give an unreasonable price. Suppose $r = 0, u = 2, d = \frac{1}{2}$, $S_0 = K = 1$ and $p = \frac{1}{2}$, for which the above formula gives $\widehat{C}_0 = \frac{1}{2}$.

Then as a trader we could do the following:

- At time 0, write 4 options and sell them for 2 €, borrow one additional Euro to buy three shares. Note that the net investment is zero.
- Now wait until time 1. Then

  - If the share price has gone up, the shares are worth 6. We sell them to get 6 € in Cash. Since the share price S_T (which is 2) is bigger than the strike K (which is 1), the options will be executed, costing us 4 € and we have to pay 1 € back. Thus our strategy leaves us with a net profit of 1 €.
  - If the share price has gone down, the options have become worthless. We sell the shares, giving us $\frac{3}{2}$ € and thereby, after paying 1 € back, leaving us with a profit of $\frac{1}{2}$ €.

Thus, whatever happens, we are left with a strictly positive profit without having taken any risk and even without initial investment of capital. Such a situation is called an *arbitrage opportunity* and it is usually assumed that such opportunities do not exist in a viable market for the following reason: if you can make a riskless profit of at least $\frac{1}{2}$ €, without initial investment, then you can as well make a riskless profit of a million Euros by simply scaling up the above strategy, that is, by writing 8 million options and so on. In fact, in our model, the riskless profit is not limited at all. Obviously, a market with such a feature is useless[1].

On the other hand, it can easily be shown that there is one and only one option price in this model that does not allow for arbitrage, namely,

$$C_0 = \frac{1}{1+r}\left(\widetilde{p}\max(S_0u - K, 0) + (1 - \widetilde{p})\max(S_0d - K, 0)\right),$$

where $\widetilde{p} = \frac{1+r-d}{u-d}$. The distinctive feature of $\widetilde{p}$ is that $\frac{1}{1+r}(\widetilde{p}S_0u + (1-\widetilde{p})S_0d) = S_0$, that is, the stock price process is a *martingale* with respect to this new probability. We do not give a precise definition for this. Intuitively, a martingale is a process X such that the conditional expectation of X_{t+s} given the evolution of X up to time s is X_t. Thus a martingale is a model for the gain process of a player in a fair game.

To summarize, the arbitrage-free price of an option can be written as the discounted expected value of its payoff, but the expectation has to be taken under a suitable probability measure equivalent—but in general not equal—to the original one.

---

[1] Note, however, that we have tacitly assumed that the strategy scales indefinitely and that, in particular, our buying of shares does not change the market price.

Remark 7.1 No-arbitrage arguments usually make rather strong assumptions:

- Bonds can be bought and sold at the same rate and for every positive maturity. In particular, an investor can lend and borrow money at the same rate.

- There is no restriction on short-selling. That is, an investor may borrow a share and sell it in order to gain from falling prices. At a later time she or he has to buy the share back at the market price to be able to return it.

- There are no transaction costs.

- There are no liquidity effects: an investor can buy or sell unlimited numbers of securities at the market price without changing the price. In particular, the bid and ask prices coincide.

- All investors have the same information about the market.

- In continuous-time models, an investor may trade continuously at every point in time.

7.3 The Black-Scholes Model

The simple model in the preceding section can be extended to an n-step setup. Again, the unique arbitrage-free price for a contingent claim is given as a discounted expected value, under some equivalent probability measure, of the claim's payoff at expiry. It is now tempting to let n go to infinity to obtain a continuous-time model. Indeed, this can be done in rigorous fashion so that we arrive at a model of the form:

$$
\begin{aligned}
B_t &= B_0 \exp(rt), \\
S_t &= S_0 \exp(\mu t + \sigma W_t),
\end{aligned}
\tag{7.1}
$$

$t \in [0, T]$, where W is a Brownian motion (sometimes also called "Wiener process," after N. Wiener, the Mathematician who first described the modern mathematical model for this process), that is, a continuous-time stochastic process with specific properties. The exact mathematical definition of Brownian motion will be given in Sect. 8.2, and we invite the reader who is not already familiar with this concept to skip ahead and have a look at Definition 8.15. The model described by (7.1) is often called *the* Black-Scholes model.

As in the one-step model, there exists a probability measure $\widetilde{\mathbb{P}}$, equivalent to the original measure $\mathbb{P}$, such that $t \mapsto B_t^{-1} S_t$ becomes a martingale. Under this new probability measure

$$
S_t = S_0 \exp\left(\left(r - \frac{\sigma^2}{2}\right) t + \sigma \widetilde{W}_t\right),
$$

where $\widetilde{W}_t = W_t - \frac{1}{\sigma}(r - \mu - \frac{\sigma^2}{2})t$ is a Brownian motion under $\widetilde{\mathbb{P}}$. The measure $\widetilde{\mathbb{P}}$ can in fact be explicitly given in terms of the Radon-Nikodym derivative, which is given by $\frac{\mathrm{d}\widetilde{\mathbb{P}}}{\mathrm{d}\mathbb{P}} = \exp(\nu W_T - \frac{\nu^2}{2}T)$, with $\nu = \frac{1}{\sigma}(r - \mu - \frac{\sigma^2}{2})$. That means that the expected value $\widetilde{\mathbb{E}}[X]$ of any bounded random variable measurable with respect to $\{W_s : 0 \le s \le T\}$ can be computed by the formula:

$$\widetilde{\mathbb{E}}[X] = \mathbb{E}\left[\exp\left(\nu W_T - \frac{\nu^2}{2}T\right)X\right],$$

where $\widetilde{\mathbb{E}}$ denotes expectation with respect to $\widetilde{\mathbb{P}}$.

This new probability measure is now used to price derivatives in this model: if C is some European contingent claim, that is, a derivative whose payoff C_T at time T is a function of $S_t, 0 \le t \le T$, then its arbitrage-free price at time 0 is given by

$$C_0 = \widetilde{\mathbb{E}}[B_T^{-1} C_T]. \tag{7.2}$$

When C_T depends only on finitely many S_{t_j}, $j = 1, \ldots, m$, then the expectation in (7.2) can be written as an m-dimensional integral, which is where QMC enters the game. The details of this will be given in Sect. 8.2.

In our continuous-time model, we assume that the option can be traded at any time prior to its maturity T. For this, the time t analog of (7.2) is

$$B_t^{-1} C_t = \widetilde{\mathbb{E}}[B_T^{-1} C_T], \tag{7.3}$$

or $C_t = B_t \widetilde{\mathbb{E}}[B_T^{-1} C_T]$.

Example 7.2 A European Call option on a share with price process $(S_t)_{t \ge 0}$ and with strike K and maturity T has payoff $C_T = \max(S_T - K, 0)$. The pricing equation (7.2) therefore gives the option price in the Black-Scholes model at time t as

$$C_0 = \exp(-rT)\widetilde{\mathbb{E}}[\max(S_T - K, 0)].$$

Since

$$S_T = S_0 \exp\left(\left(r - \frac{\sigma^2}{2}\right)T + \sigma \widetilde{W}_T\right),$$

and since $(r - \frac{\sigma^2}{2})T + \sigma \widetilde{W}_T$ is a normal random variable with mean $(r - \frac{\sigma^2}{2})T$ and variance $\sigma^2 T$, we get

$$C_0 = \frac{\exp(-rT)}{\sqrt{2\pi\sigma^2 T}} \int_{-\infty}^{\infty} \max(S_0 \exp(x) - K, 0)) \exp\left(-\frac{(x - (r - \frac{\sigma^2}{2})T)^2}{2\sigma^2 T}\right) \mathrm{d}x$$

$$= \frac{\exp(-rT)}{\sqrt{2\pi\sigma^2 T}} \int_{\log(\frac{K}{S_0})}^{\infty} (S_0 \exp(x) - K) \exp\left(-\frac{(x - (r - \frac{\sigma^2}{2})T)^2}{2\sigma^2 T}\right) \mathrm{d}x.$$

It is a nice exercise to compute this integral. It turns out that its value is given by the famous *Black-Scholes option pricing formula*

$$C_0 = S_0 \Phi(d_1) - \exp(-rT)K \Phi(d_2), \tag{7.4}$$

where $\Phi(x) = \frac{1}{\sqrt{2\pi}} \int_{-\infty}^{x} \exp(-t^2/2)\,dt$ and

$$d_1 = \frac{\log \frac{S_0}{K} + (r + \frac{\sigma^2}{2})T}{\sigma\sqrt{T}} \quad\text{and}\quad d_2 = \frac{\log \frac{S_0}{K} + (r - \frac{\sigma^2}{2})T}{\sigma\sqrt{T}}. \tag{7.5}$$

So in this case we get a closed-form formula and there is no need to apply simulation techniques. The price C_t for $0 \leq t \leq T$ can be obtained from Equations (7.4) and (7.5) simply by substituting S_t for S_0 and $T - t$ for T.

Another class of examples for which there often exist closed formulas are barrier and lookback options, where the payoff depends on the maximum or minimum of the price over a given interval.

We move on to a somewhat harder example.

Example 7.3 The payoff of an *Asian option* written on a share with price process $(S_t)_{t\in[0,T]}$ depends on the average price over some interval $[T_0, T]$, $T_0 < T$, where T is the expiry date of the option. The payoff of a so-called *fixed strike* Asian call option is given by

$$C_T^{\text{fix}} = \max\left(\frac{1}{T - T_0} \int_{T_0}^{T} S_\tau\,d\tau - K, 0\right),$$

and the payoff of a so-called *floating strike* Asian call option is given by

$$C_T^{\text{flt}} = \max\left(\frac{1}{T - T_0} \int_{T_0}^{T} S_\tau\,d\tau - S_T, 0\right).$$

Up to now, nobody has found an explicit formula for either Asian option, but there are rather efficient methods using partial differential equations (PDE) to compute the value. Nevertheless, this example is a nice benchmark for simulation methods.

Because of its simplicity, the Black-Scholes model does not provide us with many interesting examples for simulation. One step toward demanding problems is to look at the m-dimensional Black-Scholes model.

Consider m shares $S^1, \ldots, S^m$ whose price processes are given by

$$S_t^j = S_0^j \exp\left(\mu_j t + \sum_{l=1}^{k} \sigma_{jl} W_t^l\right)$$

$t \in [0, T]$, where $W^1, \ldots, W^k$ are k independent Brownian motions and $\sigma = (\sigma_{jl})_{jl}$ is a $m \times k$ matrix. In this model neither the existence nor the uniqueness of a probability measure that makes each process $(\exp(-rt) S_t^j)_{t\in[0,T]}$ a martingale is granted. In fact, every solution $\boldsymbol{v} = (v_1, \ldots, v_k)^\top$ with $v_j \in \mathbb{R}$ of the linear system

$$\sigma \boldsymbol{v} = r\mathbf{1} - \boldsymbol{\mu} - \frac{1}{2}\text{diag}(\sigma\sigma^\top) \tag{7.6}$$

gives rise to such a measure, where $\mathbf{1}$ is the m-dimensional column vector with all entries equal to 1 and $\boldsymbol{\mu} = (\mu_1, \ldots, \mu_m)^\top$. To see this, consider that, if such a solution exists, we can write $\widetilde{W}_t^j = W_t^j - \nu_j t$ and we can construct a measure equivalent to $\mathbb{P}$ such that $\widetilde{W}^j$ is a Brownian motion under the new measure, just as in the one-dimensional case. In fact we can find one measure $\widetilde{\mathbb{P}}$ such that $\widetilde{W}^1, \ldots, \widetilde{W}^k$ are k independent Brownian motions. The Radon-Nikodym derivative is given by

$$\frac{d\widetilde{\mathbb{P}}}{d\mathbb{P}} = \prod_{j=1}^{k} \exp\left(\nu_j W_T^k - \frac{\nu_j^2}{2} T \right).$$

By substitution we see that the price processes now take on the form

$$S_t^j = S_0^j \exp\left(\left(r - \frac{1}{2}(\sigma\sigma^\top)_{jj} \right) t + \sum_{l=1}^{k} \sigma_{jl} \widetilde{W}_t^l \right),$$

and we see that the discounted price processes remain constant on average, i.e., they are martingales. It can also be shown that no equivalent martingale measure exists if (7.6) has no solution, but we will not attempt this here.

The m-dimensional Black-Scholes model is interesting from the point of view of (optimal) portfolio selection, but it also provides us with practical high-dimensional integration problems through derivative pricing. An arbitrage-free price for a derivative with payoff C_T can again be given as

$$C_0 = \exp(-rT)\widetilde{\mathbb{E}}\left[C_T\right],$$

where $\widetilde{\mathbb{E}}$ denotes expectation under a suitable equivalent martingale measure $\widetilde{\mathbb{P}}$. As mentioned above, every solution ν of (7.6) gives rise to such a measure, and in general different measures will give different prices.

In practice, one has to make a reasonable choice for the pricing measure, and usually this is done by *calibrating* the model to market data. However, one must not forget that only one of many possible arbitrage-free prices is computed and that a different price in the market need not imply the existence of arbitrage.

Example 7.4 A classical basket option on shares with prices $S^1, \ldots, S^d$ and weights $w^1, \ldots, w^d \in \mathbb{R}$ and strike K has payoff

$$\max\left(w^1 S_T^1 + \cdots + w^d S_T^d - K, 0 \right).$$

More complicated dependencies on the price processes can be encountered in practice. In particular, the payoff may depend on the time-averages of the price processes. Then the option also has some Asian characteristics. For basket options on several shares the PDE method for pricing soon becomes intractable. Here, we really have to use simulation.

7.4 SDE Models

In many models from financial mathematics, the share price process is not given explicitly but is described via a *stochastic differential equation*, in short *SDE*.

For example, also the one-dimensional Black-Scholes model can be described by an SDE:

$$\mathrm{d}S_t = \widehat{\mu} S_t \, \mathrm{d}t + \sigma S_t \, \mathrm{d}W_t,$$

$$S_0 = s_0.$$

The term $\mathrm{d}W_t$ is interpreted as an infinitesimal increment of a Brownian motion, where a finite increment over the time interval $[t, t + h]$ of Brownian motion is a Gaussian variable with mean 0 and variance h (and thus with standard deviation $\sqrt{h}$). The almost surely unique solution[2] to this SDE with initial value s_0 is

$$S_t = s_0 \exp\left(\widehat{\mu} t + \sigma W_t - \frac{\sigma^2}{2} t \right),$$

such that for $\widehat{\mu} = \mu + \frac{\sigma^2}{2}$ we recover the price process from (7.1).

More generally, a model could be defined by an $m + 1$-dimensional SDE

$$\mathrm{d}S_t = \mu(t, S_t) \, \mathrm{d}t + \sigma(t, S_t) \, \mathrm{d}W_t$$
$$S_0 = s_0, \tag{7.7}$$

where $S = (S^0, \ldots, S^m)$ is an $m + 1$-dimensional stochastic process and $s_0 = (s^0, \ldots, s^m) \in \mathbb{R}^{m+1}$. In this general model not all the components need to correspond to share prices or indeed to prices at all. For example, a coordinate process might be a model for an exchange rate or an interest rate. It is assumed that one coordinate is the price of an asset that can function as a numeraire in that it is always positive. Without loss we may assume that S^0 is the numeraire. Discounting now means dividing by S^0, that is, the discounted process is the m-dimensional process $\widetilde{S}$ given by

$$t \mapsto (S_t^0)^{-1}(S_t^1, \ldots, S_t^m).$$

Only those components of $\widetilde{S}$ that correspond to price processes of traded assets need to be martingales under the equivalent martingale measure.

---

[2] This is a consequence of the famous Itô formula from stochastic analysis. In short, the Itô formula states that for a function f which is C^1 in the first variable and C^2 in the second variable, we have

$$\mathrm{d}f(t, W_t) = \frac{\partial f}{\partial t}(t, W_t) \, \mathrm{d}t + \frac{\partial f}{\partial W}(t, W_t) \, \mathrm{d}W_t + \frac{1}{2} \frac{\partial^2 f}{\partial W^2}(t, W_t) \, \mathrm{d}t.$$

Example 7.5 Consider the so-called Heston model (already under an equivalent martingale measure):

$$\mathrm{d}B_t = r\,B_t\,\mathrm{d}t$$
$$\mathrm{d}S_t = r\,S_t\,\mathrm{d}t + \sqrt{V_t}\,S_t\,(\rho\,\mathrm{d}W_t^1 + \sqrt{1-\rho^2}\,\mathrm{d}W_t^2)$$
$$\mathrm{d}V_t = \kappa(\theta - V_t)\,\mathrm{d}t + \xi\sqrt{V_t}\,\mathrm{d}W_t^1$$
$$(B_0, S_0, V_0) = (b_0, s_0, v_0)\,.$$

Here, r, κ, θ, ξ are positive constants, μ is a real constant, and $-1 < \rho < 1$ is a correlation coefficient. The third component of our process, V, is the so-called volatility of the share price and is not a traded asset. It has the form of a so-called Cox-Ingersoll-Ross process (CIR process). The first process B models a bond price and thus serves as the numeraire. It is worth mentioning that, despite there not being an explicit solution known for the SDE, there is a semi-exact formula for the price of a European call option in the Heston model using Laplace inversion.

We do not concern ourselves with the theory of SDEs since this topic would easily fill another book and indeed there are many excellent books available on SDEs and many of them cover applications in finance.

From the point of view of MC and QMC simulation, it is mainly of interest to know that under suitable regularity requirements on the coefficients of the SDE there exists a unique solution and that under slightly stronger conditions this solution can be approximated by discrete algorithms.

Let S_T be the solution to the SDE at time T and let $\widehat{S}_N$ be some approximation to S_T computed on the time grid $0 = t_0 < t_1 < \ldots < t_N = T$ with fineness $\delta = \max_{1 \leq k \leq N}(t_k - t_{k-1})$. We say that $\widehat{S}_N$ converges to S_T in the *strong sense* with order γ, if $\mathbb{E}[|S_T - \widehat{S}_N|] = O(\delta^\gamma)$.

Sometimes it is enough to compute some characteristics of the solution like $\mathbb{E}[f(S_T)]$ for a function f belonging to some class C. This question is linked to the concept of *weak convergence* of numerical schemes. The benefit is that the weak order of an approximation scheme is usually higher than the strong order of the same scheme.

The most straightforward solution method for SDEs is the *Euler-Maruyama method*: given (7.7), we compute an approximate solution $\widehat{S}$ on the time nodes $0, h, \ldots, nh = T$ via

$$\widehat{S}_0 = S_0$$
$$\widehat{S}_{k+1} = \widehat{S}_k + \mu(kh, \widehat{S}_k)h + \sigma(kh, \widehat{S}_k)\Delta W_{k+1}, \qquad (7.8)$$

where $\Delta W_{k+1} := W_{(k+1)h} - W_{kh}$. It follows from the definition of Brownian motion, Definition 8.15, that ΔW_{k+1} is a normal random vector with expectation 0 and covariance matrix $\sqrt{h}I_{m+1}$, where I_{m+1} is the $(m+1) \times (m+1)$ identity matrix. Frequently, (7.8) is therefore stated in the form:

$$\widehat{S}_{k+1} = \widehat{S}_k + \mu(kh, \widehat{S}_k)h + \sigma(kh, \widehat{S}_k)\sqrt{h}Z_{k+1},$$

where $Z_1, Z_2, \ldots$ is a sequence of independent standard normal vectors. However, we will prefer the original form when using QMC.

Under suitable regularity conditions (Lipschitz continuous in the second variable, sublinear growth with first variable, sufficient smoothness) on the coefficient functions μ, σ of the SDE, the Euler-Maruyama scheme converges in the strong sense with order $\frac{1}{2}$ and in the weak sense with order 1, such that, for sufficiently regular f, $\mathbb{E}[f(\widehat{S}_{nh})]$ is a decent approximation to $\mathbb{E}[f(S_T)]$, for sufficiently small h.

Note that the coefficients of the Heston model, Example 7.5, are not Lipschitz. Thus we cannot rely on the Euler-Maruyama scheme to converge. It can be shown that the SDE for the CIR process V has a solution and is almost surely non-negative. Moreover, if the so-called Feller condition $2\kappa\theta \geq \xi^2$ is satisfied, then V is almost surely strictly positive. Nevertheless, it may easily happen that the Euler-Maruyama approximation of V becomes negative, so one has to modify the equation for V to force it to remain non-negative.

We report two other schemes for solving autonomous SDEs numerically, which under appropriate conditions on the coefficients converge in the strong sense with order 1. The first is the *Milstein scheme*,

$$\widehat{S}_{k+1} = \widehat{S}_k + \mu(\widehat{S}_k)h + \sigma(\widehat{S}_k)\Delta W_{k+1} + \frac{1}{2}\sigma(\widehat{S}_k)\sigma'(\widehat{S}_k)(\Delta W_{k+1}^2 - h),$$

where σ' is the derivative of σ. The second is an example for a *Runge-Kutta scheme*, with the advantage of not requiring a derivative:

$$\widehat{S}_{k+1} = \widehat{S}_k + \mu(\widehat{S}_k)h + \sigma(\widehat{S}_k)\Delta W_{k+1} + \frac{1}{2}(\sigma(Y_k) - \sigma(\widehat{S}_k))(\Delta W_{k+1}^2 - h)\frac{1}{\sqrt{h}},$$

where the supporting value Y_k is given by $Y_k = \widehat{S}_k + \sigma(\widehat{S}_k)\sqrt{h}$.

A problem that can occur in practice is that the simulated path can leave the domain of definition while the exact solution does not. For example, the approximate stock price and/or the volatility process may become negative, even if the theoretical process is always non-negative. Besides choosing small discretization intervals, there are sophisticated methods for treating this problem, but for which we have to refer the reader to the further reading section.

7.5 Lévy Models

Lévy processes are generalizations of Brownian motion. The mathematical definition will be given in Sect. 8.4. The additional feature of general Lévy processes, which makes them particularly interesting for financial modeling, is that they allow for jumps. In analogy to the Gaussian models, i.e., models built on Brownian motion, Lévy models come in two flavors. There are explicit models where the stock price is exponential Lévy motion:

$$S_t = S_0 \exp(L_t),$$

where L is a Lévy process with $\mathbb{E}[\exp(L_t)] < \infty$. Alternatively, the stock price might again be given by an SDE, i.e.:

$$dS_t = f(t, S_{t-}) \, dL_t.$$

Here, S_{t-} is short for the left-hand limit of S at time t, $S_{t-} = \lim_{h \to 0-} S_{t+h}$. If it is possible to sample from the increments of L, then the Euler-Maruyama scheme still allows us to simulate a discrete approximation to the solution S,

$$\widehat{S}_{k+1} = \widehat{S}_k + f(kh, \widehat{S}_k)(L_{(k+1)h} - L_{kh}).$$

From the point of view of option pricing, it is important that the market is arbitrage-free so that it admits an equivalent probability measure under which the discounted prices of tradable assets are martingales. This can be achieved, for example, with the so-called Esscher transform, a change of measure under which the process L is again a Lévy process.

7.6 Price Sensitivities (Greeks)

The arbitrage-free value of a financial derivative is not the only characteristic that is of interest to investors. Almost equally important is the question about sensitivity of the value to input parameters. The interest is two-fold. For one thing, some of those parameters have to be estimated by one method or the other, usually by matching market prices of derivatives that are liquidly traded. Now the sensitivities tell us how errors in the estimates of the parameters propagate.

But for investors it is also of interest to know how the price changes with respect to market conditions, like interest rates or volatility, which may be interpreted as a measure of (lack of) confidence in price stability.

Sensitivities can be also used for *hedging*. For example, if we know how the option price changes with respect to the share price, we can use this information to insure our option position locally against price changes by adding a suitable (maybe negative) number of shares to our portfolio. Or, vice versa, options can be used to locally insure against the loss in value of a share.

Another application is risk assessment for portfolios of derivatives. Summing up, the Greeks of a whole portfolio give the sensitivity of the portfolio with respect to the same parameter. The marginal change in value of the portfolio with respect to the parameter thus roughly equals the Greek of the portfolio times the change.

The commonly used symbols for the sensitivities of options in the Black-Scholes model are Greek letters and are therefore commonly called the "Greeks" of the option. We have stated in Sect. 7.3 that the price of a European call option in the Black-Scholes model can be written as a smooth function

$$C = C(S, K, \sigma, r, T - t).$$

The most commonly used price sensitivities are "Delta" $= \frac{\partial}{\partial S} C$, "Gamma" $= \frac{\partial^2}{\partial S^2} C$, "Rho" $= \frac{\partial}{\partial r} C$, "Theta" $= \frac{\partial}{\partial t} C = -\frac{\partial}{\partial T} C$, "Vega" $= \frac{\partial}{\partial \sigma} C$ (actually, "Vega" is not a Greek letter).

Consider the dependence of the option price on some parameter ξ. If the value of the option at expiry is given by $C_T(\xi)$, then, according to (7.3), the price sensitivity at time t with respect to ξ is

$$\frac{\partial}{\partial \xi} C_t(\xi) = \lim_{h \to 0} \exp(-r(T-t)) \widetilde{\mathbb{E}} \left[h^{-1}(C_T(\xi+h) - C_T(\xi)) \right].$$

If we can exchange the order of limit and integration, then we get

$$\frac{\partial}{\partial \xi} C_t(\xi) = \exp(-r(T-t)) \widetilde{\mathbb{E}} \left[\frac{\partial}{\partial \xi} C_T(\xi) \right],$$

and computation of the sensitivity is conceptually the same as computation of the price.

A counterexample is provided by the European digital call option. Its payoff is differentiable almost everywhere with derivative zero. However, there is one point where the differential quotient tends to infinity. So here, changing the order of differentiation and expectation is obviously not allowed.

Usually the expectation smoothens the payoff, so that the price becomes a smooth function. But if we use simulation techniques to compute the price, then the integral is approximated by a finite sum which inherits the non-smoothness from the payoff function. Thus one has to find different ways of computing the Greeks. We will return to this problem after we have discussed simulation methods for pricing.

Further Reading

The use of Brownian motion as a model for stock price evolution dates back to the dissertation thesis of Louis Bachelier, "Théorie de la Spéculation," from 1900. Bachelier is widely considered to be the founding father of financial mathematics.

The Black-Scholes pricing equation for European call options has been proven by Black, Scholes, and Merton in the early 1970s, [13,106]. They did not use the notions of martingales and equivalent martingale measures, an idea proposed in the late 1970s by Harrison and Kreps [51], but rather derived a partial differential equation (PDE) for the price by finding a replicating strategy for the option. The corresponding PDE and its modern generalizations provide an alternative framework for valuing derivatives that is usually preferred for low dimensions. For a comprehensive treatment of general arbitrage theory see Delbaen and Schachermayer [22]. A nice introduction to quantitative financial mathematics is provided by the book by Albrecher, Binder, Lautscham, and Mayer [3]. An easily understandable introduction to the fundamental concepts of financial mathematics and financial engineering is also provided in the trilogy [92–94] by Larcher.

The standard reference for numerical treatment of SDEs is the book by Kloeden and Platen [80]. This book also covers stochastic analysis for continuous processes and the theory of SDEs. The regularity conditions needed for strong and weak convergence of the schemes provided in the text together with corresponding proofs can also be found there.

For the problem of possible negativity of the volatility of stock price process in the Heston model see again [80] and also [4]. It has been shown in [52] that if the Feller condition is not satisfied, i.e., if $2\kappa\theta < \sigma^2$, then the strong order of convergence of any discretization scheme is at most $\frac{2\kappa\theta}{\sigma^2}$.

A very good reference for Lévy models in finance is the book by Cont and Tankov [20].

Exercises

7.1 A share can be viewed as a derivative on itself with maturity $T > 0$ and payoff equal to S_T. A bond with $B_0 = 1$ can be viewed as a derivative on a share with deterministic payoff $B_T = \exp(rT)$ at time T. The payoff of a portfolio of derivatives on a share with the same maturity is the sum of the single payoffs.

Draw the payoffs of the following derivatives/portfolios:

(a) 1 share;

(b) 1 share and $-K$ bonds;

(c) 1 share and -1 call option with strike K;

(d) 1 call option and -1 put option, both with strike K;

(e) 1 share, $-K$ bonds, -1 call option, 1 put option, both options with strike K.

7.2 Prove the so-called *Put-Call parity*: Assume a market consisting of a share with price process $(S_t)_{t\in[0,T]}$, European put and call options on the share with the same strike K and same expiry T, and a bond with deterministic price process $(\exp(rt))_{t\in[0,T]}$, for some $r \in (0, \infty)$. Let $(P_t)_{t\in[0,T]}$ and $(C_t)_{t\in[0,T]}$ be the price processes of the put and call options, respectively.

Show that if for some $t \in [0, T)$ we have $S_t - \exp(-r(T - t))K \neq C_t - P_t$, then there exists an arbitrage opportunity in this market that can be realized using a simple buy-and-hold strategy. In other words: in an arbitrage-free market we have the put-call parity

$$S_t - \exp(-r(T - t))K = C_t - P_t \quad \text{for all } t \in [0, T].$$

7.3 A forward contract gives its holder the right and the obligation to buy one share at time T in the future at a pre-arranged price, the "forward price."

(a) Draw the payoff of a forward contract.

(b) Show that the forward price must equal $\exp(rT)S_0$ by constructing a buy-and-hold arbitrage otherwise.

7.4 Using the definition $\Phi(x) := \frac{1}{\sqrt{2\pi}} \int_{-\infty}^{x} \exp(-y^2/2)\,\mathrm{d}y$, and $\Phi(\infty) = 1$ compute the integrals

$$I_1 = \frac{1}{\sqrt{2\pi}} \int_{-\infty}^{x} \exp(\theta y) \exp(-y^2/2)\,\mathrm{d}y$$

$$I_2 = \frac{1}{\sqrt{2\pi}} \int_{x}^{\infty} \exp(\theta y) \exp(-y^2/2)\,\mathrm{d}y$$

$$I_3 = \frac{1}{\sqrt{2\pi}} \int_{-\infty}^{\infty} \exp(\theta y) \exp(-y^2/2)\,\mathrm{d}y.$$

Hint: Complete the square.

7.5 Let X be a standard normal variable, i.e., X has probability density function $f(x) = \frac{1}{\sqrt{2\pi}} \exp(-x^2/2)$. Show that for any $\theta \in \mathbb{R}$ we have

$$\mathbb{E}[\exp(\theta X)] = \exp(\theta^2/2).$$

Hint: Use Exercise 7.4.

7.6 Deduce from Exercise 7.5 that in the Black-Scholes model

$$\mathbb{E}\left[S_{t_2}|S_{t_1}\right] = \exp(r(t_2 - t_1))S_{t_1}.$$

Here, $\mathbb{E}[X|Y]$ denotes conditional expectation of the random variable X with respect to the random variable Y, which informally means that in evaluating the expectation Y can be treated like a constant.

7.7 Verify the Black-Scholes option pricing formula Equation (7.4). *Hint:* Use Exercise 7.4.

Monte Carlo and Quasi-Monte Carlo Simulation

8

In this chapter we will learn the basics of pricing derivatives using simulation methods. We will consider both Monte Carlo and quasi-Monte Carlo but—of course—with a special emphasis on the latter. The aim of our exposition is not to provide a large toolbox for the quantitative analyst, but to help getting started with the topic. QMC pricing is an active area of research by its own and the reader is encouraged to consult the specialized literature. We will, however, take a look at some popular examples that frequently serve as benchmarks for refined simulation techniques.

8.1 Non-uniform Random Number Generation

Most random variables encountered in practical models are not uniformly distributed. We are therefore interested in methods for generating pseudo- or quasi-random numbers with a given distribution from their uniform counterparts.

The most straightforward method is the so-called inversion method which will be presented in the first subsection.

We are also going to present the class of acceptance-rejection methods for generating random numbers with a given distribution. These methods have the reputation of being generally inapplicable for QMC. We try to give a more differentiated view on that topic.

Inversion Method

We introduce this method for the special case of invertible cumulative distribution functions and defer the general method to the exercises.

Consider a real random variable X. Its *cumulative distribution function (CDF)* is defined as the function $F : \mathbb{R} \to [0, 1]$, $F(x) = \mathbb{P}[X \leq x]$ for all $x \in \mathbb{R}$. Let

© The Author(s), under exclusive license to Springer Nature Switzerland AG 2026
G. Leobacher and F. Pillichshammer, *Introduction to Quasi-Monte Carlo Integration and Applications*, Compact Textbooks in Mathematics,
https://doi.org/10.1007/978-3-032-05446-3_8

us assume that F is invertible, that is, there exists a function $G : (0, 1) \to \mathbb{R}$ with $G(F(x)) = x$ for all $x \in \mathbb{R}$ and $F(G(u)) = u$ for all $u \in (0, 1)$.

Suppose now that the random variable U is uniformly distributed on $(0, 1)$ and define a real random variable $Y = G(U)$. Then Y has the same distribution as X. To see this, let $y \in \mathbb{R}$. Then

$$\mathbb{P}[Y \leq y] = \mathbb{P}[G(U) \leq y] = \mathbb{P}\left[F(G(U)) \leq F(y)\right] = \mathbb{P}[U \leq F(y)] = F(y),$$

as claimed.

A sufficient condition for a CDF to be invertible is that it has a positive probability density function (PDF) on $\mathbb{R}$ (see Exercise 8.1).

For some distributions the inversion of the CDF is almost trivial. For example, consider a Cauchy-distributed random variable X with parameters (x_0, γ), which has PDF

$$f(x) = \frac{1}{\pi} \frac{\gamma}{(x - x_0)^2 + \gamma^2} \quad \text{for } x \in \mathbb{R}.$$

The CDF is then given by

$$F(x) = \mathbb{P}[X \leq x] = \int_{-\infty}^{x} f(t) \, dt = \frac{1}{\pi} \arctan\left(\frac{x - x_0}{\gamma}\right) + \frac{1}{2}.$$

F is obviously invertible, with inverse $F^{-1}(u) = x_0 + \gamma \tan(\pi(u - \frac{1}{2}))$. Thus if U is uniformly distributed on $(0, 1)$, then $x_0 + \gamma \tan(\pi(U - \frac{1}{2}))$ is Cauchy distributed with parameter (x_0, γ).

For other distributions the inversion procedure can only be done numerically.

Definition 8.1 A random variable X is *normally distributed* with parameters μ and $\sigma > 0$ if it has PDF

$$f_X(x) = \frac{1}{\sqrt{2\pi\sigma^2}} \exp\left(-\frac{(x - \mu)^2}{2\sigma^2}\right).$$

We write $X \sim N(\mu, \sigma^2)$. If, in addition, $\mu = 0$ and $\sigma = 1$, then we call X a *standard normal variable*.

The inverse of a normal CDF cannot be given by a formula using only algebraic combinations of elementary functions. However, there are highly accurate methods for approximating the standard normal CDF by rational functions, and implementations are available in every sufficiently popular programming language.

If Z is a standard normal variable and μ, σ are real numbers with $\sigma > 0$, then $X = \sigma Z + \mu$ is a normal random variable with parameters μ and σ. So being able to sample from the standard normal distribution is all that is needed for sampling from general normals.

Acceptance-Rejection Method

Inverting a CDF numerically can be computationally expensive. A very versatile and cheap alternative method for generating a random variable with prescribed PDF f is the *acceptance-rejection method*. For its implementation we need another distribution for which it is cheap to sample from, e.g., via the inversion method. Let g be the PDF of this distribution. Moreover, we need that, for some $0 < c \le 1$, $cf(x) \le g(x)$ for all $x \in \mathbb{R}$.

The algorithm is as follows:

Algorithm 8.2 (Acceptance-rejection sampling)

1. Generate a sample Y with PDF g and an independent uniform random variable U.
2. If $Ug(Y) \le cf(Y)$, set $X = Y$ else go back to step 1.

We call f the target PDF and g the proposal PDF. We give a proof that the algorithm indeed gives a random variable with the desired distribution, and that c should be as big as possible so that the algorithm stops after as few steps as possible.

Note that under our assumption on f and g it is obvious that $g(x) = 0$ implies $f(x) = 0$ for all $x \in \mathbb{R}$. Moreover, note that $\mathbb{P}[g(Y) > 0] = 1$ if Y has density g.

Theorem 8.3 *Let X be the random variable generated by Algorithm 8.2 and let N be number of repetitions of Algorithm 8.2 before giving X. Then X has PDF f, and N has geometric distribution with parameter c, i.e., $\mathbb{P}[N = k] = c(1-c)^{k-1}$ for all $k \in \mathbb{N}$. In particular, the algorithm needs on average $\frac{1}{c}$ repetitions to complete.*

Proof Let $Y_1, U_1, Y_2, U_2, \ldots$ be a sequence of independent random variables, where each Y_j is distributed according to PDF g and each U_j is uniformly distributed on $[0, 1]$. Then for any measurable set $A \subseteq \mathbb{R}$

$$
\mathbb{P}\left[Y_k \in A \text{ and } U_k g(Y_k) \le cf(Y_k)\right]
$$
$$
= \int_A \mathbb{P}\left[U_k g(Y_k) \le cf(Y_k)\,\Big|\, Y_k = y\right] g(y)\,\mathrm{d}y
$$
$$
= \int_A \frac{cf(y)}{g(y)} g(y) \chi_{(0,\infty)}(g(y))\,\mathrm{d}y
$$
$$
= c \int_A f(y) \chi_{(0,\infty)}(g(y))\,\mathrm{d}y
$$
$$
= c \int_A f(y)\,\mathrm{d}y.
$$

In particular,

$$
\mathbb{P}\left[U_k g(Y_k) \le cf(Y_k)\right] = \mathbb{P}\left[Y_k \in \mathbb{R} \text{ and } U_k g(Y_k) \le cf(Y_k)\right] = c \int_\mathbb{R} f(y)\,\mathrm{d}y = c
$$

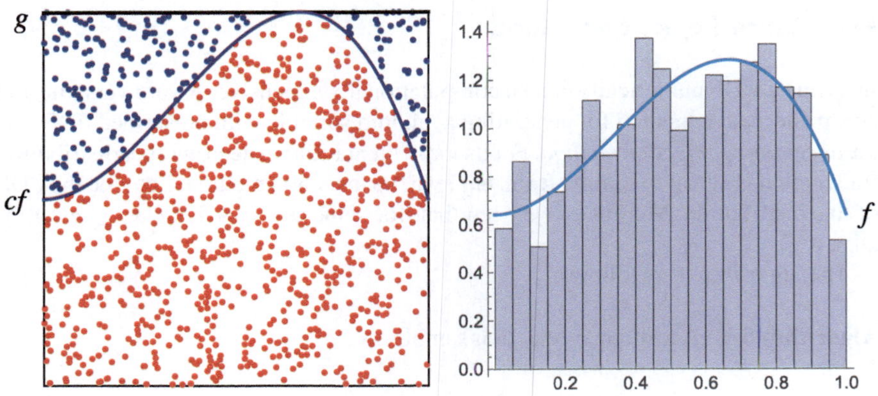

Fig. 8.1 Illustration of the acceptance-rejection algorithm where $g \equiv 1$ and where 1000 random points in the unit square have been used. On the right-hand side we see the histogram of the random points so generated, compared with the probability PDF f

for every k, so that the random variable $N := \min\{k \in \mathbb{N} : U_k g(Y_k) \leq cf(Y_k)\}$ has geometric distribution with parameter c.

According to Algorithm 8.2 we have $X = Y_N$. Therefore,

$$\mathbb{P}[Y_k \in A \text{ and } N = k]$$
$$= \mathbb{P}[Y_k \in A \text{ and } U_1 g(Y_1) > cf(Y_1) \text{ and } \ldots$$
$$\ldots \text{ and } U_{k-1} g(Y_{k-1}) > cf(Y_{k-1}) \text{ and } U_k g(Y_k) \leq cf(Y_k)]$$
$$= \mathbb{P}[Y_k \in A \text{ and } U_k g(Y_k) \leq cf(Y_k)](1 - c)^{k-1},$$

thanks to the independence of $U_1, \ldots, U_k, Y_1, \ldots, Y_k$. Now

$$\mathbb{P}[X \in A] = \sum_{k=1}^{\infty} \mathbb{P}[Y_k \in A \text{ and } N = k]$$

$$= \sum_{k=1}^{\infty} c \int_A f(y)\,\mathrm{d}y \,(1 - c)^{k-1} = \int_A f(y)\,\mathrm{d}y,$$

simply by computing the geometric sum.

The average number of repetitions is $\mathbb{E}[N] = \sum_{k=1}^{\infty} kc(1-c)^{k-1} = \frac{1}{c}$, which is a simple exercise. $\qquad\square$

The method can be illustrated by a picture like the left-hand side of Fig. 8.1. Here $g \equiv 1$ on the unit interval and c is a constant such that $cf(u) \leq 1 = g(u)$ for all $u \in [0, 1]$. Random points are sampled in the unit square and those points which lie below the graph of cf are projected onto the x-axis. The right-hand side of Fig. 8.1 shows a normalized histogram of the projected points together with the PDF f.

Example 8.4 The PDF of a gamma distributed random variable X with parameters (a, b) is given by

$$f(x) = \begin{cases} \frac{1}{\Gamma(a)} b^a x^{a-1} \exp(-bx) & \text{if } x > 0, \\ 0 & \text{if } x \le 0, \end{cases}$$

where Γ is Euler's gamma function,

$$\Gamma(a) := \int_0^\infty x^{a-1} \exp(-x)\, dx.$$

If we want to sample from a gamma distribution there is no loss in restricting to the case $b = 1$, since if X is gamma distributed with parameters $(a, 1)$ and $b > 0$, then $b^{-1}X$ is gamma distributed with parameters (a, b).

Suppose now that $a > 1$. Our idea is to sample from the exponential distribution, which is easy (Exercise 8.8). Thus we want to find some constants $\beta > 0$ and $0 < c \le 1$ such that $cf(x) \le \beta \exp(-\beta x)$ and take $g(x) = \beta \exp(-\beta x)$ in Algorithm 8.2.

We need $\frac{c}{\Gamma(a)} x^{a-1} \exp(-x) \le \beta \exp(-\beta x)$ for all $x \ge 0$. For that, we maximize the function $h(x) := \frac{x^{a-1} \exp(-x)}{\Gamma(a)\beta \exp(-\beta x)}$ using first-order conditions. We then set $c = h(x_0)^{-1}$, where x_0 is the maximizer of h. It can be shown that $\beta = a$ gives the lowest rejection rate, i.e., the biggest c.

The case where $a < 1$ is slightly more complicated and we leave it for the exercises. The choice of g in the case $a > 1$ is not optimal in that the constant c may be comparatively small. There are refined methods available for choosing good sampling densities for acceptance rejection algorithms.

Box-Muller Method and Marsaglia-Bray Algorithm

In this section we will present two popular methods for generating standard normal vectors. But first recall the definition of a normal (or Gaussian) random vector:

Definition 8.5 A random vector $X = (X_1, \ldots, X_s)$ is said to be *normally distributed* with mean $\mu \in \mathbb{R}^s$ and covariance matrix $\Sigma > 0$ if it has joint PDF

$$f_X(x) = \frac{1}{\sqrt{(2\pi)^s \det(\Sigma)}} \exp\left(-\frac{(x - \mu)\Sigma^{-1}(x - \mu)^\top}{2}\right) \quad \text{for } x \in \mathbb{R}^s$$

that is, for every measurable set $A \subseteq \mathbb{R}^s$

$$\mathbb{P}[X \in A] = \int_{\mathbb{R}^s} \chi_A(x) f_X(x)\, dx.$$

Here, $\Sigma > 0$ means that Σ has to be positive definite, i.e., $x\Sigma x^\top > 0$ for all $x \in \mathbb{R}^s \setminus \{0\}$.

Consider a two-dimensional standard normal vector (X, Y), i.e., $\mu = 0$ and $\Sigma = I_2$, the identity matrix. Then we have

$$
\mathbb{P}\left[\sqrt{X^2 + Y^2} \le r\right] = \frac{1}{2\pi} \int_{-r}^{r} \int_{-\sqrt{r^2-y^2}}^{\sqrt{r^2-y^2}} \exp\left(-\frac{x^2 + y^2}{2}\right) dx \, dy
$$

$$
= \frac{1}{2\pi} \int_{0}^{r} \int_{0}^{2\pi} \rho \exp\left(-\frac{\rho^2}{2}\right) d\varphi \, d\rho
$$

$$
= 1 - \exp\left(-\frac{r^2}{2}\right).
$$

It follows that the modulus of (X, Y) has CDF $F_R(r) = 1 - \exp(-r^2/2)$. But that means that we can generate a random radius by inversion of F_R, $F_R^{-1}(u) = \sqrt{-2\log(1 - u)}$. Since the PDF has rotational symmetry, this gives us a method for generating independent pairs of standard normal variables.

Algorithm 8.6 (Box-Muller)

1. Generate two independent $U[0, 1)$ random samples U, V.
2. Let $R = \sqrt{-2\log(1 - U)}$.
3. Let $X = R\cos(2\pi V)$ and $Y = R\sin(2\pi V)$.
4. Return (X, Y).

There is an acceptance-rejection-type variant of the Box-Muller method which is known as Marsaglia-Bray algorithm:

Algorithm 8.7 (Marsaglia-Bray)

1. Generate two independent $U[0, 1)$ random samples U, V.
2. Let $U_1 = 2U - 1$ and $V_1 = 2V - 1$.
3. If $U_1^2 + V_1^2 \ge 1$ reject (U, V) and start from the beginning.
4. Else let $S = U_1^2 + V_1^2$.
5. If $S = 0$ set $(X, Y) = (0, 0)$.
6. Else set $X = U_1\sqrt{-2\log(S)/S}$ and $Y = V_1\sqrt{-2\log(S)/S}$.
7. Return (X, Y).

Theorem 8.8 *The vector (X, Y) generated by either of Algorithms 8.6 or 8.7 has standard normal distribution.*

We leave the proof that (X, Y) are independent standard normal variables to the reader.

Both the Box-Muller method and the Marsaglia-Bray method are very appealing due to their elegance. We will argue later that using acceptance-rejection methods with QMC has its pitfalls. The Box-Muller method is a special kind of inversion method, but it does not map one coordinate to one coordinate; rather, it generates two coordinates at a time.

Though the approximation of the inverse CDF by rational functions is probably less elegant, and the corresponding algorithm is certainly harder to memorize, it is computationally only marginally more demanding.

Importance Sampling

For some densities it is very hard—if not impossible—to invert the CDF exactly, and frequently it is very expensive to do so numerically.

On the other hand, it is not always necessary to generate exactly from the given distribution but rather one samples from a distribution that is close (in some sense that remains to be made precise) to it and adjusts for the error made. This method is called *importance sampling* or, in the present context, *smooth rejection*.

We present the idea in a one-dimensional setup, the general case is straightforward. Consider a random variable X with PDF f_X and suppose we want to compute $\mathbb{E}[h(X)]$ for some function h. Let F_X denote the corresponding CDF, $F_X(x) = \int_{-\infty}^{x} f_X(t)\, dt$. Normally, we would compute

$$\mathbb{E}[h(X)] \approx \frac{1}{N} \sum_{n=1}^{N} h(F_X^{-1}(U_n))$$

using the inversion method, where $U_1, \ldots, U_N$ is a uniform pseudo-random sequence or a low-discrepancy sequence.

Suppose now that we do not know how to (cheaply) invert F_X. However, assume that there is another PDF g for which the corresponding CDF G, given by $G(x) = \int_{-\infty}^{x} g(t)\, dt$, is easily inverted. Assume, for simplicity, that $g(t) > 0$ for all t. Then

$$\mathbb{E}[h(X)] = \int_{-\infty}^{\infty} h(x) f_X(x)\, dx = \int_{-\infty}^{\infty} h(x) \frac{f_X(x)}{g(x)} g(x)\, dx = \mathbb{E}\left[h(Y) \frac{f_X(Y)}{g(Y)} \right],$$

where Y is a random variable with PDF g. Now the last expected value can be computed by sampling from the PDF g using the inversion method:

$$\mathbb{E}\left[h(Y) \frac{f_X(Y)}{g(Y)} \right] \approx \frac{1}{N} \sum_{n=1}^{N} h\left(G^{-1}(U_n) \right) \frac{f_X(G^{-1}(U_n))}{g(G^{-1}(U_n))}.$$

When choosing the PDF g, one must take care not to make the integrand less regular. For example, suppose we want to compute $\mathbb{E}[\sin(X)]$, where X has a hyperbolic distribution:

$$f_X(x) = \left(\int_{\mathbb{R}} \exp\left(-\sqrt{1 + t^2}\right)\, dt \right)^{-1} \exp\left(-\sqrt{1 + x^2}\right).$$

Let Y have standard normal distribution, $g(y) = \frac{1}{\sqrt{2\pi}} \exp(-y^2/2)$. Then

$$\mathbb{E}[\sin(X)] = \mathbb{E}\left[\sin(Y)\frac{f_X(Y)}{g(Y)}\right],$$

but $f_X(y)/g(y)$ grows like $\exp(y^2/2)$ as $|y| \to \infty$. Thus we have sacrificed boundedness of our integrand. Taking a double exponential PDF instead, i.e., $g(x) = \frac{1}{2}\exp(|x|)$ preserves boundedness. In turn, that particular g introduces a kink into the integrand, i.e., a place where the integrand is not differentiable. Usually that does less harm than unboundedness. If need arises, the kink can be dealt with by using a PDF made of three different parts.

Remark 8.9 When using Monte Carlo, one may also sample from the PDF g using the rejection method. The goal of importance sampling is then to reduce the variance of the integrand to speed up convergence.

Remark 8.10 Importance sampling is particularly useful for sampling from a random vector whose components have a complicated correlation structure.

Whether or Not—and How—To Use Acceptance-Rejection With QMC

We already mentioned that using acceptance-rejection algorithms with QMC is not straightforward. Indeed, the view that those algorithms should not (or cannot) be combined at all is not uncommon.

The question of whether or not acceptance-rejection algorithms can be used together with QMC depends, of course, on the context. Before we approach an answer to that question, we first discuss how the algorithm can be applied and—equally important—how it cannot.

But first consider MC simulation. We are given a pseudo-random number generator that gives us a sequence $(U_n)_{n\in\mathbb{N}_0}$ of numbers in $[0, 1)$ which are, ideally, indistinguishable from a truly random sequence of independent random variables with uniform distribution on $[0, 1)$. From the sequence $(U_n)_{n\in\mathbb{N}_0}$ we now compute a sequence $(X_n)_{n\in\mathbb{N}_0}$ of independent random variables with given distributions by using the acceptance-rejection algorithm with $cf \leq g$. According to Theorem 8.3 for $N \in \mathbb{N}$ we need on average N/c elements of the sequence $(U_n)_{n\in\mathbb{N}_0}$ to generate $X_0, \ldots, X_{N-1}$. Which of the uniform distributions will be used in the generation of a particular X_n cannot be known in advance (and is not important for MC). If in the simulation we need random d-dimensional vectors, this can be done simply by grouping $X_0, X_1, \ldots$ into d-tuples.

For QMC the situation is quite different. We have a low-discrepancy sequence $(u_n)_{n\in\mathbb{N}_0}$ in the s-dimensional unit cube. This sequence has inherently an s-dimensional structure, and when generating a d-dimensional vector with given distribution and independent components from it, we need to take care not to loose that structure. We provide different approaches as examples.

Example 8.11 The most straightforward transfer of the corresponding MC scheme would be to use a one-dimensional low-discrepancy sequence, just like we use a one-dimensional random number generator for MC. This will give very bad results: consider, for example, the case where we want to construct two independent uniforms. Then the densities f and g in the acceptance-rejection algorithms coincide with $c = 1$.

Thus, no pair is rejected. Hence, applying the acceptance-rejection algorithm to the one-dimensional sequence $(x_n)_{n \in \mathbb{N}_0}$ will result in the two-dimensional sequence $((x_0, x_2), (x_4, x_6), (x_8, x_{10}), \ldots)$.

For example, if $(x_n)_{n \in \mathbb{N}_0}$ is the van der Corput sequence in base 2, then $x_{4k} \in [0, \frac{1}{4})$ and $x_{4k+2} \in [\frac{1}{4}, \frac{1}{2})$ for $k = 0, 1, 2, \ldots$. Thus all points of the two-dimensional sequence will lie in one rectangle of volume 2^{-4} and consequently are not even uniformly distributed.

Example 8.12 Another straightforward transfer of the corresponding MC scheme suggests to use a $s = 2d$-dimensional low-discrepancy sequence, and check for every pair of coordinates whether it is rejected. If no pair is rejected, we get a vector with the desired distribution.

We provide a simple example. Let f be the PDF of the gamma distribution with parameter a, $f(x) = x^{a-1} \exp(-x)/\Gamma(a)$ and let g be the PDF of the exponential distribution with parameter b, $g(x) = b \exp(-bx)$. If $a = 1.2$ and $b = 0.85$, then $bf(x) \leq g(x)$. We apply the rejection algorithm to some lattice rule in dimension 4, that is, the first two components are used to generate the first gamma variable while the last two components will be used to generate the second one. If rejection occurs in generating either of the components, the whole four-dimensional point is rejected.

The resulting sequence $(x_n)_{n \in \mathbb{N}_0}$ will have the distribution of two independent $\Gamma(1.2, 0.85)$ variables, so applying the corresponding CDF to the components gives a sequence $(u_n)_{n \in \mathbb{N}_0}$ which is uniform in the unit square. However there is no reason why it should have any additional structure, like having low-discrepancy or being a $(t, 4)$-sequence. Figure 8.2 compares $(u_n)_{n \in \mathbb{N}_0}$ with the first and third components of the original lattice. Of course, the whole number of points in the lattice must be greater than the number plotted so we can show an equal number of points in both plots.

It can be seen that, while the points on the left still bear some similarities to the lattice on the right, they also show some characteristics typical for random numbers, like the presence of clusters and holes. Nevertheless, we may have the hope that the result is not worse than when using MC, especially when c is close to 1.

What else could we do? A key observation is that the acceptance-rejection algorithm does not require the densities f and g to be one-dimensional.

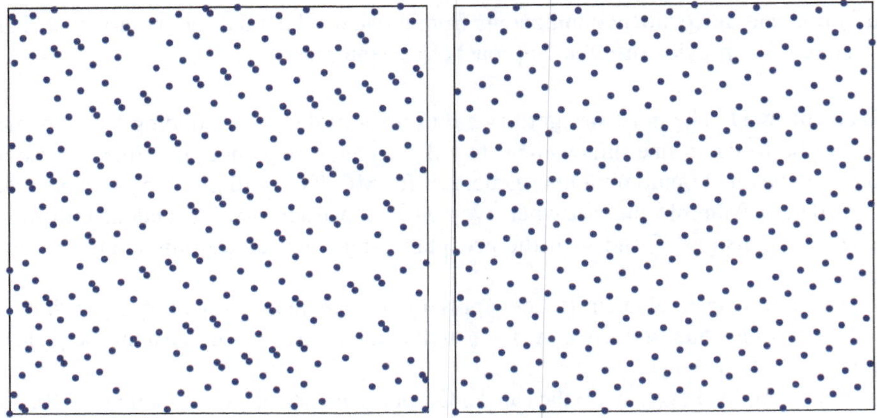

Fig. 8.2 Comparison of rejected and original lattice points

Example 8.13 Suppose we want to generate a d-dimensional vector with independent components, each distributed according to the one-dimensional PDF f. So, the vector is distributed according to the d-dimensional PDF $f(x) = \prod_{k=1}^{d} f(x_k)$, where $x = (x_1, \ldots, x_d)$. As usual, we assume that there exists a one-dimensional PDF g from which it is easy to sample and for which there exists c with $cf \leq g$. Then $g(x) = \prod_{k=1}^{d} g(x_k)$ is a d-dimensional PDF from which it is easy to sample (just do it coordinate-wise) and for which $c^s f \leq g$. By "easy to sample from" we now explicitly mean that we have a function H such that $H(U)$ is distributed according to g for uniform U.

Thus we may apply Algorithm 8.2 in the following way: take a low-discrepancy sequence $(x_n)_{n \in \mathbb{N}_0}$ in the $s = (d+1)$-dimensional unit cube, compute $Y_n = (H(x_{n,1}), \ldots, H(x_{n,s}))$ for every n, and use only those points for which

$$x_{n,s} \, g(Y_n) \leq c^d f(Y_n),$$

and discard all others.

If we map the random vector obtained in this way back to the unit cube the picture is similar to Fig. 8.2 (for $d = 2$). The benefit is that we need only a $d+1$-dimensional low-discrepancy sequence instead of a $2d$-dimensional one.

Figure 8.3 shows graphs similar to those from Fig. 8.1, but with random points replaced by points from the Faure sequence. We see that the normalized histogram matches the PDF very well compared to the same histogram for random points. When the figures are compared, dismissing the use of QMC with acceptance-rejection altogether does not seem justified.

An issue with the acceptance-rejection method is that it sometimes makes the dependence of the result of an MC simulation on the model parameters less smooth. It is clear that the result of a true MC simulation is by definition stochastic. If one

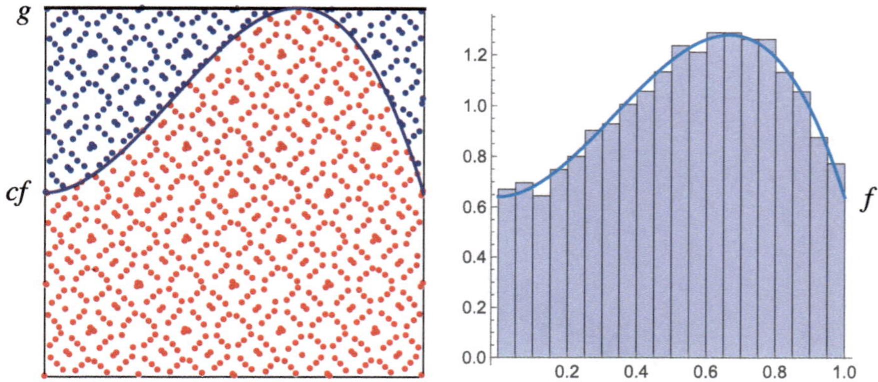

Fig. 8.3 Illustration of the acceptance-rejection algorithm where $g \equiv 1$ and where 1000 points from the Faure sequence have been used instead of random ones. On the right-hand side we see the histogram of the points so generated, compared with the probability PDF function f. The histogram should be compared to that in Fig. 8.1

searches for model parameters which minimize (a function of) the integral that is computed, then this has the practical drawback that, for example, Newton's method cannot be used. In practice, it is therefore common to fix the random sequence for the MC simulation, i.e., the random generator is started afresh for each set of parameters. In this sense, the MC method becomes similar to QMC in that the point set is now deterministic.

However, if acceptance-rejection is used for the generation of random variables, then the integral as a function of the model parameters can still become noisy.

Example 8.14 Let $(X_j^\lambda)_{j=1,\dots,s}$ be a sequence of i.i.d. Gamma$(\lambda, 1)$ random variables and $Y^\lambda := \sum_{j=1}^{s} X_j^\lambda$. Let further $\overline{\lambda} \in \mathbb{R}$ be fixed and $f(y) := y - \overline{\lambda} s$.
   We want to approximate

$$\alpha(\lambda) = \mathbb{E}\left[f(Y^\lambda) \right]$$

by the estimator

$$\widehat{\alpha}_N(\lambda) = \frac{1}{N} \sum_{j=0}^{N-1} f(Y_j^\lambda)$$

for different values of $\lambda \in (\overline{\lambda} - \epsilon, \overline{\lambda} + \epsilon)$. We compare the following scenarios:

1. We use an MC method and acceptance-rejection with a suitable exponential distribution as dominating function. The pseudo-random number generator is restarted for every choice of λ, so that in fact we use the same sequence for every integral evaluation.
2. We use a low-discrepancy QMC sequence (here: a Sobol' sequence) together with the inverse transform method.

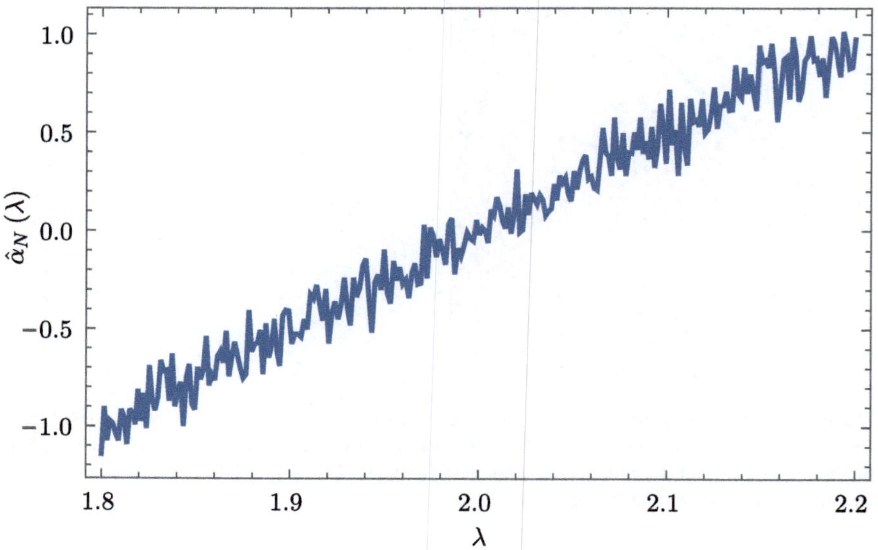

Fig. 8.4 Acceptance-rejection method with a fixed MC point set

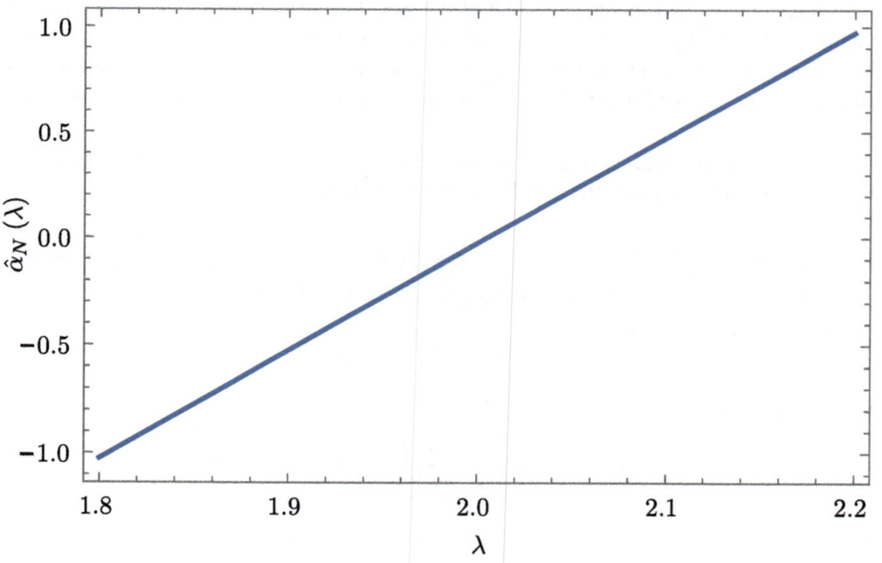

Fig. 8.5 Inverse transform method with a QMC point set

We plot the function $\widehat{\alpha}_N$ with $s = 5$, $N = 1024$, $\overline{\lambda} = 2$ and $\epsilon = 0.2$, where λ changes in steps of 0.001. In Fig. 8.4 one can see quite some noise while in Fig. 8.5 the graph is very smooth.

Smoothness is of importance if, for example, one wants to minimize $\alpha(\lambda)$. An application would be calibration of a financial model to market data.

Of course, this introduction of noise need not happen in any case. For example, when one generates $N(\mu, \sigma^2)$ random variables using Marsaglia-Bray, one in fact generates standard normal variables and transforms them into the required variables afterward. Thus, different parameters μ, σ do not result in different input variables being rejected.

The observant reader will have noticed that the above reservation about the use of acceptance-rejection algorithms is not confined to QMC, but holds for MC as well.

So what is the answer to the questions posed in the title? Acceptance-rejection algorithms destroy some of the structure of low-discrepancy sets/sequences. There is no reason to believe that the resulting point set still has a particularly low discrepancy. So the recommendation is to avoid acceptance-rejection if the random variables can be generated by the inversion method with reasonable effort, or if importance sampling can be used. A situation where this may fail to be the case is when the PDF f is not explicitly known, or when f is only known up to a scalar factor. In many cases there is still a good chance that QMC performs better than MC.

8.2   Generation of Brownian Paths

Many problems from finance, but also from physics, encompass phenomena which are modeled by Brownian motion. In this section we give the basic definition and describe some methods for sampling from Brownian motion.

Brownian Motion: Definition and Properties

Definition 8.15 A *standard Brownian motion* W in $\mathbb{R}^d$ is a stochastic process in continuous time, defined on some probability space $(\Omega, \mathcal{F}, \mathbb{P})$, having the following properties:

1. $W_0 = 0$ almost surely;
2. W has *stationary increments*, that is, for any $t_2 \geq t_1 \geq 0$ the random variables $W_{t_2} - W_{t_1}$ and $W_{t_2 - t_1}$ have the same distribution;
3. W has *independent increments*, that is, for any $n \in \mathbb{N}$ and any $t_1, \ldots, t_n \in [0, \infty)$ with $t_0 := 0 < t_1 < t_2 < \ldots < t_n$, the random variables $W_{t_1} - W_{t_0}, W_{t_2} - W_{t_1}, \ldots, W_{t_n} - W_{t_{n-1}}$ are independent;
4. $\sqrt{1/t}\, W_t$ is a standard normal $\mathbb{R}^d$-valued random variable for every $t > 0$;
5. W has *continuous paths*, that is, for each $\omega \in \Omega$ the mapping $t \mapsto W_t(\omega)$ is continuous.

For applications we usually only need to evaluate the Brownian path at finitely many nodes $t_1, \ldots, t_d$. We therefore define a *discrete Brownian path* with discretization $0 < t_1 < \ldots < t_d$ as a Gaussian vector $(W_{t_1}, \ldots, W_{t_d})$ with mean zero and covariance matrix

$$
\left(\min(t_j, t_k) \right)_{j,k=1}^{d} = \begin{pmatrix} t_1 & t_1 & t_1 & \ldots & t_1 \\ t_1 & t_2 & t_2 & \ldots & t_2 \\ t_1 & t_2 & t_3 & \ldots & t_3 \\ \vdots & \vdots & \vdots & \ddots & \vdots \\ t_1 & t_2 & t_3 & \ldots & t_d \end{pmatrix}.
$$

Classical Constructions

There are three classical constructions of discrete Brownian paths:

- The *forward method*, also known as *step-by-step method* or *piecewise method*.
- The *Brownian bridge construction* or *Lévy-Ciesielski construction*.
- The *principal component analysis construction* (PCA construction).

The forward method is also the most straightforward one: given a standard normal vector $X = (X_1, \ldots, X_d)$ the discrete Brownian path is computed inductively by

$$
W_{t_1} = \sqrt{t_1}\, X_1 \quad \text{and} \quad W_{t_{k+1}} = W_{t_k} + \sqrt{t_{k+1} - t_k}\, X_{k+1}.
$$

Using that $\mathbb{E}[X_j X_k] = \delta_{jk}$, it is easy to see that $(W_{t_1}, \ldots, W_{t_d})$ has the required correlation matrix. Besides its simplicity, the main attractivity of the forward method lies in the fact that it is very efficient: given that the values $\sqrt{t_{k+1} - t_k}$ are pre-computed, generation of a path takes only generation of the standard normal vector plus d multiplications and $d - 1$ additions.

An alternative construction is the Brownian bridge construction, which allows the values $W_{t_1}, \ldots, W_{t_d}$ to be computed in any given order. The main observation that makes this possible is the following lemma, the proof of which is left to the reader (Fig. 8.6).

Lemma 8.16 *Let W be a Brownian motion and let $t_1 < t_2 < t_3$. Then the conditional distribution of W_{t_2} given W_{t_1}, W_{t_3} is $N(\mu, \sigma^2)$ with*

$$
\mu = \frac{t_3 - t_2}{t_3 - t_1} W_{t_1} + \frac{t_2 - t_1}{t_3 - t_1} W_{t_3} \quad \text{and} \quad \sigma^2 = \frac{(t_3 - t_2)(t_2 - t_1)}{t_3 - t_1}.
$$

Suppose the elements of $(W_{t_1}, \ldots, W_{t_d})$ should be computed in the order $W_{t_{\pi(1)}}, W_{t_{\pi(2)}}, \ldots, W_{t_{\pi(d)}}$ for some permutation π of d elements. In computing $W_{t_{\pi(j)}}$ we need to take into account the previously computed elements, and at most two of those are of relevance, the one next to $\pi(j)$ on the left and the one next to $\pi(j)$ on

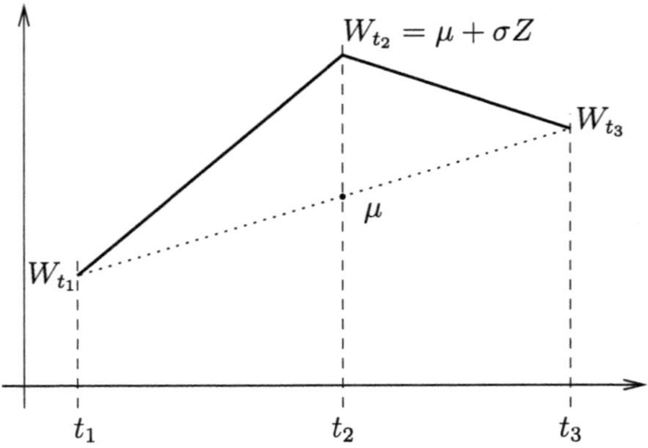

Fig. 8.6 Illustration of Lemma 8.16. Z is a standard normal variable

the right: define for every $j \in \{1, \ldots, d\}$ two sets,

$$L(j) := \{k : k < \pi(j) \text{ and } \pi^{-1}(k) < j\}$$
$$R(j) := \{k : k > \pi(j) \text{ and } \pi^{-1}(k) < j\}.$$

Thus L contains all the indices k that are smaller than $\pi(j)$ and for which W_{t_k} has already been constructed and R contains all the indices k that are greater than $\pi(j)$ and for which W_{t_k} has already been constructed. Now define

$$l(j) := \begin{cases} 0 & \text{if } L(j) = \emptyset, \\ \max L(j) & \text{if } L(j) \neq \emptyset, \end{cases}$$

$$r(j) := \begin{cases} \infty & \text{if } R(j) = \emptyset, \\ \min R(j) & \text{if } R(j) \neq \emptyset, \end{cases}$$

and set $W_{t_0} = 0$,

$$W_{t_{\pi(j)}} := \begin{cases} W_{t_{l(j)}} + \sqrt{t_{\pi(j)} - t_{l(j)}}\, X_j & \text{if } r(j) = \infty, \\[2ex] \dfrac{t_{r(j)} - t_{\pi(j)}}{t_{r(j)} - t_{l(j)}} W_{t_{l(j)}} + \dfrac{t_{\pi(j)} - t_{l(j)}}{t_{r(j)} - t_{l(j)}} W_{t_{r(j)}} \\[1ex] \quad + \sqrt{\dfrac{(t_{\pi(j)} - t_{l(j)})(t_{r(j)} - t_{\pi(j)})}{t_{r(j)} - t_{l(j)}}}\, X_j & \text{if } r(j) < \infty, \end{cases}$$

where $X = (X_1, \ldots, X_d)$ is a standard normal random vector.

It is easy to check that the vector $(W_{t_1}, \ldots, W_{t_d})$ constructed in that way has again covariance matrix $(\min(t_j, t_k))_{j,k}$. The functions l and r, as well as the coefficients of $W_{t_{l(j)}}, W_{t_{r(j)}}, X_j$, do not depend on the random vector X so they can be pre-computed. In some special cases, the functions l and r can be given explicitly, for example, if the $\pi(t_j)$ are the first n elements of the van der Corput sequence. Therefore the Brownian bridge construction is also very efficient: besides the generation of the vector X, computation of one sample uses at most $2d$ additions and $3d$ multiplications.

Moreover, we see that the forward construction is a special case of the Brownian bridge construction where π is the identical permutation.

The PCA construction exploits the fact that the correlation matrix of $(W_{t_1}, \ldots, W_{t_d})$ is positive definite and can therefore be written in the form $V D V^{-1}$ for a diagonal matrix D with positive entries and an orthogonal matrix V. D can be written as $D = D^{\frac{1}{2}} D^{\frac{1}{2}}$, where $D^{\frac{1}{2}}$ is the element-wise positive square root of D. Now the PCA construction from a standard normal random vector X is given by

$$(W_{t_1}, \ldots, W_{t_d})^\top = V D^{\frac{1}{2}} X^\top.$$

The proof that $(W_{t_1}, \ldots, W_{t_d})$ is a discrete Brownian path on the nodes $t_1, \ldots, t_d$ is an easy consequence of the following lemma:

Lemma 8.17 *Let A, Σ be any $d \times d$ matrices with $A A^\top = \Sigma$ and let X be a standard normal vector. Then $Y^\top = A X^\top$ is a normal vector with covariance matrix Σ, i.e., $\mathbb{E}[Y_j Y_k] = \Sigma_{jk}$.*

Proof Since every linear combination of independent normal random variables is still normal, $A X^\top$ is normal. We compute the covariance matrix:

$$\mathbb{E}\left[(A X^\top)_j (A X^\top)_k\right] = \mathbb{E}\left[\sum_{l=1}^{d} A_{jl} X_l \sum_{m=1}^{d} A_{km} X_m\right]$$

$$= \sum_{l=1}^{d} \sum_{m=1}^{d} A_{jl} A_{km} \mathbb{E}[X_l X_m]$$

$$= \sum_{l=1}^{d} A_{jl} A_{kl} = (A A^\top)_{jk} = \Sigma_{jk}.$$

$\square$

The disadvantage of the PCA for high-dimensional problems is that the matrix-vector multiplication, having computational complexity $O(d^2)$, becomes comparatively costly.

What is Wrong About the Forward Construction?

We have provided three different constructions of Brownian paths with one standing apart in that it is clearly the most simple one. So why not use the forward construction for every application?

The answer is that theory predicts a big integration error for QMC if dimensions are big and the number of integration nodes is of realistic order, like a couple of millions only. But one may have the hope that if only a limited number of input

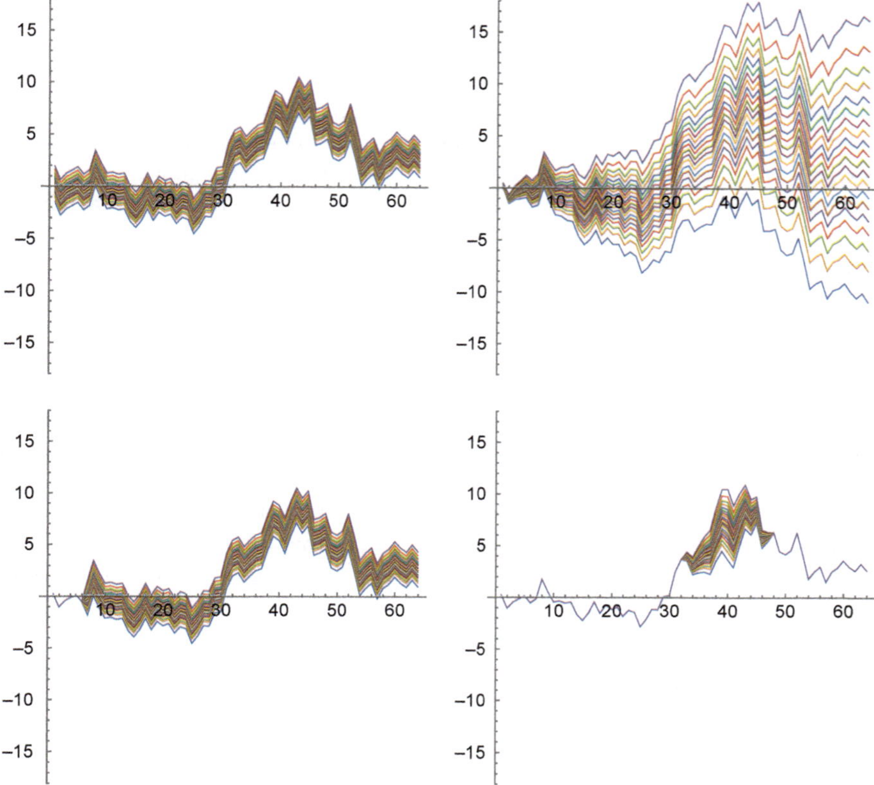

Fig. 8.7 Paths of Brownian motion constructed with the forward construction (left) and the Brownian bridge construction (right). All but one parameters are fixed

parameters have significant importance for the result, then QMC might behave very similar to a low-dimensional integration problem.

Figure 8.7 shows the influence of input parameters on the whole discrete path. We compare the forward construction on the left with the Brownian bridge construction on the right. In the two upper plots all but the first input variables are held fixed. We see that the influence of the first variable on the overall behavior of the path (in an informal sense) is bigger for the Brownian bridge construction.

In the two lower plots all but the seventh input variables are held fixed. We see that in the forward construction only values of the path after the seventh node are influenced, but the overall influence is only slightly smaller than that of the first variable. In contrast, the influence of the seventh variable in the Brownian bridge construction is restricted to the third quarter and is much smaller than that of the first variable.

The above notion of "behaving like a low-dimensional problem" can be made precise using the notion of *effective dimension*. It must be added though, that, despite its popularity the concept of effective dimension alone does not fully explain the success of the alternative constructions. An alternative approach is to use the concept of weighted spaces as presented in Chap. 4. One can argue that dimensions with higher index have a much smaller influence on the integrand than earlier dimensions, thus the norm of the integrand in a suitable weighted space (and thus the error of integration) is small. The problem is now to provide such a weighted space, together with suitable weights. The Korobov spaces described in Chap. 4 are spaces of periodic functions on the unit cube, while most problems in finance lead to non-periodic integrands that are either unbounded or defined on all of $\mathbb{R}^s$. We cannot provide such a space and there is a great number of authors who investigated this problem and it remains largely unsolved at the present. So there is a gap in showing error bounds like those provided for weighted Korobov spaces for financial problems. Numerical experiments do however support the conjecture that similar bounds could be proven.

To answer the question posed in the header: there is nothing wrong with the forward construction, but for some classes of problems other constructions achieve lower errors, at least empirically. For other problems the forward construction may be just fine, presumably for the kind of problems for which the input variables already have different influence on the integrand. Yet some other problems may be such that they are intrinsically high-dimensional, i.e., any construction will yield an integrand that is ill-suited for QMC integration.

Evenly Spaced Discretization Nodes

The case where the $t_1, \ldots, t_d$ are evenly spaced, i.e., where $t_j = \frac{jT}{d}$, $j = 1, \ldots, d$, is of special interest. First, for many applications even spacing is natural. Second, as we will see below, we have a broader range of efficient constructions at our disposal for even spacing. There is no loss in assuming that $T = 1$.

In this setup the covariance matrix equals

$$\left(\frac{1}{d} \min(j, k) \right)_{j,k=1}^{d} = \frac{1}{d} \begin{pmatrix} 1 & 1 & 1 & \ldots & 1 \\ 1 & 2 & 2 & \ldots & 2 \\ 1 & 2 & 3 & \ldots & 3 \\ \vdots & \vdots & \vdots & \ddots & \vdots \\ 1 & 2 & 3 & \ldots & d \end{pmatrix}.$$

We will denote this matrix by $\Sigma^{(d)}$ or, if there is no danger of confusion, simply by Σ.

Note that we can compute the Cholesky decomposition of Σ rather easily: $\Sigma^{(d)} = SS^\top$, where

$$S = S^{(d)} := \frac{1}{\sqrt{d}} \begin{pmatrix} 1 & 0 & 0 & \ldots & 0 \\ 1 & 1 & 0 & \ldots & 0 \\ 1 & 1 & 1 & \ldots & 0 \\ \vdots & \vdots & \vdots & \ddots & \vdots \\ 1 & 1 & 1 & \ldots & 1 \end{pmatrix}.$$

Note that if $\mathbf{y} = (y_1, \ldots, y_d)$ is a vector in $\mathbb{R}^d$, then $S\mathbf{y}^\top$ is the cumulative sum over y divided by $\sqrt{d}$,

$$S\mathbf{y}^\top = \frac{1}{\sqrt{d}}(y_1, y_1 + y_2, \ldots, y_1 + \cdots + y_d)^\top.$$

We have the following two easy lemmas:

Lemma 8.18 *Let A be any $d \times d$ matrix with $AA^\top = \Sigma$ and let X be a standard normal vector. Then $W^\top = AX^\top$ is a discrete Brownian path with discretization $\frac{1}{d}, \frac{2}{d}, \ldots, \frac{d-1}{d}, 1$.*

Proof This is a special case of Lemma 8.17. $\qquad\qquad\qquad\qquad\qquad\qquad\qquad\square$

Lemma 8.19 *Let A be any $d \times d$ matrix with $AA^\top = \Sigma$. Then there is an orthogonal $d \times d$ matrix V with $A = SV$. Conversely, $SV(SV)^\top = \Sigma$ for every orthogonal $d \times d$ matrix V.*

Proof Suppose $AA^\top = \Sigma$, such that $AA^\top = SS^\top$. Note that S is invertible and define $V = S^{-1}A$. Then

$$VV^\top = S^{-1}AA^\top(S^{-1})^\top = S^{-1}SS^\top(S^{-1})^\top = I_d,$$

showing that V is orthogonal. The converse follows from the fact that for orthogonal V we have $V^\top = V^{-1}$. $\qquad\qquad\qquad\qquad\qquad\qquad\qquad\qquad\qquad\qquad\qquad\qquad\square$

For evenly spaced discretization nodes the orthogonal matrices corresponding to the classical matrices can often be given explicitly. The orthogonal transform corresponding to the forward method is the identical mapping on the $\mathbb{R}^d$. For $d = 2^k$, the orthogonal transform corresponding to the Brownian bridge construction where W is computed in the order $W_1, W_{\frac{1}{2}}, W_{\frac{1}{4}}, W_{\frac{3}{4}}, W_{\frac{1}{8}}, W_{\frac{3}{8}}, W_{\frac{5}{8}}, \ldots$ is given by the inverse Haar transform. For the PCA, the orthogonal transform can be computed using the fast discrete sine transform and thus the computation cost is $O(d \log d)$. The advantage of the representation of A in Lemma 8.19 is that there are many orthogonal matrices that allow for fast matrix-vector multiplication, that is, a path of length d can be computed using $O(d \log d)$ operations. Examples include the Walsh transform, discrete sine/cosine transform, Hilbert transform, and others.

In the case where the discrete Brownian path is used for solving an SDE, there is one more thing to be noted: only the increments of the Brownian path are needed. But for a discrete Brownian path W on the nodes $\frac{1}{d}, \ldots, \frac{d}{d}$, computing the increments means multiplication by S^{-1}. Thus, if W is generated using orthogonal transform V, i.e.:

$$(W_{\frac{1}{d}}, \ldots, W_{\frac{d}{d}})^{\top} = SV(X_1, \ldots, X_n)^{\top},$$

then

$$(W_{\frac{1}{d}}, W_{\frac{2}{d}} - W_{\frac{1}{d}} \ldots, W_{\frac{d}{d}} - W_{\frac{d-1}{d}})^{\top} = V(X_1, \ldots, X_n)^{\top}.$$

Example 8.20 Consider the problem of valuating an Asian option, i.e., an option with payoff function $\max\left(\frac{1}{T-T_0} \int_{T_0}^{T} S_t \, dt - K, 0\right)$, in the Heston model (cf. Example 7.5):

$$dB_t = r B_t \, dt,$$
$$dS_t = r S_t \, dt + \sqrt{V_t} S_t (\rho \, dW_t^{(1)} + \sqrt{1 - \rho^2} \, dW_t^{(2)}),$$
$$dV_t = \kappa (\theta - V_t) \, dt + \xi \sqrt{V_t} \, dW_t^{(1)},$$
$$(B_0, S_0, V_0) = (b_0, s_0, v_0),$$

where $r, \rho, \kappa, \theta, \xi$ are positive constants and $W^{(1)}, W^{(2)}$ are independent standard Brownian motions. We solve the SDE using the simple Euler-Maruyama method given in Equation (7.8), with $h = 2T/s$ ($s \in \mathbb{N}$ will be the dimension of the resulting integration problem, and we take s to be even so we can use half of the coordinates for each Brownian path): $\widehat{B}_0 = b_0, \widehat{S}_0 = s_0, \widehat{V}_0 = v_0$ and

$$\widehat{B}_{(k+1)h} = \widehat{B}_{kh} + r\widehat{B}_{kh}h = b_0(1 + rh)^{k+1},$$
$$\widehat{S}_{(k+1)h} = \widehat{S}_{kh} + r\widehat{S}_{kh}h + \sqrt{\widehat{V}_{kh}}\widehat{S}_{kh}\left(\rho\Delta W_{(k+1)h}^{(1)} + \sqrt{1 - \rho^2}\Delta W_{(k+1)h}^{(2)}\right),$$
$$\widehat{V}_{(k+1)h} = \widehat{V}_{kh} + \kappa(\theta - \widehat{V}_{kh})h + \xi\sqrt{\widehat{V}_{kh}}\Delta W_{(k+1)h}^{(1)},$$

for $k = 0, \ldots, s/2 - 1$, where $\Delta W_{(k+1)h}^{(j)} = W_{(k+1)h}^{(j)} - W_{kh}^{(j)}$, $j = 1, 2$. The payoff of the Asian option is approximated by replacing the integral by a corresponding sum, i.e.:

$$\max\left(\frac{1}{T - T_0} \int_{T_0}^{T} S_t \, dt - K, 0\right) \approx \max\left(\frac{h}{T - T_0} \sum_{k=\lfloor \frac{T_0}{T}\frac{s}{2}\rfloor+1}^{s/2} \widehat{S}_{kh} - K, 0\right).$$

Next we need $\Delta W_{kh}^{(j)}$ for $k = 1, \ldots, \frac{s}{2}$, $j = 1, 2$, the increments of the two Brownian motions. First we apply the inverse standard normal CDF component-wise to every point of an s-dimensional low-discrepancy sequence, giving us an s-dimensional sequence $(X_n)_{n\in\mathbb{N}_0}$. Now we can either partition each X_n into two $\frac{s}{2}$-dimensional vectors and use each to generate the increments of one of the Brownian motions using some orthogonal transform. For example, one could use the Brownian bridge construction to generate the first Brownian motion and the PCA to generate the second one. Alternatively, and more generally, we can use one s-dimensional orthogonal transform on X_n and use the first $\frac{s}{2}$ components as the increments of the first Brownian motion and the remaining ones for the second one.

We test the different approaches numerically for a fixed set of parameters: let $s_0 = 100$, $v_0 = 0.3$, $r = 0.03$, $\rho = 0.2$, $\kappa = 2$, $\theta = 0.3$, $\xi = 0.5$ be the model parameters and let the option parameters be $K = 100$, $T_0 = 0$, $T = 1$. Note that the approximate payoff simplifies to

$$\max\left(\frac{2}{s}\sum_{k=1}^{s/2}\widehat{S}_{kh} - K, 0\right).$$

For performing the integration, we use the classical 64-dimensional Sobol' sequence with a random 64-dimensional shift added. We plot the $\log_2$ of the standard deviation over 64 integral evaluations each using 2^m points of the sequence, $m = 2, \ldots, 10$.

The left-hand graph of Fig. 8.8 shows the $\log_2$ of the standard deviation along m for four different transforms: the identity, "Forward," the orthogonal transform corresponding to the Brownian bridge (i.e., the inverse Haar transform), "BB," the one corresponding to PCA and the Brownian bridge applied separately to the inputs of the two Brownian paths, "BB2." On the x-axis we plot the $\log_2$ of the number of integration points, i.e., m, while along the y-axis we plot the $\log_2$ of the standard deviation of the result over 64 runs.

We can see that, as in many practical examples, the PCA performs best. Maybe surprisingly, the idea of using two independent Brownian bridge constructions performs worse than the two combined transforms, but still much better than the identical transform.

We complement this graph with the corresponding one for a "ratchet option", that is, an option with payoff

$$f(S_{\frac{T}{d}}, S_{\frac{2T}{d}}, \ldots, S_T) = \frac{1}{d}\sum_{j=1}^{d}\chi_{[0,\infty)}\left(S_{\frac{jT}{d}} - S_{\frac{(j-1)T}{d}}\right)S_{\frac{jT}{d}}.$$

The errors are plotted on the right-hand side of Fig. 8.8. We can see that the orthogonal transforms that were so successful in the case of an Asian option now perform worse than the identity.

Thus the choice of a suitable orthogonal transform for generation of the Brownian paths depends on the payoff function. Exactly how this transform should be chosen, and for which types of payoffs it accelerates convergence, is still subject to research.

Coming back to the general case of unevenly spaced discretization nodes we note the following: suppose you have nodes $0 < t_1 < \ldots < t_d$. We may compute

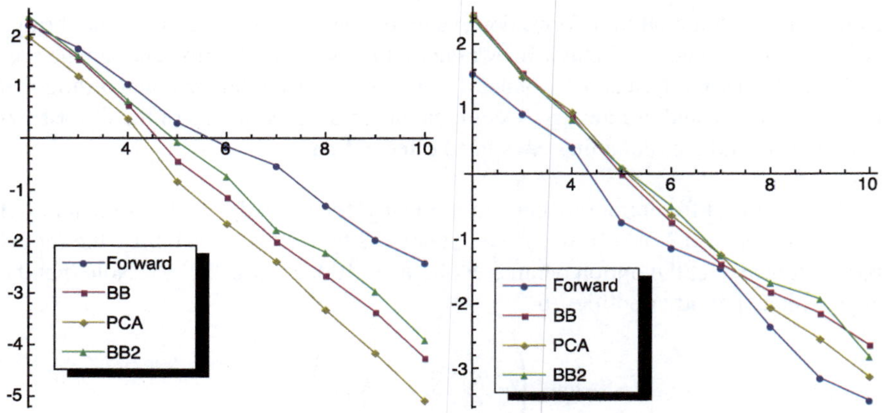

Fig. 8.8 Left: Convergence of the price of an Asian option under different transforms. Right: same graph for the ratchet option

an evenly spaced path $(W_{\frac{1}{d}}, W_{\frac{2}{d}}, \ldots, W_{\frac{d}{d}})$ using our favorite orthogonal transform, then compute

$$\widetilde{W} = \sqrt{d}\left(\sqrt{t_1}\,W_{\frac{1}{d}},\ \sqrt{t_2 - t_1}(W_{\frac{2}{d}} - W_{\frac{1}{d}}),\ \ldots,\ \sqrt{t_d - t_{d-1}}(W_{\frac{d}{d}} - W_{\frac{d-1}{d}})\right).$$

Then $\widetilde{W}$ is a discrete Brownian path with discretization $0 < t_1 < \ldots < t_d$.

8.3 Computation of Price Sensitivities

In Sect. 7.6, we not only learned that it is important to be able to compute price sensitivities, but also that the most straightforward method to compute them, namely, the interchange of differentiation and integration, is not always feasible. This is because usually the payoff function is not differentiable or even discontinuous. We will briefly discuss alternatives to that approach.

Consider a European put option, i.e., an option with payoff $\max(K - S_T, 0)$ and arbitrage-free price

$$P = \exp(-rT)\mathbb{E}[\max(K - S_T, 0)]. \tag{8.1}$$

We are interested in the Delta, i.e., in $\frac{\partial P}{\partial S_0}$. The variable S_0 enters (8.1) via $S_T = S_0 \exp((r - \frac{\sigma^2}{2})T + \sigma W_T)$. We cannot simply differentiate under the expectation because the function $s \mapsto \max(K - s, 0)$ is not everywhere differentiable.

But we may note that $\log S_T$ is a normal random variable with mean $(r - \frac{\sigma^2}{2})T + \log S_0$ and variance $\sigma^2 T$, such that

$$P = \frac{\exp(-rT)}{\sqrt{2\pi}} \int_{\mathbb{R}} \max(K - \exp(y), 0) \exp\left(-\frac{(y - \log S_0 - (r - \frac{\sigma^2}{2})T)^2}{2\sigma^2 T}\right) \, dy.$$

Now the variable S_0 appears in an expression that *is* differentiable and now we may differentiate under the integral:

$$\frac{\partial P}{\partial S_0} = \frac{\exp(-rT)}{\sqrt{2\pi}} \int_{\mathbb{R}} \max(K - \exp(y), 0) \frac{\partial}{\partial S_0} \exp\left(-\frac{(y - \log S_0 - (r - \frac{\sigma^2}{2})T)^2}{2\sigma^2 T}\right) \, dy$$

$$= \frac{\exp(-rT)}{\sqrt{2\pi}} \int_{\mathbb{R}} \max(K - \exp(y), 0)$$

$$\times \frac{y - \log S_0 - (r - \frac{\sigma^2}{2})T}{S_0 \sigma^2 T} \exp\left(-\frac{(y - \log S_0 - (r - \frac{\sigma^2}{2})T)^2}{2\sigma^2 T}\right) \, dy.$$

Writing $\frac{\partial P}{\partial S_0}$ again as an expectation, we get

$$\frac{\partial P}{\partial S_0} = \mathbb{E}\left[\frac{\log(S_T/S_0) - (r - \frac{\sigma^2}{2})T}{S_0 \sigma^2 T} \max(K - S_T, 0)\right].$$

We formulate this discovery in a more general notation: given a random payoff Z_θ depending on a Brownian path and a parameter θ, we want to compute $\frac{\partial}{\partial \theta}\mathbb{E}[Z_\theta]$. The expression $\mathbb{E}[\frac{\partial}{\partial \theta} Z_\theta]$ may not exist, but sometimes it is possible to write

$$\frac{\partial}{\partial \theta}\mathbb{E}[Z_\theta] = \mathbb{E}[Z_\theta \Lambda]$$

for some suitable random variable Λ. This random variable is also called the *likelihood ratio* for the corresponding sensitivity.

In the above example, it was very easy to compute the random variable Λ, and corresponding random variables can be found for sensitivities with respect to the other parameters. It can also be seen from the above argument that the likelihood ratio is largely independent of the payoff of the option: we may replace the payoff of the European put by any other bounded payoff which depends only on S_T. For more complicated dependence on the path, for example, if the option is of Asian type, the likelihood ratio changes.

There is an elegant theory for systematic computation of likelihood ratios using Malliavin calculus. Presentation of that method is beyond the scope of this book, but the further reading section cites an interesting source.

We conclude this section with a useful lemma, of which the above calculation method of the Delta of a European put option is a special case. This result is useful for many practical examples and does not have the theoretical overhead of Malliavin calculus.

Lemma 8.21 *Let $(Z_\theta)_\theta$ be a family of random variables with densities $z \mapsto \rho(z, \theta)$, i.e.:*

$$\mathbb{P}[Z_\theta \in A] = \int_A \rho(z, \theta)\, dz$$

for every θ. Suppose that the partial derivative $\frac{\partial \rho}{\partial \theta}(z, \theta)$ exists for all θ and all $z \in \mathbb{R}$, and that there exists an integrable function $K : \mathbb{R} \to \mathbb{R}$ with $|\frac{\partial \rho}{\partial \theta}(z, \theta)| \leq K(z)$ for all θ and all $z \in \mathbb{R}$. Then for every bounded and measurable function $f : \mathbb{R} \to \mathbb{R}$

$$\frac{\partial}{\partial \theta}\mathbb{E}[f(Z_\theta)] = \mathbb{E}\left[f(Z_\theta)\frac{\partial \log(\rho(Z_\theta, \theta))}{\partial \theta} \right].$$

Proof We write the expectation as integral and we may exchange differentiation and integration under the assumptions on ρ and f. This way we obtain

$$\frac{\partial}{\partial \theta}\mathbb{E}[f(Z_\theta)] = \frac{\partial}{\partial \theta}\int_\mathbb{R} f(z)\rho(z, \theta)\, dz = \int_\mathbb{R} f(z)\frac{\partial \rho(z, \theta)}{\partial \theta}\, dz$$

$$= \int_\mathbb{R} f(z)\frac{\frac{\partial \rho(z, \theta)}{\partial \theta}}{\rho(z, \theta)}\rho(z, \theta)\, dz = \int_\mathbb{R} f(z)\frac{\partial \log(\rho(z, \theta))}{\partial \theta}\rho(z, \theta)\, dz$$

$$= \mathbb{E}\left[f(Z_\theta)\frac{\partial \log(\rho(Z_\theta, \theta))}{\partial \theta} \right],$$

where we have used the chain rule and written the integral as an expectation, again.

$\square$

8.4   Generation of Lévy Paths

To begin with, we give the formal definition of a Lévy process.

Definition 8.22 A *Lévy process* L in $\mathbb{R}^d$ is a stochastic process in continuous time, defined on some probability space $(\Omega, \mathcal{F}, \mathbb{P})$, having the following properties:

1. $L_0 = 0$ almost surely.
2. L has *stationary increments*, that is, for any $s, t \geq 0$, the random variables $L_{s+t} - L_t$ and L_s have the same distribution.
3. L has *independent increments*, that is, for any $n \in \mathbb{N}$ and any $t_1, \ldots, t_n \in [0, \infty)$ with $t_0 := 0 < t_1 < t_2 < \ldots < t_n$, the random variables $L_{t_1} - L_{t_0}, \ldots, L_{t_n} - L_{t_{n-1}}$ are independent.
4. L is *continuous in probability*, i.e., for all $t \geq 0$ and $c > 0$,

$$\lim_{h \to 0} \mathbb{P}[|L_{t+h} - L_t| > c] = 0.$$

5. L has *càdlàg paths*, that is, for each $\omega \in \Omega$ the mapping $t \mapsto L_t(\omega)$ is right-continuous with limits from the left.

We will concentrate on simulating discrete paths, i.e., path values at only finitely many times, and therefore properties 4. and 5. are of minor importance for our purpose. One property that follows from the above definition is that a Lévy process is already completely characterized by the distribution of L_1. Examples are provided by the Poisson process, where L_1 has Poisson distribution[1] and by Brownian motion, where $L_1 \sim N(0, 1)$. The so-called Lévy-Khintchine formula states that for any Lévy process there are numbers $b, \sigma \in \mathbb{R}$ and a measure v on $\mathbb{R}\backslash\{0\}$ with $\int_{|x|<1} x^2 v(dx) < \infty$ such that the characteristic function of L_t is given by

$$\phi_{L_t}(u) := \mathbb{E}\left[\exp(iuL_t)\right] = \exp(t\,\psi(u))$$

with

$$\psi(u) = ibu - \frac{\sigma^2}{2}u^2 + \int_{|x|\geq 1} (\exp(iux) - 1)v(dx) + \int_{|x|<1} (\exp(iux) - 1 - iux)v(dx).$$

Thus, the distribution of L_t (and therefore of an increment $L_{t+s}-L_t$) can be computed via Fourier inversion. For some distributions like the normal, Poisson, and gamma distribution, the PDF of the increment can be given explicitly.

It is actually straightforward to construct a discrete Lévy path on a given set of nodes $0 < t_1 < \ldots < t_d$: let F_t^{-1} denote the inverse of the distribution function of L_t. Let $U_1, \ldots, U_d$ be independent $U(0, 1)$ random variables. Define

$$L_{t_1} := F_{t_1}^{-1}(U_1)$$
$$L_{t_k} := L_{t_{k-1}} + F_{t_k - t_{k-1}}^{-1}(U_k)\,.$$

That is, the forward method works immediately. The other constructions have no direct generalizations to Lévy processes, except for special cases for which the conditional distribution of L_m given L_l, L_r for $l < m < r$ can be computed.

However, there is a simple trick that recovers some of the qualitative features of those transforms: we may rewrite the forward construction of the discrete Lévy path as

$$L_{t_1} := F_{t_1}^{-1}(\Phi(Y_1))$$
$$L_{t_k} := L_{t_{k-1}} + F_{t_k - t_{k-1}}^{-1}(\Phi(Y_k))\,,$$

where $Y_1, \ldots, Y_d$ are independent standard normal variables and Φ is the standard normal CDF. The orthogonal transform is now employed simply in that the $Y_1, \ldots, Y_d$ are generated from our input variables $X_1, \ldots, X_d$ by multiplication with the orthogonal matrix, i.e., $Y^\top = VX^\top$.

---

[1] A random variable X has Poisson distribution with parameter $\lambda > 0$, if $\mathbb{P}[X = k] = \exp(-\lambda)\lambda^k/k!$ for all $k = 0, 1, 2, \ldots$.

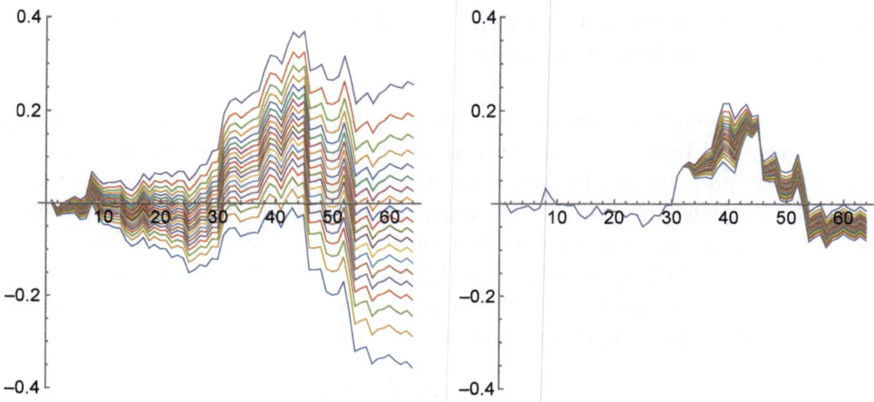

Fig. 8.9 NIG process paths constructed with Brownian bridge orthogonal transform. Left figure: all but the first variables held fixed. Right figure: all but the seventh variable held fixed

Figure 8.9 illustrates the effect of this method on the construction of discrete normal inverse Gaussian[2] (NIG) Lévy paths. The figure on the right shows the effect of the seventh input variable. In comparison to the corresponding Brownian motion example from Fig. 8.7 we see that the effect is less localized, but still the seventh variable mostly influences the behavior of the path on the interval $[\frac{1}{2}, \frac{3}{4}]$. Note that the plots are slightly misleading since they interpolate linearly between the discretization points and thus look like continuous functions. In reality, the paths of an NIG process are (with probability 1) discontinuous with infinitely many jumps in every non-empty open interval. It is important to keep this in mind if, for example, some characteristic of the first entry time of the path into some set is to be computed, as is the case, e.g., for barrier options.

8.5 Multilevel (quasi-)Monte Carlo

Suppose we want to approximate $\mathbb{E}[Y]$ for some random variable Y which has finite expectation. Suppose further that we have a sequence of sufficiently regular functions $f^\ell : \mathbb{R}^{d_\ell} \to \mathbb{R}$ such that

$$\lim_{\ell \to \infty} \mathbb{E}[f^\ell(X^\ell)] = \mathbb{E}[Y], \tag{8.2}$$

---

[2] The PDF of the NIG distribution with parameters $\alpha, \delta, \gamma, \mu$ is given by $f(x) = \frac{\alpha\delta K_1\left(\alpha\sqrt{\delta^2+(x-\mu)^2}\right)}{\pi\sqrt{\delta^2+(x-\mu)^2}} \exp(\delta\gamma + \beta(x - \mu))$, where K_j denotes a modified Bessel function of the second kind. Here the increments have been sampled from the NIG distribution for simplicity. In general, sampling from L_t for $t \neq 1$ requires Fourier inversion.

where for each $\ell \in \mathbb{N}_0$, X^ℓ denotes a d_ℓ-dimensional standard normal vector. In most cases the f^ℓ will be the discrete versions of a function defined on the Brownian paths with d_ℓ discretization nodes, and typically $d_\ell = 2^\ell$. A standard example is provided by the fixed strike Asian option, which has payoff

$$f(W) := \max\left(\frac{1}{T}\int_0^T S_0 \exp\left(\sigma\sqrt{T}W_{t/T} + \left(r - \frac{\sigma^2}{2}\right)t\right)\,dt - K, 0\right),$$

where W is a standard Brownian motion, S_0 is the stock price at time 0, K is the strike of the option, σ is the volatility, and r is the interest rate. W will be approximated by a discrete path of the form $SV^\ell X^\ell$ where, for example, V^ℓ is the orthogonal transform corresponding to $d_\ell = 2^\ell$-dimensional PCA.

Equation (8.2) states that there exists a sequence of algorithms which approximate $\mathbb{E}[Y]$ with increasing accuracy. For example, if $f^\ell(X^\ell)$ has finite variance, we can approximate $\mathbb{E}[Y]$ by $\frac{1}{N}\sum_{k=0}^{N-1} f^\ell(X_k^\ell)$ using sufficiently large ℓ and N, where $(X_k^\ell)_{k\in\mathbb{N}_0}$ is a sequence of independent standard normal vectors.

Usually, evaluation of $f^\ell(X_k^\ell)$ becomes more costly with increasing ℓ. Multilevel methods sometimes help to save significant proportions of computation time by computing more samples for the coarser approximations, which need less computation time but have higher variance.

We have, for large L,

$$\begin{aligned}
\mathbb{E}[Y] &\approx \mathbb{E}\left[f^L(X^L)\right] \\[2mm]
&= \mathbb{E}\left[f^0(X^0)\right] + \sum_{\ell=1}^{L}\left(\mathbb{E}\left[f^\ell(X^\ell)\right] - \mathbb{E}\left[f^{\ell-1}(X^{\ell-1})\right]\right) \\[2mm]
&= \mathbb{E}\left[f^0(X^0)\right] + \sum_{\ell=1}^{L}\left(\mathbb{E}\left[f^\ell(X^\ell)\right] - \mathbb{E}\left[f_c^{\ell-1}(X^\ell)\right]\right) \\[2mm]
&= \mathbb{E}\left[f^0(X^0)\right] + \sum_{\ell=1}^{L}\left(\mathbb{E}\left[f^\ell(X^\ell) - f_c^{\ell-1}(X^\ell)\right]\right),
\end{aligned} \qquad (8.3)$$

where $(f_c^\ell)_{\ell\in\mathbb{N}_0}$ is an arbitrary sequence of functions $f_c^\ell : \mathbb{R}^{d_{\ell+1}} \to \mathbb{R}$ with $\mathbb{E}[f_c^{\ell-1}(X^\ell)] = \mathbb{E}[f^\ell(X^\ell)]$. The "$c$" in f_c^ℓ stands for "coarse level."

The most basic example for f_c^ℓ is given for $d_\ell = m^\ell$ by $f_c^\ell = f^\ell \circ C_{m,\ell}$, where $C_{m,\ell}$ is the linear map defined by the matrix

$$(C_{m,\ell})_{i,j} := \begin{cases} \frac{1}{\sqrt{m}} & \text{if } (i-1)m + 1 \le j \le im,\ 1 \le i \le m^\ell, \\ 0 & \text{else.} \end{cases}$$

For example,

$$
C_{2,\ell} := \begin{pmatrix}
\frac{1}{\sqrt{2}} & \frac{1}{\sqrt{2}} & 0 & 0 & 0 & \cdots & 0 & 0 \\
0 & 0 & \frac{1}{\sqrt{2}} & \frac{1}{\sqrt{2}} & 0 & \cdots & 0 & 0 \\
\vdots & \vdots & \vdots & \vdots & \vdots & & \vdots & \vdots \\
0 & 0 & 0 & 0 & 0 & \cdots & \frac{1}{\sqrt{2}} & \frac{1}{\sqrt{2}}
\end{pmatrix}.
$$

In general, f_c^ℓ is chosen in a way to get small variance for $f^\ell(X^\ell) - f_c^{\ell-1}(X^\ell)$.

Equation (8.3) becomes useful if, as is often the case in practice, the expectation $\mathbb{E}\left[f^\ell(X^\ell) - f^{\ell-1}(C_{m,\ell}X^\ell)\right]$ can be approximated to the required level of accuracy using less function evaluations N_ℓ for bigger ℓ while the costs c_ℓ per function evaluation increase. Suppose the error of approximation of $\mathbb{E}\left[f^\ell(X^\ell) - f_c^{\ell-1}(X^\ell)\right]$ using N_ℓ points is $e_\ell(N_\ell)$. We choose $N_0, \ldots, N_L$ so that

$$
e_0(N_0) + \cdots + e_L(N_L) \leq \varepsilon
$$

while minimizing the total cost

$$
c = c_0 N_0 + \cdots + c_L N_L.
$$

In that way the total computation cost is typically much lower than it would be if $\mathbb{E}[f^L(X^L)]$ would be computed directly.

We conclude with an example in which multilevel MC is combined with orthogonal transforms and QMC.

Example 8.23 We compare multilevel MC with three multilevel QMC methods, forward, PCA, and the regression algorithm from [75] numerically. For that we choose the parameters in a Black-Scholes model as $r = 0.04$, $\sigma = 0.3$, $S_0 = 100$, and we aim to value an Asian call option with parameters $K = 100$ and $T = 1$. At the finest level we choose 2^{10} discretization points and at each coarser level the number of points is divided in by 2, i.e., $L = 10$ and $m = 2$. The number of sample points are doubled at each level starting with N_L sample points at the finest level L. For the QMC approaches we take a Sobol' sequence with a random shift. In Table 8.1 we compare for different values N_L both the average and the standard deviation of the price of the Asian call option based on 1000 independent runs. Moreover, the average computing time for one run is given in brackets. As we can see, the QMC methods generally give better results in terms of accuracy in that simple example. In terms of computation times, the PCA suffers from its slightly more costly path construction and is not better than the forward method with double number of samples. The results for the regression method show that there is room for improvement beyond PCA.

Table 8.1 Multilevel (Q)MC using 2^{10} time steps ($L = 10$). The average and the standard deviation of the option price are based on 1000 runs. The average computing time is given in brackets

| | Multilevel | | | | Multilevel QMC | | | | |
| | Monte Carlo | | Forward | | PCA | | Regression | |
N_L	Average	Stddev	Average	Stddev	Average	Stddev	Average	Stddev
2	7.717	0.41	7.735	0.19×10^{-1}	7.736	0.16×10^{-1}	7.739	0.10×10^{-1}
	(0.0057 s)		(0.0057 s)		(0.0088 s)		(0.0069 s)	
4	7.738	0.19	7.734	0.71×10^{-2}	7.736	0.44×10^{-2}	7.738	0.29×10^{-2}
	(0.0074 s)		(0.0074 s)		(0.0118 s)		(0.0091 s)	
8	7.748	0.54×10^{-1}	7.737	0.30×10^{-2}	7.737	0.14×10^{-2}	7.736	0.10×10^{-2}
	(0.0101 s)		(0.0100 s)		(0.0165 s)		(0.0124 s)	
16	7.746	0.40×10^{-1}	7.736	0.11×10^{-2}	7.737	0.69×10^{-3}	7.736	0.30×10^{-3}
	(0.0157 s)		(0.0157 s)		(0.0279 s)		(0.0194 s)	
32	7.728	0.31×10^{-1}	7.736	0.49×10^{-3}	7.737	0.21×10^{-3}	7.736	0.10×10^{-3}
	(0.0266 s)		(0.0265 s)		(0.0585 s)		(0.0326 s)	
64	7.739	0.81×10^{-2}	7.736	0.20×10^{-3}	7.737	0.69×10^{-4}	7.737	0.32×10^{-4}
	(0.0486 s)		(0.0484 s)		(0.1202 s)		(0.0583 s)	

Further Reading

The standard algorithm for sampling from the inverse normal CDF can be found in the article by Wichura [146], and the corresponding source code in FORTRAN can be found online. Translations into several other programming languages exist.

The standard reference for MC pricing of financial instruments is the book by Glasserman [47], which also contains some material on QMC simulation.

The topic of non-uniform random variate generation is treated in quite some detail in the book by Hörmann, Leydold, and Derflinger [70]. There it is shown how the proposal PDF for a given target PDF can be computed automatically with very small rejection rate.

Dick and Zhu [38] study the combination of a special class of acceptance-rejection algorithms and QMC. They provide estimates for the discrepancy of the generated point set with respect to the target PDF and they find that it is of lower order than that of random points.

Example 8.14 is taken from the article [41] by Eichler, Leobacher, and Zellinger. There the problem of calibrating (financial) models is considered in some detail.

We noted in the text that the cost for generating Brownian paths using PCA is costly due to the general matrix-vector multiplication involved. Keiner and Waterhouse [79] describe an approximate PCA for which the cost of matrix-vector multiplication is $O(d \log d)$. Another efficient approach by means of a so-called "fast QMC matrix-vector multiplication" was developed in [28] (see also [26, Chap. 16]).

The notion of effective dimension was first formalized by Caflisch, Morokoff, and Owen in [16] using the concept of ANOVA decomposition. The latter has some interesting applications in complexity analysis, see, for example, [118].

In the article [99], Leobacher reviews a number of orthogonal transforms that can be computed using $O(d \log d)$ operations. For example, one can find a detailed algorithm for the use of the fast discrete sine transform for fast computation of the PCA, which has originally been proposed by Scheicher, see [129].

The example for which the Brownian bridge and other transforms perform worse than the identical one, i.e., the ratchet option, was first considered in the context of QMC pricing by Papageorgiou in [125]. Papageorgiou concludes that the effectivity of a transform is intimately tied to the payoff structure. His results were formalized and generalized by Wang and Sloan [144]. In the articles [74,75] by Leobacher and Irrgeher, it is tried to choose the orthogonal transform in a way that puts as much variance as possible into the dependence of the first input variable. To this end the payoff is approximated by a linear function g ("regression") and an orthogonal (Householder-) transform V is computed such that $g \circ V$ only depends on X_1. This V is taken as the orthogonal transform for the original problem. See also Imai and Tan [72] for a different approach.

Examples of Lévy processes for which a bridge construction is possible are provided by Avramidis and L'Ecuyer in [7]. They show how to obtain such a construction for the gamma process, the Lévy process for which L_t has gamma distribution with parameters (ta, b), $a, b \in (0, \infty)$, that is:

$$\mathbb{P}[L_t \leq z] = \int_0^z \frac{b^{ta} x^{ta-1}}{\Gamma(ta)} \exp(-bx)\, \mathrm{d}x,$$

and also for the variance-gamma process, which is a Lévy process of the form $t \mapsto W_{L_t}$, where L is a gamma process and W is Brownian motion.

The idea to use orthogonal transforms on normals to obtain constructions for Lévy paths similar to the Brownian bridge was first used in [98] for the Brownian bridge and later, but independently, in [73] for general orthogonal transforms.

Fournier et al. provide an introduction to using Malliavin calculus for computing Greeks using MC [45].

Multilevel MC is a technique for speeding up MC simulation, especially for SDE models. It has gained a lot of recognition over the last couple of years, starting with the pioneering work by Giles [46] and Heinrich [53].

Exercises

8.1 Let $f : \mathbb{R} \to \mathbb{R}$ be a positive PDF, i.e.:

- f is measurable;
- $f(x) > 0$ for all $x \in \mathbb{R}$;
- $\int_{\mathbb{R}} f(x)\,dx = 1$.

Show that $F : \mathbb{R} \to (0, 1)$, $F(x) := \int_{-\infty}^{x} f(y)\,dy$ is bijective.

8.2 Suppose f is a continuous positive PDF and $F(x) := \int_{-\infty}^{x} f(y)\,dy$ for all $x \in \mathbb{R}$. Show that the inverse function of F satisfies the ordinary differential equation $G'(u) = 1/f(G(u))$.

8.3 Let $F : \mathbb{R} \to [0, 1]$ be a CDF, that is, F is monotone, right-continuous with $\lim_{x \to -\infty} F(x) = 0$ and $\lim_{x \to \infty} F(x) = 1$. Define the *quasi-inverse* of F by

$$F^{-}(u) := \inf\{x \in \mathbb{R} : F(x) \ge u\} \quad \text{for all } u \in (0, 1).$$

Show that

   a. F^{-} is non-decreasing.

   b. $F(F^{-}(u)) \ge u$ and $F^{-}(F(x)) \le x$ for all $u \in (0, 1)$, and all $x \in \mathbb{R}$.

   c. Conclude that for all $u \in (0, 1)$, and all $x \in \mathbb{R}$, we have $F^{-}(u) \le x$ if and only if $F(x) \ge u$.

   d. Conclude further that for any random variable U which is uniformly distributed on $(0, 1)$ we have that $F^{-}(U)$ is a real random variable with CDF F, i.e., $\mathbb{P}[F^{-}(U) \le x] = F(x)$.

   e. Show that F^{-} coincides with F^{-1} if F is invertible.

8.4 Another special case of great practical interest is where X takes values in $\mathbb{Z}$. What does the quasi-inverse look like in that case? How can one write a computer program to sample from the distribution of X?

8.5 Write a simple computer program that generates a geometric random variable with parameter q using a variant of the inversion method.

8.6 A doubly exponential random variable X with parameters α, β has PDF

$$f(x) := \frac{1}{\frac{1}{\alpha} + \frac{1}{\beta}} \times \begin{cases} \exp(\alpha x) & \text{if } \alpha < 0, \\ \exp(-\beta x) & \text{if } \beta \geq 0. \end{cases}$$

    a. Show that f is a PDF, i.e., f is measurable, $f \geq 0$ almost everywhere and $\int_{\mathbb{R}} f(x)\,dx = 1$.

    b. Show that the CDF of X is invertible.

    c. Compute the CDF of X.

8.7 Let X be a real random variable with values in I where I is either $-(\infty, b]$, or $[a, b]$ or $[a, \infty)$. Suppose that the restriction of F to I is invertible. Show that the inversion method works analog to the case considered in the text.

8.8 The PDF of an exponential random variable X is given by $f(x) = \lambda \exp(-\lambda x)$ for $x \geq 0$, $f(x) = 0$ for $x \leq 0$. Show that the CDF of X is invertible on $[0, \infty)$ and compute its inverse.

8.9 The PDF of the arcsine distribution is given by $f(x) = \frac{1}{\pi \sqrt{x(1-x)}}$ for $x \in (0, 1)$ and $f(x) = 0$ for $x \notin (0, 1)$. Compute the corresponding CDF, show that it is invertible on $[0, 1]$ and compute the inverse.

8.10 Show that the Marsaglia-Bray algorithm indeed generates a two-dimensional standard vector.

8.11 Find a continuous PDF for sampling from a gamma distribution with parameter $a < 1$ using the acceptance-rejection method. *Hint*: First find a function $g_1 : (0, 1] \to \mathbb{R}$ which is greater than the target PDF in its domain, then find a function $g_2 : [1, \infty) \to \mathbb{R}$ for the other part. $c^{-1}g$ equals g_1 on $(0, 1]$ and g_2 on $(0, \infty)$, and c has to be computed by integration.

8.12 Show how to construct a discrete Brownian path on the nodes $t_j = \frac{jT}{d}$ from a path on the nodes $\frac{1}{d}, \ldots, \frac{d}{d}$.

8.13 Prove Lemma 8.16.

References

1. C. Aistleitner, Covering numbers, dyadic chaining and discrepancy. J. Complex. **27**, 531–540 (2011)
2. Ch. Aistleitner, Tractability results for the weighted star-discrepancy. J. Complex. **30**(4), 381–391 (2014)
3. H. Albrecher, A. Binder, V. Lautscham, P. Mayer, *Introduction to Quantitative Methods for Financial Markets* (Birkhäuser, Basel, 2013)
4. L. Andersen, Simple and efficient simulation of the heston stochastic volatility model. J. Comput. Finance **11**, 1–42 (2008)
5. N. Aronszajn, Theory of reproducing kernels. Trans. Am. Math. Soc. **68**, 337–404 (1950)
6. E.I. Atanassov, On the discrepancy of the Halton sequences. Math. Balkanica (N.S.) **18**, 15–32 (2004)
7. A.N. Avramidis, P. L'Ecuyer, Efficient Monte Carlo and quasi-Monte Carlo option pricing under the variance-gamma model. Manage. Sci. **52**, 1930–1944 (2006)
8. N.S. Bakhvalov, Approximate computation of multiple integrals. Vestnik Moskov. Univ. Ser. Mat. Meh. Astr. Fiz. Him. **4**, 3–18 (1959). (Russian)
9. N.S. Bakhvalov, On the approximate calculation of multiple integrals. J. Complex. **31**(4), 502–516 (2015)
10. J. Beck, W.W.L. Chen, *Irregularities of distribution* (Cambridge University Press, Cambridge, 1987)
11. A. Berlinet, C. Thomas-Agnan, *Reproducing Kernel Hilbert spaces in probability and statistics* (Kluwer Academic, Boston, 2004)
12. D. Bilyk, M. Lacey, The supremum norm of the discrepancy function: recent results and connections in *Monte Carlo and Quasi-Monte Carlo Methods 2012*, volume 65 of *Springer Proceedings in Mathematics and statistics*, ed. by J. Dick, F.Y. Kuo, G.W. Peters, I.H. Sloan (Springer, Heidelberg, 2013), pp. 23–38
13. F. Black, M. Scholes, The pricing of options and corporate liabilities. J. Polit. Econ. **81**, 637–654 (1973)

G. Leobacher and F. Pillichshammer, *Introduction to Quasi-Monte Carlo Integration and Applications*, Compact Textbooks in Mathematics,
https://doi.org/10.1007/978-3-032-05446-3

14. H. Brass, K. Petras. *Quadrature Theory. The Theory of Numerical Integration on a Compact Interval*, volume 178 of *Mathematical Surveys and Monographs* (American Mathematical Society, Providence, RI, 2011)

15. V.A. Bykovskii, The discrepancy of the Korobov lattice points. Izv. Math. **76**, 446–465 (2012)

16. R.E. Caflisch, W. Morokoff, A. Owen, Valuation of mortgage-backed securities using Brownian bridges to reduce effective dimension. J. Comput. Finance **1**, 27–46 (1997)

17. B. Chazelle, *The discrepancy method* (Randomness and Complexity. Cambridge University Press, Cambridge, 2000)

18. W.W.L. Chen, M.M. Skriganov, Explicit constructions in the classical mean squares problem in irregularities of point distribution. J. Reine Angew. Math. **545**, 67–95 (2002)

19. T. Cochrane, Trigonometric approximation and uniform distribution modulo one. Proc. Am. Math. Soc. **103**, 695–702 (1988)

20. R. Cont, P. Tankov, *Financial modelling with jump processes* (Chapman and Hall/CRC, 2004)

21. N.G. de Bruijn, K.A. Post, A remark on uniformly distributed sequences and Riemann integrability. Indag. Math. **30**, 149–150 (1968). Nederl. Akad. Wetensch. Proc. Ser. A **71**

22. F. Delbaen, W. Schachermayer, *The mathematics of arbitrage* (Springer, Berlin, 2006)

23. J. Dick, A. Hinrichs, L. Markhasin, F. Pillichshammer, Optimal L_p-discrepancy bounds for second order digital sequences. Israel J. Math. **221**(1), 489–510 (2017)

24. J. Dick, A. Hinrichs, F. Pillichshammer, Proof techniques in quasi-Monte Carlo theory. J. Complex. **31**, 327–371 (2015)

25. J. Dick, P. Kritzer, G. Leobacher, F. Pillichshammer, Constructions of general polynomial lattice rules based on the weighted star discrepancy. Finite Fields Appl. **13**(4), 1045–1070 (2007)

26. J. Dick, P. Kritzer, F. Pillichshammer, *Lattice rules—Numerical integration, approximation, and discrepancy*. Springer Series in Computational Mathematics, vol. 58 (Springer, Cham, 2022)

27. J. Dick, F.Y. Kuo, F. Pillichshammer, I.H. Sloan, Construction algorithms for polynomial lattice rules for multivariate integration. Math. Comput. **74**(252), 1895–1921 (2005)

28. J. Dick, F.Y. Kuo, Q.T. Le Gia, C. Schwab, Fast QMC matrix-vector multiplication. SIAM J. Sci. Comput. **37**(3), A1436–A1450 (2015)

29. J. Dick, F.Y. Kuo, I.H. Sloan, High dimensional integration—the quasi-Monte Carlo way. Acta Numer. **22**, 133–288 (2013)

30. J. Dick, G. Leobacher, F. Pillichshammer, Construction algorithms for digital nets with low weighted star discrepancy. SIAM J. Numer. Anal. **43**(1), 76–95 (2005)

31. J. Dick, H. Niederreiter, On the exact t-value of Niederreiter and Sobol' sequences. J. Complex. **24**, 572–581 (2008)

32. J. Dick, D. Nuyens, F. Pillichshammer, Lattice rules for nonperiodic smooth integrands. Numer. Math. **126**, 259–291 (2014)

33. J. Dick, F. Pillichshammer, *Digital nets and sequences* (Discrepancy Theory and Quasi-Monte Carlo Integration. Cambridge University Press, Cambridge, 2010)

34. J. Dick, F. Pillichshammer, Discrepancy theory and quasi-Monte Carlo integration, in *Panorama of Discrepancy Theory*, volume 2107 of *Lecture Notes in Mathematics*, ed. by W.W.L. Chen, A. Srivastav, G. Travaglini (Springer, Cham, 2014), pp. 539–619

35. J. Dick, F. Pillichshammer, Explicit constructions of point sets and sequences with low discrepancy, in *Uniform Distribution and Quasi-Monte Carlo Methods*. ed. by P. Kritzer, H. Niederreiter, F. Pillichshammer, A. Winterhof. Discrepancy, Integration and Applications, volume 15 of Radon Series on Computational and Applied Mathematics. (De Gruyter, Berlin, 2014), pp.63–86

36. J. Dick, F. Pillichshammer, Optimal $\mathcal{L}_2$ discrepancy bounds for higher order digital sequences over the finite field $\mathbb{F}_2$. Acta Arith. **162**(1), 65–99 (2014)

37. J. Dick, F. Pillichshammer, The weighted star discrepancy of Korobov's p-sets. Proc. Am. Math. Soc. **143**(12), 5043–5057 (2015)

38. J. Dick, H. Zhu, A discrepancy bound for a deterministic acceptance-rejection sampler. Research report, The University of New South Wales, 2013

39. C. Doerr, M. Gnewuch, M. Wahlström, Calculation of discrepancy measures and applications, in *Panorama of Discrepancy Theory*, volume 2107 of *Lecture Notes in Mathematics*, ed. by W.W.L. Chen, A. Srivastav, G. Travaglini (Springer, Cham, 2014), pp. 621–678

40. M. Drmota, R.F. Tichy, *Sequences, discrepancies and applications*. Lecture Notes in Mathematics, vol. 1651 (Springer, Berlin, 1997)

41. A. Eichler, G. Leobacher, H. Zellinger, Calibration of financial models using quasi-Monte Carlo. Monte-Carlo Methods Appl. **17**, 99–131 (2011)

42. H. Faure. Discrépance de suites associées à un système de numération (en dimension s). Acta Arith. **41**, 337–351 (1982). (French)

43. H. Faure, P. Kritzer, New star discrepancy bounds for (t, m, s)-nets and (t, s)-sequences. Monatsh. Math. **172**, 55–75 (2013)

44. H. Faure, P. Kritzer, F. Pillichshammer, From van der Corput to modern constructions of sequences for quasi-Monte Carlo rules. Indag. Math. **26**(5), 760–822 (2015)

45. E. Fournier, J. Lasry, J. Lebouchoux, P. Lions, N. Touzi, Applications of Malliavin calculus to Monte Carlo methods in finance. Finance Stochast. **3**, 391–412 (1999)

46. M.B. Giles, Multilevel Monte Carlo path simulation. Oper. Res. **56**, 607–617 (2008)

47. P. Glasserman, *Monte Carlo methods in financial engineering* (Springer, New York, 2004)

48. M. Gnewuch, H. Pasing, Ch. Weiss, A generalized Faulhaber inequality, improved bracketing covers, and applications to discrepancy. Math. Comput. **90**(332), 2873–2898 (2021)

49. T. Goda, K. Suzuki, T. Yoshiki, Lattice rules in non-periodic subspaces of Sobolev spaces. Numer. Math. **141**(2), 399–427 (2019)

50. G.H. Hardy, E.M. Wright, *An introduction to the theory of numbers*, 10th edn. (Clarendon Press, Oxford, 1979)

51. J.M. Harrison, D.M. Kreps, Martingales and arbitrage in multiperiod securities markets. J. Econ. Theory **20**, 381–408 (1979)

52. M. Hefter, A. Jentzen, On arbitrarily slow convergence rates for strong numerical approximations of Cox-Ingersoll-Ross processes and squared Bessel processes. Finance Stochast. **23**(1), 139–172 (2019)

53. S. Heinrich, Multilevel Monte Carlo methods. Lect. Notes Comput. Sci. **2179**, 58–67 (2001)

54. S. Heinrich, E. Novak, G.W. Wasilkowski, H. Woźniakowski, The inverse of the star discrepancy depends linearly on the dimension. Acta Arith. **96**, 279–302 (2001)

55. P. Hellekalek, General discrepancy estimates: the Walsh function system. Acta Arith. **67**(3), 209–218 (1994)

56. P. Hellekalek, G. Larcher (eds.), *Random and Quasi-Random point sets*. Lecture Notes in Statistics, vol. 138 (Springer, New York, 1998)

57. F.J. Hickernell, Quadrature error bounds with applications to lattice rules. SIAM J. Numer. Anal. **33**, 1995–2016. Erratum: SIAM. J. Numer. Anal. **34**(853–866), 1997 (1996)

58. F.J. Hickernell, Lattice rules: how well do they measure up? in *Random and quasi-random point sets*, volume 138 of *Lecture Notes for Statistics* (Springer, New York, 1998), pp. 109–166

59. F.J. Hickernell, Obtaining $O(n^{-2+\epsilon})$ convergence for lattice quadrature rules, in *Monte Carlo and Quasi-Monte Carlo methods 2000*. ed. by K.T. Fang, F.J. Hickernell, H. Niederreiter. pp. (Springer, Berlin, 2002), pp.274–289

60. F.J. Hickernell, H. Woźniakowski, Tractability of multivariate integration for periodic functions. J. Complex. **17**, 660–682 (2001)

61. A. Hinrichs, Covering numbers, Vapnik-Červonenkis classes and bounds for the star-discrepancy. J. Complex. **20**, 477–483 (2004)

62. A. Hinrichs, E. Novak, M. Ullrich, H. Woźniakowski, The curse of dimensionality for numerical integration Zhn of smooth functions. Math. Comput. **83**(290), 2853–2863 (2014)

63. A. Hinrichs, E. Novak, M. Ullrich, H. Woźniakowski, The curse of dimensionality for numerical integration of smooth functions II. J. Complex. **30**(2), 117–143 (2014)

64. A. Hinrichs, E. Novak, H. Woźniakowski, The curse of dimensionality for monotone and convex functions of many variables. J. Approx. Theory **27**, 955–965 (2011)

65. A. Hinrichs, F. Pillichshammer, W.C. Schmid, Tractability properties of the weighted star discrepancy. J. Complex. **24**(2), 134–143 (2008)

66. A. Hinrichs, F. Pillichshammer, S. Tezuka, Tractability properties of the weighted star discrepancy of the Halton sequence. J. Comput. Appl. Math. **350**, 46–54 (2019)

67. E. Hlawka, Funktionen von beschränkter Variation in der Theorie der Gleichverteilung. Ann. Mat. Pura Appl. **54**, 325–333 (1961). ((German))

68. E. Hlawka, *Theorie der Gleichverteilung* (Bibliographisches Institut, Mannheim, 1979)

69. W. Hoeffding, Probability inequalities for sums of bounded random variables. J. Am. Statist. Assoc. **58**, 13–30 (1963)

70. W. Hörmann, J. Leydold, G. Derflinger, *Automatic nonuniform random variate generation* (Springer, Berlin, 2004)

71. L.K. Hua, Y. Wang, *Applications of number theory to numerical analysis* (Springer, Berlin, Heidelberg, New York, 1981)

72. J. Imai, K.S. Tan, A general dimension reduction technique for derivative pricing. J. Comput. Finance **10**, 129–155 (2007)

73. J. Imai, K.S. Tan, An accelerating quasi-Monte Carlo method for option pricing under the generalized hyperbolic Lévy process. SIAM J. Sci. Comput. **31**, 2282–2302 (2009)

74. C. Irrgeher, G. Leobacher, Fast orthogonal transforms for pricing derivatives with quasi-Monte Carlo, in *Proceedings of the 2012 Winter Simulation Conference*, ed. by C. Laroque et al. (2012)

75. C. Irrgeher, G. Leobacher, Fast orthogonal transforms for multilevel quasi-Monte Carlo simulation in computational finance, in *Handelingen Contactforum Actuarial and Financial Mathematics Conference, Interplay between Finance and Insurance*, Brussels, ed. by W. Vanmaele et al. (2013)

76. S. Joe, Component by component construction of rank-1 lattice rules having $O(n^{-1}(\ln n))^d)$ star-discrepancy, in *Monte Carlo and Quasi-Monte Carlo methods 2002*. ed. by H. Niederreiter (Springer, Berlin, 2004), pp.293–298

77. S. Joe, Constructions of good rank-1 lattice rules based on the weighted star discrepancy, in *Monte Carlo and Quasi-Monte Carlo methods 2004*. ed. by H. Niederreiter, D. Talay (Springer, Berlin, 2006), pp.181–196

78. M.A. Kalos, P.A. Whitlock, *Monte Carlo methods*, 2nd edn. (Weinheim, Wiley-Blackwell, 2008)

79. J. Keiner, B.J. Waterhouse, Fast principal components analysis method for finance problems with unequal time steps, in *Monte Carlo and Quasi-Monte Carlo methods 2008*. ed. by P. L'Ecuyer, A.B. Owen (Springer, Berlin, 2009), pp.455–465

80. P.E. Kloeden, E. Platen, *Numerical solution of stochastic differential equations* (Springer, Berlin, Heidelberg, New York, 1992)

81. D.E. Knuth, *The Art of Computer Programming, vol. 2: Seminumerical Algorithms* (Addison-Wesley Publishing Co., Reading, Mass.-London-Don Mills, Ont, 1969)

82. J.F. Koksma, Some integrals in the theory of uniform distribution modulo 1. Mathematica, Zutphen. B. **11**, 49–52 (1942)

83. N.M. Korobov, *Number-theoretic methods in approximate analysis* (Gosudarstv. Izdat. Fiz.-Mat. Lit, Moscow, 1963). ((Russian))

84. P. Kritzer, F. Pillichshammer, Low discrepancy polynomial lattice point sets. J. Number Theory **132**(11), 2510–2534 (2012)

85. L. Kuipers, H. Niederreiter, *Uniform distribution of sequences* (Wiley, New York, 1974)

86. F.Y. Kuo, D. Nuyens, Application of quasi-Monte Carlo methods to PDEs with random coefficients–an overview and tutorial, in *Monte Carlo and Quasi-Monte Carlo methods*, volume 241 of *Springer Proceedings in Mathematics and statistics*, ed. by A.B. Owen, P.W. Glynn (Springer, Cham, 2018), pp. 53–71

87. F.Y. Kuo, D. Nuyens, Hot new directions for quasi-Monte Carlo research in step with applications, in *Monte Carlo and Quasi-Monte Carlo Methods*, volume 241 of *Springer Proceedings in Mathematics and Statistics*, ed. by A.B. Owen, P.W. Glynn (Springer, Cham, 2018), pp. 123–144

88. G. Larcher, On the distribution of sequences connected with good lattice points. Monatsh. Math. **101**, 135–150 (1986)

89. G. Larcher, A best lower bound for good lattice points. Monatsh. Math. **104**, 45–51 (1987)

90. G. Larcher, Nets obtained from rational functions over finite fields. Acta Arith. **63**(1), 1–13 (1993)

91. G. Larcher, Digital point sets: analysis and application, in *Random and Quasi-Random Point Sets*, volume 138 of *Lecture Notes in Statistics*, ed. by P. Hellekalek, G. Larcher (Springer, New York, 1998). In [56]

92. G. Larcher, *The Art of Quantitative Finance vol. 1. Trading, Derivatives and Basic Concepts*. Springer Texts Business and Economics (Springer, Cham, 2023)

93. G. Larcher. *The Art of Quantitative Finance vol. 2. Volatilities, Stochastic Analysis and Valuation Tools*. Springer Texts Business and Economics (Springer, Cham, 2023)

94. G. Larcher, *The Art of Quantitative Finance vol. 3. Risk, Optimal Portfolios, and Case Studies*. Springer Texts Business and Economics (Springer, Cham, 2023)

95. P. L'Ecuyer, P. Hellekalek, Random number generators: selection criteria and testing, in *Random and Quasi-Random Point Sets*, volume 138 of *Lecture Notes in Statistics*, ed. by P. Hellekalek, G. Larcher (Springer, New York, 1998). In [56]

96. P. L'Ecuyer, P. Marion, M. Godin, F. Puchhammer, A tool for custom construction of QMC and RQMC point sets, in *Monte Carlo and Quasi-Monte Carlo methods*, ed. by A. Keller. Springer Proceedings in Mathematics and Statistics, vol. 387 (Springer, Cham, 2022), pp.51–70

97. Ch. Lemieux, *Monte-Carlo and Quasi-Monte Carlo sampling* Springer Series in Statistics. (Springer, New York, 2009)

98. G. Leobacher, Stratified sampling and quasi-Monte Carlo simulation of Lévy processes. Monte-Carlo Methods Appl. **12**, 231–238 (2006)

99. G. Leobacher, Fast orthogonal transforms and generation of Brownian paths. J. Complex. **28**, 278–302 (2012)

100. G. Leobacher, F. Pillichshammer, Bounds for the weighted L^p discrepancy and tractability of integration. J. Complex. **19**, 529–547 (2003)

101. M.B. Levin, On the lower bound of the discrepancy of Halton's sequence I. C. R. Math. Acad. Sci. Paris **354**(5), 445–448 (2016)

102. M.B. Levin, On the upper bound of the L_2-discrepancy of Halton's sequence. Acta Arith. **202**(3), 205–225 (2022)

103. M.B. Levin, On the lower bound of the discrepancy of (t, s)-sequences: II. Online J. Anal. Comb. (12), Paper No. 3, 74 (2017)

104. R. Lidl, H. Niederreiter, *Introduction to finite fields and their applications* (Cambridge University Press, Cambridge, revised edition, 1994)

105. J. Matoušek, *Geometric Discrepancy* (Springer, Berlin, 1999)

106. R.C. Merton, Theory of rational option pricing. Bell J. Econ. Manage. Sci. **4**, 141–183 (1973)

107. T. Müller-Gronbach, E. Novak, K. Ritter, *Monte-Carlo Algorithmen* (Springer, Berlin, Heidelberg, 2012)

108. H. Niederreiter, Point sets and sequences with small discrepancy. Monatsh. Math. **104**, 273–337 (1987)

109. H. Niederreiter. Low-discrepancy point sets obtained by digital constructions over finite fields. Czechoslovak Math. J. **42**(117)(1), 143–166 (1992)

110. H. Niederreiter, *Random Number Generation and Quasi-Monte Carlo Methods*. CBMS-NSF Series in Applied Mathematics, vol. 63 (SIAM, Philadelphia, 1992)

111. H. Niederreiter, C.P. Xing, Low-discrepancy sequences and global function fields with many rational places. Finite Fields Appl. **2**, 241–273 (1996)

112. H. Niederreiter, C.P. Xing, Quasirandom points and global function fields, in *Finite Fields and Applications (Glasgow, 1995)*, volume 233 of London Mathematical Society Lecture Note Series (Cambridge University Press, Cambridge, 1996), pp. 269–296

113. H. Niederreiter, C.P. Xing, *Rational Points on Curves over Finite Fields: Theory and Applications*. London Mathematical Society Lecture Note Series, vol. 285 (Cambridge University Press, Cambridge, 2001)

114. E. Novak, Numerische Verfahren für Hochdimensionale Probleme und der Fluch der Dimension. Jahresber. Deutsch. Math.-Verein. **101**, 151–177 (1999)

115. E. Novak, Some results on the complexity of numerical integration, in *Monte Carlo and quasi-Monte Carlo methods*, volume 163 of Springer Proceedings in Mathematics and statistics (Springer, [Cham], 2016), pp. 161–183

116. E. Novak, F. Pillichshammer, Conditions for tractability of the weighted L_p-discrepancy and integration in non-homogeneous tensor product spaces. Monatsh. Math. **206**(2), 449–470 (2025)

117. E. Novak, F. Pillichshammer, The L_p-discrepancy for finite $p>1$ suffers from the curse of dimensionality. Proc. Am. Math. Soc. **153**(4), 1447–1459 (2025)

118. E. Novak, H. Woźniakowski, *Tractability of Multivariate Problems, Volume I: Linear Information* (EMS, Zurich, 2008)

119. E. Novak, H. Woźniakowski, *Tractability of Multivariate Problems, Volume II: Standard Information for Functionals* (EMS, Zurich, 2010)

120. E. Novak, H. Woźniakowski, *Tractability of Multivariate Problems, Volume III: Standard Information for Operators* (EMS, Zurich, 2012)

121. D. Nuyens, *Fast Construction of Good Lattice Rules*. Ph.D. thesis, Katholieke Universiteit Leuven, Leuven (2007)

122. D. Nuyens, The construction of good lattice rules and polynomial lattice rules. In *Uniform distribution and quasi-Monte Carlo methods*, volume 15 of *Radon Series on Computational and Applied Mathematics* (De Gruyter, Berlin, 2014), pp. 223–255

123. D. Nuyens, R. Cools, Fast algorithms for component-by-component construction of rank-1 lattice rules in shift-invariant reproducing kernel Hilbert spaces. Math. Comput. **75**, 903–920 (2006)

124. D. Nuyens, R. Cools, Fast component-by-component construction, a reprise for different kernels, in *Monte Carlo and Quasi-Monte Carlo Methods 2004*. ed. by H. Niederreiter, D. Talay (Springer, Berlin, 2006), pp.373–387

125. A. Papageorgiou, The Brownian bridge does not offer a consistent advantage in quasi-Monte Carlo integration. J. Complex. **18**, 171–186 (2002)

126. S.H. Paskov, J.F. Traub, Faster evaluation of financial derivatives. J. Portfolio Manag. **22**, 113–120 (1995)

127. F. Pillichshammer, Polynomial lattice point sets, in *Monte Carlo and quasi-Monte Carlo methods 2010*, volume 23 of *Springer Proceedings in Mathematics and statistics* (Springer, Heidelberg, 2012), pp. 189–210

128. W. Rudin, *Real and Complex Analysis* (McGraw-Hill Book Co., New York, 1987)

129. K. Scheicher, Complexity and effective dimension of discrete Lévy areas. J. Complex. **23**, 152–168 (2007)

130. W.C. Schmid, Projections of digital nets and sequences. Math. Comput. Simul. **55**, 239–247 (2001)

131. W.M. Schmidt, *Irregularities of Distribution X* (Academic Press, New York, 1977), pp. 311–329

132. R. Schürer, W.C. Schmid, MinT: a database for optimal net parameters, in *Monte Carlo and quasi-Monte Carlo methods 2004* (Springer, Berlin, 2006), pp. 457–469

133. R. Schürer, W.C. Schmid, MinT–architecture and applications of the (t, m, s)-net and OOA database. Math. Comput. Simul. **80**(6), 1124–1132 (2010)

134. M.M. Skriganov, Harmonic analysis on totally disconnected groups and irregularities of point distributions. J. Reine Angew. Math. **600**, 25–49 (2006)
135. I.H. Sloan, S. Joe, *Lattice Methods for Multiple Integration* (Oxford University Press, New York and Oxford, 1994)
136. I.H. Sloan, A.V. Reztsov, Component-by-component construction of good lattice rules. Math. Comput. **71**, 263–273 (2002)
137. I.H. Sloan, H. Woźniakowski, When are quasi-Monte Carlo algorithms efficient for high-dimensional integrals? J. Complex. **14**, 1–33 (1998)
138. I.H. Sloan, H. Woźniakowski, Tractability of multivariate integration for weighted Korobov classes. J. Complex. **17**, 697–721 (2001)
139. I.M. Sobol', Distribution of points in a cube and approximate evaluation of integrals. Ž. Vyčisl. Mat. i Mat. Fiz. **7**, 784–802 (1967)
140. V.N. Temlyakov, Cubature formulas, discrepancy, and nonlinear approximation. J. Complex. **19**, 352–391 (2003)
141. S. Tezuka, *Uniform Random Numbers: Theory and Practice*. The Kluwer International Series in Engineering and Computer Science, vol. 315 (Kluwer Academic Publishers, Dordrecht, 1995)
142. A. Topuzoğlu, A. Winterhof, Pseudorandom sequences, in *Topics in Geometry, Coding Theory and Cryptography*, volume 6 of *Algebra and Its Applications*, ed. by A. Garcia, H. Stichtenoth (Springer, Dordrecht, 2007), pp. 135–166
143. J.F. Traub, A.G. Werschulz, *Complexity and Information* (Cambridge University Press, Cambridge, 1998)
144. X. Wang, I.H. Sloan, Quasi-Monte Carlo methods in financial engineering: an equivalence principle and dimension reduction. Oper. Res. **59**, 80–95 (2011)
145. H. Weyl, Über die Gleichverteilung mod. Eins. Math. Ann. **77**, 313–352 (1916)
146. M.J. Wichura, Algorithm AS 241: the percentage points of the normal distribution. J. R. Stat. Soc. Ser. C (Applied Statistics) **37**, 477–484 (1988)

Index

A
Absolutely continuous function, 61
Acceptance-rejection method, 203
ANOVA Sobolev space, 125
Arbitrage
   -free price, 189, 190, 193
   opportunity, 189
Arcsine distribution, 232
Asian option, 192

B
Basket option, 193
Black-Scholes
   model, 190, 191, 194
   option pricing formula, 191
Bond, 185
Bounded functional, 63
Box-Muller method, 206
Brownian
   bridge construction, 214
   motion, 213
   path, 213

C
Cauchy distribution, 202
CBC algorithm, 84, 89, 97, 118, 170
Characteristic function, 14
Circulant matrix, 90
Combined L_p discrepancy, 71
Contingent claim, 186

Coordinate weights, 72
Cumulative Distribution Function (CDF), 201
Curse of dimensionality, 2, 111, 130, 178, 183, 184

D
Digital
   method, 149
   net, 149
   sequence, 154
   (t, m, s)-net, 149
   (t, s)-sequence, 154
Digitally shifted point set, 174
Direction numbers, 161
Discrepancy, 21
   combined L_p, 71
   extreme, 21
   function, 27
   L_p, 27
   Roth's lower bound, 28
   star, 21
   triangle inequality, 54
   weighted L_p, 73
   weighted star, 73
Discrete Brownian path, 213
Discrete exponential valuation, 157
Distribution
   arcsine, 232
   exponential, 205
   gamma, 205

© The Editor(s) (if applicable) and The Author(s), under exclusive license to Springer Nature Switzerland AG 2026   241
G. Leobacher and F. Pillichshammer, *Introduction to Quasi-Monte Carlo Integration and Applications*, Compact Textbooks in Mathematics,
https://doi.org/10.1007/978-3-032-05446-3